Water Science and Technology Library

Volume 112

The aim of the *Water Science and Technology Library* is to provide a forum for dissemination of the state-of-the-art of topics of current interest in the area of water science and technology. This is accomplished through publication of reference books and monographs, authored or edited. Occasionally also proceedings volumes are accepted for publication in the series. *Water Science and Technology Library* encompasses a wide range of topics dealing with science as well as socio-economic aspects of water, environment, and ecology. Both the water quantity and quality issues are relevant and are embraced by *Water Science and Technology Library*. The emphasis may be on either the scientific content, or techniques of solution, or both. There is increasing emphasis these days on processes and *Water Science and Technology Library* is committed to promoting this emphasis by publishing books emphasizing scientific discussions of physical, chemical, and/or biological aspects of water resources. Likewise, current or emerging solution techniques receive high priority. Interdisciplinary coverage is encouraged. Case studies contributing to our knowledge of water science and technology are also embraced by the series. Innovative ideas and novel techniques are of particular interest.

Comments or suggestions for future volumes are welcomed.

Vijay P. Singh, Department of Biological and Agricultural Engineering & Zachry Department of Civil and Environment Engineering, Texas A&M University, USA
Email: vsingh@tamu.edu

All contributions to an edited volume should undergo standard peer review to ensure high scientific quality, while monographs should also be reviewed by at least two experts in the field.

Manuscripts that have undergone successful review should then be prepared according to the Publisher's guidelines manuscripts: https://www.springer.com/gp/authors-editors/book-authors-editors/book-manuscript-guidelines

More information about this series at http://www.springer.com/series/6689

Sourav Das · Abhra Chanda · Tuhin Ghosh
Editors

Pond Ecosystems of the Indian Sundarbans

An Overview

Springer

Editors
Sourav Das
School of Oceanographic Studies
Jadavpur University
Kolkata, West Bengal, India

Abhra Chanda
School of Oceanographic Studies
Jadavpur University
Kolkata, West Bengal, India

Tuhin Ghosh
School of Oceanographic Studies
Jadavpur University
Kolkata, West Bengal, India

ISSN 0921-092X ISSN 1872-4663 (electronic)
Water Science and Technology Library
ISBN 978-3-030-86785-0 ISBN 978-3-030-86786-7 (eBook)
https://doi.org/10.1007/978-3-030-86786-7

This Springer imprint is published by the registered company Springer Nature Switzerland AG
The registered company address is: Gewerbestrasse 11, 6330 Cham, Switzerland

Dedicated to the people of Indian Sundarbans who are incessantly struggling with climatic disasters

Preface

Inland aquatic bodies play a crucial role in regulating the atmospheric gaseous concentrations (especially greenhouse gases) and providing a habitat for various flora and fauna and other ecosystem services. The lakes and reservoirs all through the world have received substantial attention from the global scientific community. Several pieces of research focused on their biogeochemical dynamics, role in present-day society, and the implication in earning a livelihood. The comparatively smaller water bodies, i.e., the ponds, have received much less attention. In the recent past, several pieces of research indicated that these small aquatic bodies are capable of governing climate phenomena on a synoptic scale. Ponds are a common landscape feature of rural sectors in various tropical countries, and India is no exception. The Sundarbans (shared by India and Bangladesh) is renowned as the largest mangrove forest in the world. This unique ecosystem shelters a spectrum of both floral and faunal biodiversity. Besides, this region is an abode for more than 4.4 million people. Situated in the lower end of the world's largest delta (the Ganga-Brahmaputra-Meghna delta), this region shelters a marginalized section of society in the majority. The ponds are essential in carrying out several daily life activities and are available in almost every household in this part of the world. Lately, the uncertainty in capture fisheries and the increasing demand for fish jointly led to the flourishing of aquaculture ponds in this region (as also observed in many coastal sectors throughout the globe). The purpose of most of the household and community ponds is to meet the freshwater requirements. It has been a tradition in the Indian Sundarbans to dig a pond in the household plots, which meets several purposes in day-to-day activities. Bathing, washing utensils, feeding the agricultural lands during the dry season, homestead gardening, and feeding the cattle are some of the most common activities related to ponds in the Indian Sundarbans. Though these ponds are in plenty through the Indian Sundarban Biosphere Reserve, their biogeochemistry, greenhouse gas emission potential, productivity rates, nutrient, and pollutant dynamics are scarcely studied. This book is perhaps an endeavor of the first of its kind, which tried to cover all these aspects of the ponds of Indian Sundarbans, under one umbrella. This timely book can act as the foundational basis for limnologists and hydro-geologists to have firsthand baseline information on these crucial lentic ecosystems. Estuarine and marine scientists, ecologists, biogeochemists, environmentalists, and social scientists, whose interest lies in the use and performance of lentic ecosystems of the coastal sector of India and other parts of the world, would find interest in the present title. Intermediate to advanced level students can be beneficial by going through this book.

The book opens with an introductory chapter on the ponds of the Indian Sundarbans, which presented a basic overview of these lentic ecosystems. The second chapter dealt with the land use/land cover dynamics (we considered four administrative blocks as representatives for the entire region). We emphasized the size class of the ponds and their distribution pattern. The third and fourth chapters studied the impact of water quality on the livelihood of local people and the fish diversity in the lentic ecosystems of this region. The fifth chapter discussed the role of iron fertilization to enhance the fish yield from these ponds. The role of these ponds in greenhouse gas emission was discussed explicitly in the sixth to eighth chapters. The ninth chapter hinted at the possible ecosystem services of these ponds. The tenth chapter studied the relevant social aspects related to the ponds of a particular community development block of Indian Sundarbans. The eleventh to thirteen chapters covered the biogeochemical and pollutant dynamics. The observations discussed in this book can prompt future research actions from the perspective of achieving the sustainable development goals (of United Nations) like zero hunger (SDG 2) and clean water and sanitation (SDG 6). Overall, this book tried to reflect a holistic understanding of these lentic ecosystems from several viewpoints. However, many of the issues and aspects need further study. Thus, this book can act as a guide for future researchers. The findings discussed in this book indicate the aspects that require more attention.

Kolkata, India

Sourav Das
Abhra Chanda
Tuhin Ghosh

Acknowledgments

We thank Dr. Sanjibakumar Baliar Singh, Dr. Anirban Akhand, and Dr. Sudarsana Rao Pandi for their help in reviewing several chapters. We also thank the editor of Springer Nature (Margaret Deignan) for her kind help and encouragement. We specially thank Dr. Andrew C. G. Henderson (School of Geography, Politics and Sociology, Newcastle University, UK) for his kind support and encouragement throughout this book project. Moreover, we would like to acknowledge the contribution of the countless people of Indian Sundarbans who selflessly gave them their time to talk about their ponds and what it means to them.

Kolkata, India

Sourav Das
Abhra Chanda
Tuhin Ghosh

Contents

About the Editors

Dr. Sourav Das is a Post-doctoral Research Associate of the UKRI-GCRF Living Deltas Hub in the School of Oceanographic Studies at Jadavpur University, India. His research interests include biogeochemistry of marine and coastal water, ocean remote sensing, optical properties of water (inherent and apparent optical properties), and air quality monitoring. He has twelve years of research experience and has more than 60 research articles in reputed peer-reviewed journals including two books.

Dr. Abhra Chanda is an Assistant Professor, School of Oceanographic Studies at Jadavpur University, India. His research interest encompasses the biogeochemistry of lentic and lotic aquatic systems and pollutant dynamics. Blue carbon dynamics is one of the priorities of his research. He has eleven years of research experience and has more than 75 research articles in reputed peer-reviewed journals to his credit.

Prof. Tuhin Ghosh is Professor and Director of the School of Oceanographic Studies at Jadavpur University, India. He has more than two decades of research experience in coastal geomorphology, disaster management, climate change impacts, adaptation strategies, and human migration. He has more than 80 research articles in reputed peer-reviewed journals to his credit.

List of Figures

List of Tables

Chapter 1
An Introduction to the Ponds of Indian Sundarbans—An Essential Socio-Ecological System

Sourav Das, Abhra Chanda, and Tuhin Ghosh

Abstract Ponds are small lentic bodies that are abundant throughout the world. In the rural setup of many Asian countries, especially in the deltaic regions (where adjacent waters are mostly saline), ponds serve as an essential source of fresh water. The Sundarbans is renowned for being the largest mangrove forest on Earth. India and Bangladesh share this unique ecoregion. The Indian part shelters a thickly populated marginalized section of people, who exclusively rely on this forest to meet their livelihood demands. The mangroves and other land use classes of the Indian Sundarbans have received ample attention in the past. However, the millions of ponds in this setup did not receive the adequate scientific focus, which it deserves. The present book is perhaps the first attempt to furnish a holistic overview of the biogeochemical status and socio-economic importance of these ponds. Given proper management, these ponds can play a crucial role in provisioning food resources for the local inhabitants, and thus, can serve to achieve a few of the United Nations Sustainable Development Goals (SDG 2—Zero hunger; SDG 6—Clean water and sanitation). At present, these ponds remain neglected with no proper attempt of nurturing the potential ecosystem services that these aquatic ecosystems can offer. This chapter detailed the nitty–gritty of the ponds of Indian Sundarbans from all possible viewpoints and provides a foundation for the entire book.

Keywords Lentic ecosystems · Indian Sundarbans · Freshwater · Socio-economic issues · Aquatic pollution · Biogeochemistry · Gender inequality · Cultural context · Aquaculture farming · Pond management

1.1 Introduction

When biological ecosystems and the interactions within are governed and regulated by external social and economic institutions, they are defined as socio-ecological ecosystems (SESs), thus acknowledging the complex interlinking between human and natural systems (Colding and Barthel 2019). Deltas are significant SES, offering

S. Das (✉) · A. Chanda · T. Ghosh
School of Oceanographic Studies, Jadavpur University, Kolkata, West Bengal 700032, India

S. Das et al. (eds.), *Pond Ecosystems of the Indian Sundarbans*, Water Science and Technology Library 112, https://doi.org/10.1007/978-3-030-86786-7_1

a range of ecosystem services to society. They comprise only 1% of land cover worldwide but support the livelihoods of 500 million people (Ericson et al. 2006). These deltaic ecosystems provide provisional (water for irrigation and domestic purposes, food, fertilizer) and regulatory (nutrient sequestration, wastewater treatment) services. They also offer cultural (traditional knowledge systems, religious importance) and supporting (nutrient and water cycling, primary production, species diversity). Standing aquatic bodies in Indian Sundarban, such as ponds, are unlike other ecosystem service providers, such as the main river channels, mangrove forests, on account of their high economic value, multi-functionality.

The interest in pond ecosystem services (water retention) and conservation have grown over recent years. However, research efforts have been Europe-centric, focusing on the omission of ponds from the European Union Water Framework Directive (Oertli et al. 2005). There is a lack of information on the value of ponds, their water chemistry, conservation, and management practices in Southeast Asia (especially Indian Sundarban). Ponds in Indian Sundarbans are socio-ecological systems embedded in the cultural character of these areas and provide essential ecosystem services. However, they are also severely undervalued and polluted water sources. Elsewhere large concentrations of ponds, termed pondscapes, have been shown to support high biodiversity that contributes more to catchment-wide aquatic biodiversity than lakes, streams, and rivers (Davies et al. 2008). At present, no study supports the existence of pondscapes in Indian Sundarbans on account of their number and land area and their contribution to biodiversity. The present book will discuss some of these topics under one umbrella.

Ponds are ‘hotspots’ of anthropogenic activity. However, these lentic ecosystems have experienced severe environmental degradation. The pollutant loads have increased significantly in these water bodies leading to human and biodiversity health risks (e.g., eutrophication and toxic algal blooms) (Jahan et al. 2010). Moreover, these are essential natural resources to address the UN Sustainable Development Goals (SDGs), which aim to encourage strategies that improve health and education, reduce inequality, and increase economic growth while tackling global environmental change and conserving our ecosystems. This book also addresses few SDGs in terms of the pond as a resource.

The Sundarbans area is a rich ecological unit spread on the Ganges–Brahmaputra–Meghna (GBM) delta. The estuarine segments of the Rivers, Ganga, Brahmaputra, and Meghna between 21°32′N and 21°40′N and 88°05′E and 89°E (in both India and Bangladesh) shelters this unique eco-region (Spalding et al. 1997). The Indian Sundarbans have unique biodiversity, including globally threatened species, for example, the Ganges River dolphin (Platanista gangetica), the northern river terrapin (Batagur baska), the brown-winged kingfisher (Pelargopsis amauroptera), the Irrawaddy dolphin (Orcaella brevirostris), and the Royal Bengal tiger (Panthera tigris)—the only mangrove tiger on Earth (RAMSAR 2019). The mangrove ecosystem, which makes up the Indian Sundarbans, is an interconnected network of rivers, creeks, rivulets, and semi-diurnal tides with direct marine influence on the most seaward parts (Fig. 1.1). Hence, there is a range of hydrological impacts (including freshwater and coastal water) on the mangrove forest, and when combined

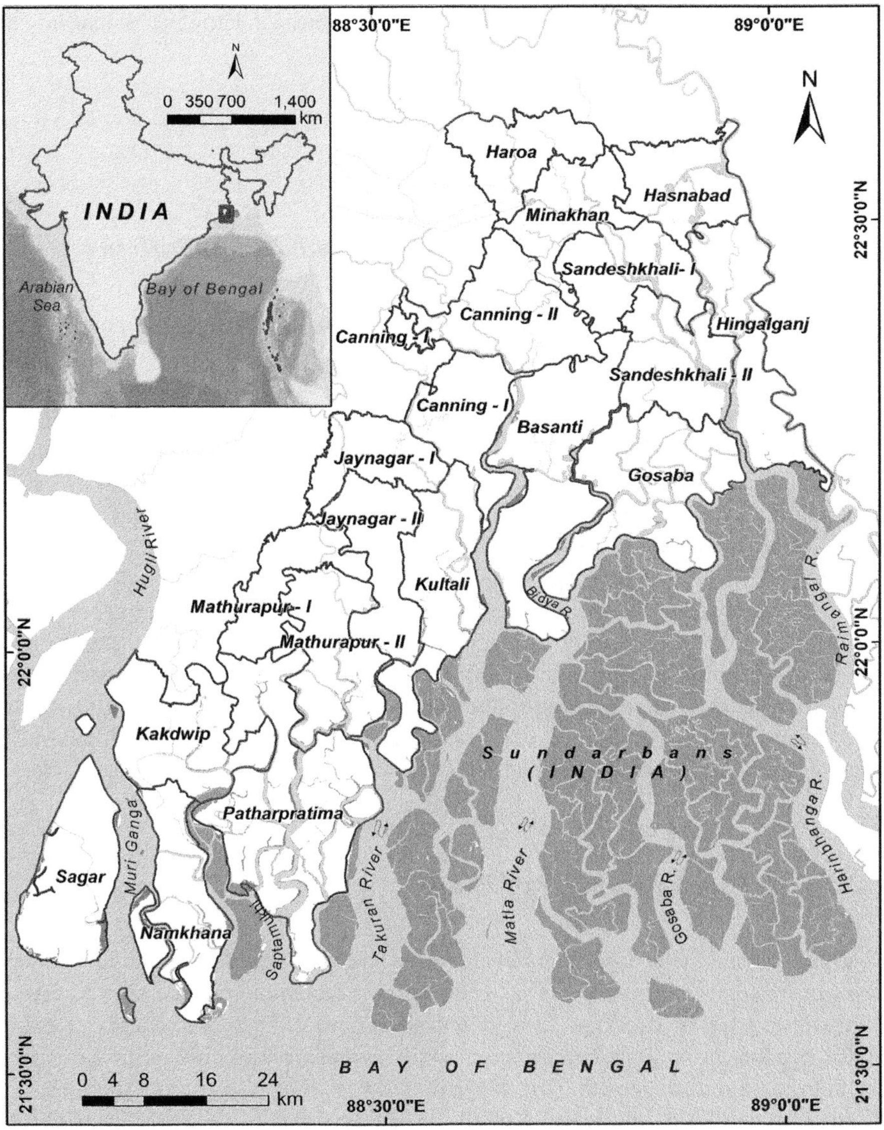

Fig. 1.1 Study area map of Indian Sundarbans consisting of nineteen community development blocks

with its topographic heterogeneity, it results in rich biodiversity (Gopal and Chauhan 2006). Because of this reason, Sundarbans mangrove forest being designated a World Heritage Site by the International Union for Conservation of Nature (IUCN) in 1987; and a wetland of international importance according to the RAMSAR convention in the year 2019; a Biosphere Reserve by United Nations Educational, Scientific and Cultural Organization (UNESCO) in 1989 (Fig. 1.1).

1.2 Ponds as Socio-Ecological Systems

1.2.1 The Nature and Uses of Ponds in Indian Sundarbans

Ponds of Indian Sundarbans have multiple functions, like irrigation, aquaculture, potable water, sanitation, bathing, and water storage. As a result, they are a common feature in the rural landscape (Roy and Nandi 2010) (Figs. 1.2, 1.3, 1.4). The majority are manually-dug with rainwater during the monsoon, groundwater, or tidal exchange (Johnston et al. 2002; Dubey et al. 2017; Kale 2017) as the primary water sources. Traditionally ponds were created for use at a domestic scale (Nhan et al. 2007), with large numbers of households in rural India (Manoj and Padhy 2015) having a pond around their home. In India, the primary function of ponds in coastal villages is for drinking water, washing, and small-scale irrigation activities (Manoj and Padhy 2015). In more recent years, the 'blue revolution' encouraged the enhancement of global food production through aquaculture, ponds and industrial aquaculture now have a high economic value in India (Ahmed 2013). Most farmers excavate the pond every six months to increase the productivity levels (Christensen et al. 2008). In that way, during the blue revolution era, pond digging will be the largest employment sector for non-landowners in Indian Sundarbans in the future (Figs. 1.2 and 1.4).

1.2.2 The Social Context of Ponds

Ponds are crucial in the day-to-day life of the Indian Sundarbans. The natural-cultural heritage of ponds in this region has followed a route from spiritual sites to multi-functioning sites of economic and domestic activities. In more recent centuries, ponds became critical spaces for community domestic and cultivation purposes, highlighting a shift in societal perception of ponds as sites of livelihood options. For instance, in the GBM delta under British colonial rule, ponds were the responsibility of the "zamindars"; local landlords and tax collectors, and who determined which ponds the communities could use for different activities; i.e., aquaculture, bathing, drinking water for cattle (Kränzlin 2000). Further, land conversion to ponds for irrigation and fish stocking led to a rent reduction (McLane 1993). Since India's independence in 1947 and the collapse of the "zamindar" system, many ponds became abandoned.

The growth of "the blue revolution" in recent decades, however, has increased the economic value of ponds again (Pucher et al. 2015). Mud crab aquaculture is quite popular in the Indian and Bangladesh Sundarbans (the world's largest mangrove

Fig. 1.2 Photographs showing some of the types, functions and management of ponds in Indian Sundarbans **a** Homestead Pond (also showing community water supply tap), **b** Agriculture Pond, **c** Crab fattening Pond, **d** Duck farming in a pond, **e** Fish Pond and **f** Increasing pond dimension before monsoon season to store more rain water (*Photos* are taken by Sourav Das)

forest, which contributes to the coastal GBM system). It can yield profits of 22,812.5 US \$ ha^{-1} $year^{-1}$ for culture (where young crabs are grown for several months until they reach a desirable size), and 30,820.8 US \$ ha^{-1} $year^{-1}$ for fattening (where soft-shelled crabs are reared for a few weeks until their exoskeleton is hardened, and typically fetch a much higher profit than "soft" crabs) (Sathiadhas and Najmudeen 2004). For rural and marginalized delta communities, however, ponds remain multi-functional entities that are central to the community and household life. The demand

Fig. 1.3 Photographs showing some of the types, functions and management of ponds in Indian Sundarbans (**a**) and (**b**) Community canal to store rain water for irrigation and other activities, **c** Fish Pond, **d** Pond beside embankment (indicating by arrow), **e** Sweet water supply (indicating by arrow) into the fish pond during summer season to maintain the water level, **f** Water supply system (indicating by arrow) for sanitation purpose from Homestead Pond (*Photos* are taken by Sourav Das)

for ponds to generate profit continues to occur alongside their need to provide local subsistence of food, water, and sanitation (Fig. 1.3).

In earlier centuries, Indian ponds grew in the vicinity of Hindu, Muslim, and Buddhist sites. In Hindu culture, water means life and pervades rituals and myths surrounding ponds as treasure keepers, sacrificial sites, and links between the underworld, spirits, and human beings (Kränzlin 2000). In Muslim culture, water is an

Fig. 1.4 Photographs showing some of the types, functions and management of ponds in Indian Sundarbans **a** Homestead Pond digging, **b** Liming of Homestead Pond, **c** Dried up Pond during summer season and **d** Community canal water (*Photos* are taken by Sourav Das)

element of purification, with the Muslim period in the Bengal delta heralding the digging of ponds to allow water for the whole community free of cost (Kränzlin 2000). Today, ponds remain at sacred sites for these religions.

Water remains an important symbol of life and rebirth in many delta cultures. However, the mightier river channels often overshadow the ponds. For instance, constitutional amendments in India and Bangladesh have granted rivers rights to life, meaning harm or damage to a river is akin to harming a living entity. Further, rivers predominate religious ceremonies. In the Ganges river, pilgrims take part in the annual Hindu holy dip for 'Gangasagar Mela' (Sinha et al. 2020). Ponds do not possess the same reverence as the delta rivers despite being central actors in daily delta life.

1.2.3 Pond Owner and User Identity

There exist hierarchal social structures between pond owners and users. In rural GBM districts, private pond ownership used to be a custom, with pond owners often more financially stable than their non-pond-owning counterparts (Belton et al. 2012). In

Bangladesh, pond size reflects social status. Wealthier households own larger ponds compared to the poorer ones who own smaller ponds. The small ponds are usually less productive and undergo intensive aquaculture (Belton et al. 2012). However, nearly a third of medium and small pond owners live below the poverty line, with more at risk of slipping below the line due to unexpected events (e.g., cyclones and ill-health) that threaten production and results in the loss of income (Belton and Azad 2012; Belton et al. 2012).

1.2.4 Ponds (Aquaculture), Migration and Loss of Livelihoods

The expansion of export-orientated aquaculture in tropical deltas has resulted in land-use changes from virgin mangrove forests to agricultural land to ponds. Indeed, government incentives and policies aimed to increase the conversion of common property such as mangrove forests to aquaculture ponds with minimal rent to alleviate rural poverty (Ahmed 1999). The displacement of traditional agricultural landscapes in the rake of growth in aquaculture has led to a surplus of agricultural workers. These laborers had to seek employment in not only other sectors but often had to migrate away from their homes (Haque and Saifuzzaman 2003). The survey of this book also revealed such types of cases in the Indian Sundarbans delta.

Aquaculture expansion has also resulted in further marginalization of communities reliant on subsistence and harvest of forestry-based products (Luttrell 2006). In the Indian Sundarbans, rural individuals who do not own ponds are typically involved in forest-based crab and prawn seed collection to supply aquaculture and have much lower incomes than those employed by the aquaculture industry directly as their income responds to local market economics compared to international markets (Chand et al. 2012). Furthermore, demand for aquaculture-wild-seed has promoted over-fishing and led to illegal fishing activity. The permits that aim to limit fishing effort restrict the collection of wild stocks to certain forest areas on specific dates. Fishers will then continue to fish in restricted forest areas, with the economic profits from selling wild stock to aquaculture overriding occasional financial penalties from illegal forest collection (Ghosh 2015).

1.2.5 Ponds and the Gender Inequality

Gender inequality is widespread in many tropical delta countries, with aquaculture highlighting this. If women get the opportunity to seek employment in this industry, they will typically earn less than their male counterparts. Women often carry out insecure, low-paid, labor-intensive works related to aquaculture, such as harvesting and packing (Gammage et al. 2006). Rural women often single-handedly run the households, which makes them vital pond users. Given the different needs of female physiology and reproductive health, this heavy pond use makes them at increased risk of health implications from poor pond water quality (Upadhyay 2005; Benneyworth et al. 2016). For instance, consumption of saline pond water, more prevalent in the dry season, has been documented to increase the risk of high blood pressure and preeclampsia in pregnant women in the Sundarbans area (Khan et al. 2011). Ponds can be an unsafe place for female users. For example, females are at greater risk of sexual harassment and physical or verbal abuse when bathing in communal ponds (Joshi et al. 2011). Menstruation remains taboo in several delta communities, which compels many women to avoid bathing when on their period (Joshi et al. 2011).

1.3 Environmental Challenges

Human activities have increased in deltaic systems like Indian Sundarbans vis-a-vis the global sea-level rise and climatic variability. Such anthropogenic and natural hazards are placing the aquatic ecosystem of Indian Sundarbans under increasing stress. Several ponds have outlets to neighboring river creeks or channels (Kränzlin 2000) due to a lack of firm river banks or embankments (Dubey et al. 2017), which allows water to be exchanged, particularly during monsoon season (heavy rainfall) and flood situations. This water exchange contaminates the surrounding water bodies (Tho et al. 2014). Dubey et al. (2016) highlighted several environmental concerns regarding the freshwater ponds of Indian Sundarbans. Saltwater inundation to the pond, disease or epidemics, storm surges, cyclones due to climate change, and uncertain rainfall lead to significant deterioration of these ponds.

1.3.1 Saltwater Inundation

Accelerated sea-level rise, increased frequency and intensity of storm surges, and the upstream withdrawal of freshwater are concerns for coastal wetlands worldwide. The upstream withdrawal patterns are often exacerbating during the pre-monsoon season due to low river discharge allowing saltwater intrusion further inland. Moreover, Chand et al. (2012) described that the saline water inundation for flooding and sea-level rise destroys the inland freshwater ecosystems. Dubey et al. (2017) revealed that more than 18% of fish farmers confirmed that these mechanisms altered more and more freshwater pond to brackish water pond day by day. Hence, increasing pond salinity may result in ponds being unusable for irrigation, bathing, and freshwater fish cultivation in the future.

1.3.2 Exchange of Dissolved Organic-Rich Waters

Indan Sundarbans receive excess organic load and nutrients from varied anthropogenic sources, i.e., household and municipal waste and surface runoffs. Due to the organic load and high temperatures, harmful algal blooms (HABs) are common in the delta pond ecosystem. Toxin producing HAB causing fish kills decreases biodiversity and increases human health risks such as headaches and skin irritations (Jahan et al. 2010). Indian delta communities witnessed several cholera outbreaks due to pond water. Mukherjee et al. (2011) reported that increasing nutrient levels and salinity of ponds enhance the longevity of V. cholerae (cholera causing bacteria) and may lead to increased future outbreaks. Pond user activities that increase the risk of such spread include mouth washing and cooking with pond water, bathing, washing utensils in ponds. Ponds used for industrial aquaculture often obtain fertilizers that are rice by-products, human and livestock waste, and crustaceans from rice grounds (Nhan et al. 2007). However, industrial and domestic discharges are more significant than ponds for nutrient loading to surrounding waters (Tho et al. 2014). However, the blend of nutrient loading and salinization of ponds has caused a drop in species diversity of phytoplankton and zooplankton (Tho et al. 2014).

1.3.3 Faecal Coliform Pollution

Open place excretion, poor sanitation, and the absence of wastewater treatment within delta systems have resulted in the substantial contamination of surface water with fecal coliform bacteria and pathogens. In India, open ponds have the maximum counts of animal and human fecal indicators compared to other water sources (Schweirer et al. 2015). Therefore, ponds act as transmitters of waterborne disease and diarrhea (Islam et al. 2000). Diarrhea is a leading cause of child mortality in India. The weakened development across the Indian Sundarbans delta points to ponds is a significant human health concern.

1.3.4 Uncertain or Irregular Rainfall Pattern

Due to the changing monsoon rainfall dynamics in tropical deltas, drought has become a typical occurrence (Kale 2017). Indian Sundarbans exclusively rely on monsoon rainfall to sustain water levels of ponds as well as groundwater. Due to low precipitation in deltas, groundwater is abstracted for irrigation and to fill up ponds (Kale 2017). Abstraction of groundwater and resulting in the lower label of the water table allows saltwater inundation.

1.3.5 The Input of Anthropogenic Pollutants

Pollution of ponds can also occur from anthropogenic pollutants such as heavy metals, pesticides, antibiotics, and insecticides. Domestic and Industrial effluents, dust, stormwater runoff, and fishing boat activities have introduced substantial amounts of these pollutants to delta regions. In India, the accretion of remnant feeds is a concern for water quality degradation, whereas 81% of farmers use such type of feed to pond in Indian Sundarbans (Dubey et al. 2017). Increasing concentrations of pollutants in aquatic ecosystems have led to environmental and ecological degradation of flora and fauna, including humans, through the bioaccumulation of toxins through the food chains. In aquaculture ponds of the Sundarbans, shrimp accumulate zinc (Zn) at higher concentrations than fish (Ghosal et al. 1997). Consuming shrimp from aquaculture ponds may therefore exceed the Provisional Maximum Tolerable Daily Intake (PMTDI) of toxic metals (Kaviraj and Guhathakurta 2004), leading to several health concerns such as diarrhea, vomiting, skin irritation, and in severe cases, cancer (Dayan and Paine 2001).

1.3.6 Extreme Climate Events

Extreme climatic events like cyclones, storm surges, and droughts alter the pond water quantity and quality. For instance, Cyclone Sidr, which hit coastal Bangladesh in 2007, affected almost 6,000 ponds by saline inundation from surging brackish tidal waters (Rabbani et al. 2010). This disaster impacted residents depending on water aid or walking long distances for safe drinking water up to 6 months following the event (Rabbani et al. 2013). Heavy rainfall events also result in ponds receiving fecally contaminated runoff and having even higher concentrations of fecal microbial contaminants compared to river waters, not only following these events but following typical monsoon periods (Islam et al. 2017). Such Cyclones (e.g., Aila 2009; Amphun 2020; Yaas 2021) also hit Indian Sundarbans and surroundings. These disasters left a catastrophic signature in the lives of rural people of Indian Sundarbans.

1.4 Pond Management Practices

1.4.1 Local Level Management

In India, the most common pond water management strategies are (1) liming (87% of farmers); (2) dewatering before restocking (53%); (3) addition of livestock waste (29%); (4) dredging (21%); (5) removal of aquatic plants (18%) (Dubey et al. 2017). Liming is a commonly used management strategy for domestic and industrial ponds to improve water quality via the following mechanisms: stabilizing pH, increasing available phosphorus, and accelerating the decomposition of organic matter (Chand et al. 2012). Dubey et al. (2017) argued that many farmers in India do not know about chemical treatments or fertilizers. Incorrect or excessive use can adversely affect water quality, primary producers, and fish. The incorrect use of management techniques may be due to some farmers being unfamiliar with the need to monitor conditions; for instance, only 61% monitor salinity and 42% monitor pH (Dubey et al. 2016). Some methods usually follow a disease outbreak. Thus, they overlook the importance of prevention. For example, 37% of farmers practiced liming, and 4% exchange water after a disease outbreak. Some farmers use potassium permanganate, copper sulfate, alum, and methylene blue, despite little evidence that the application of these chemicals aids the recovery from disease (Dubey et al. 2016). Few farmers seem to employ management techniques to prevent disease outbreaks.

Pond management to protect from extreme climate events is primarily centered on improving pond embankments, with 37% improving and heightening embankments after storms, with the principal aim to prevent livestock escaping (Dubey et al. 2017). The improvement of dikes also restricts the exchange of polluted waters to and from the surrounding environment. In India, 17% of farmers strengthened embankments with fruit trees, which will improve the wider biodiversity of the area. The proximity of trees for nesting and food sources significantly increases bird abundance and biodiversity at several managed agricultural pond sites (e.g., Lewis-Phillips et al. 2019). However, since many pond owners are not financially stable (Belton et al. 2012), 17% of farmers identified a lack of financial capability to prepare and adapt to any climatic events (Dubey et al. 2017). There is little assistance from governments for the management and maintenance of ponds at the local level.

1.4.2 National Level Pond Management Policies

Hill et al. (2018) reported that some of the countries having the pond policy exist. But there is still little information for the role and conservation status of delta ponds. Despite the environmental challenges for biodiversity and the ecosystem services of ponds, the Ramsar convention covers their protection at a basic level, but there is little national-level policy for the management and conservation of ponds. In India, there is almost no legislation regarding pond conservation. The National Water Policy 2012 encourage: (1) increasing water storage capacity including the use of ponds; (2) the use of local-level irrigation including field ponds; (3) safeguarding ponds from environmental pollutants and water diversion and (4) the restoration and maintenance of ponds. However, no routine monitoring of ponds or incentives to protect ponds and achieve the above exists. A 2010–2011 report of the Central Pollution Control Board (CPCB) showed that there are only 60 monitoring stations in India, which survey water quality in ponds. West Bengal or the Indian Sundarbans has no such stations (Manoj and Padhy 2015). The high manganese (Mn) concentrations in the river waters of the Indian Sundarbans is inappropriate for irrigation purposes. Therefore, promoting ponds for irrigation is only viable if there is no water exchange with river channels (Battachariee et al. 2015), which is not always the case, particularly following flooding events. Conserving and protecting the biodiversity of ponds of the Indian Sundarbans is missing from any national-level policy.

1.5 Ponds and the UN Sustainable Development Goals (SDGs)

The United Nations Sustainable Development Goals (SDGs) (Table 1.1) are universal goals to build a sustainable and resilient global ecosystem and human community. With the correct management, the sustainable use of ponds can address several of the UN SDGs. The pond ecosystem of Indian Sundarbans is crucial for water (SDG 6) and food security (SDG 2) because, in this context, pond use and management typically focus on provisional services.

When considering SDG 6, Indian Sundarbans ponds are often the only supply of "sweet" or freshwater for communities surrounded by brackish estuarine channels and saline or arsenic-polluted groundwater, and thus, are vital systems to help satisfy SDG 6. However, given their current water quality and management, evidence suggests that the Indian Sundarbans delta presently could not achieve SDG 6. The Sundarbans delta currently does not provide safe and accessible sanitation. Fecal and pathogenic bacteria have been recorded in high concentrations in GBM delta ponds (Sarkar et al. 2009; Islam et al. 2011, 2013), thus, pointing to the widespread problem of open defecation and poor handwashing practice in India. Further, the World Bank estimates that India spends 6.4% of its GDP due to the adverse economic impacts and costs of inadequate sanitation from poor public health (UN SDG 2019). This observation indicates that these funds tackle but do not prevent the problem. Recent sanitation programs often overlook differently gendered, ethical, caste, technical, and financial needs and discrepancies, which results in limited success (Joshi et al. 2011). Altering traditional human behaviors will require multiple level governance and extensive financial input if Indian Sundarbans pond is to provide clean water and satisfy SDG 6.

When considering SDG 2 (food security), Sundarbans deltas themselves have been the "food baskets" for centuries because of the dynamic interplay of fresh and saline water and delivery of fertile sediment during flood inundation, resulting in highly productive aquatic and terrestrial environments (Renaud et al. 2013). The easy access to wild seed and suitable biophysical conditions in these systems promoted extensive aquaculture development (Ahmed 2013). Recent observations suggest that India encompasses 94% of the global aquaculture pond area (FAO 2014). Moreover, smaller-scale operations are also crucial to local and regional food supply chains (Allison 2011). Thus, ponds act as suppliers of nutrition in deltas. It improves the welfare and economic capacity of low-income, resource- and asset-poor households and thus, satisfy the demands of SDG 2 (Ahmed and Lorica 2002; Kale 2017). Indeed there is an argument that the aquaculture industry can fulfill each of the 12 Sustainable

Table 1.1 List of united nations sustainable development goals (SDGs) and descriptions (Source: Ranängen et al. 2018)

Goal No	Goal	Goal description
1	No poverty	End poverty in all its forms everywhere
2	Zero hunger	End hunger, achieve food security and improved nutrition and promote sustainable agriculture
3	Good health and well-being	Ensure healthy lives and promote well-being for all at all ages
4	Quality education	Ensure inclusive and equitable quality education and promote lifelong learning opportunities for all
5	Gender equality	Achieve gender equality and empower all women and girls
6	Clean water and sanitation	Ensure availability and sustainable management of water and sanitation for all
7	Affordable and clean energy	Ensure access to affordable, reliable, sustainable and modern energy for all
8	Decent work and economic growth	Promote sustained, inclusive and sustainable economic growth, full and productive employment and decent work for all
9	Industry, innovation and infrastructure	Build resilient infrastructure, promote inclusive and sustainable industrialization and foster innovation
10	Reduced inequalities	Reduce inequality within and among countries
11	Sustainable cities and communities	Make cities and human settlements inclusive, safe, resilient and sustainable
12	Responsible consumption and production	Ensure sustainable consumption and production patterns
13	Climate action	Take urgent action to combat climate change and its impacts
14	Life below water	Conserve and sustainably use the oceans, seas and marine resources for sustainable development
15	Life on land	Protect, restore and promote sustainable use of terrestrial ecosystems, sustainably manage forests, combat desertification, and halt and reverse land degradation and halt biodiversity loss
16	Peace, justice and strong institutions	Promote peaceful and inclusive societies for sustainable development, provide access to justice for all and build effective, accountable and inclusive institutions at all levels

(continued)

Table 1.1 (continued)

Goal No	Goal	Goal description
17	Partnerships for the goals	Strengthen the means of implementation and revitalize the global partnership for sustainable development

Development Goals, but more so the core SDGs, which focus on improvements to nutrition, poverty alleviation, and job provision (SDGs 1, 2, and 8) (Hambrey 2017). However, economic growth (SDG 8) has occurred to the benefit of only a few, further reducing social mobility (counter to SDGs 5 and 10) and promoted intensive practices which have led to environmental degradation (counteracting SDGs 12, 13, 14, and 15) (Hambrey 2017). The present literature on the Indian Sundarbans ponds does not successfully discuss addressing the targets set out by the SDGs. Thus, this book will try to address the SDGs with different chapters.

Ponds of Indian Sundarbans will continue to face significant social and environmental challenges in the future. Delta human populations continue to expand and predicted increases in temperatures, flooding, intensity, and frequency of extreme events, threatening the already degraded condition and ability of delta ponds to provide food and water security for future delta civilians. The Covid-19 pandemic will alter how ponds are used and managed in the future. The pandemic has resulted in trade restrictions of foods, including aquaculture products, resulting in substantial economic losses and a likely continued long-term decline in the market, threatening the livelihoods of those employed in the industry (FAO 2020). For example, Bangladesh Sundarbans crab and eel farmers have already suffered financial losses following the imposed import ban to China in late January 2020 (Roy 2020). For crab farmers, China accounts for 85% of crabs export and incurred losses of USD 46.90 million by early April 2020 (Roy 2020). There is no record of the loss of Indian Sundarbans farmers. However, such bans could lead to continued deterioration of pond water quality, significant pond abandonment, conversion from intensive to more extensive local market production, land-use change to crop production and return to intensive production where financial capacity allows (in Indian Sundarban).

Additionally, disruption to food supply chains and restrictions to food market access during the pandemic (FAO 2020) may have amplified the reliance and significance of local homestead ponds in food provision. Such incidents can intensify motivation from the household, community, and broader political actors to improve pond water quality. On the other hand, loss of income from inhibited work for those already vulnerable to poverty, such as rural to urban migrants, as found in India (Ganguly 2020), will reduce financial capacity to improve pond water quality. In addition, ponds of the GBM regions impacted by Cyclone Amphan (May 2020) and Cyclone Yaas (May 2021) will be at higher risk of collapse following the social and environmental disturbance of both compounding events.

1.6 Scope of the Book

The biogeochemical properties of the pond ecosystems of Indian Sundarbans are least understood. Various pieces of research focused on the land use land cover, estuarine biogeochemistry, and mangrove cover of the Indian Sundarbans. This region is one of the most dynamic systems in the world. A significant amount of river discharge from the Hooghly River vis-à-vis the effect of monsoon and frequent depression and tropical cyclones makes the Indian Sundarbans unique from various perspectives. These extreme atmospheric events and anthropogenic input exert a considerable impact on the pond ecosystems of Indian Sundarbans. The purpose of the present work (book) is to understand the water quality parameters and their biogeochemical interaction within the pond ecosystem and different ecosystem services of lentic water bodies of the Indian Sundarbans. Therefore, it is essential to recognize the chief anthropogenic activities impacting the pond ecosystems of Indian Sundarbans, including the physicochemical variability, carbon fluxes, dissolved organic matter loading, and heavy metal pollution. These cause-effect relationships will help to develop operational management strategies to mitigate their impacts on the pond ecosystems. This book speaks about few serious global change problems on the pond ecosystem and their interactions with water quality and SDGs. The editors felt the necessity of a thorough description of the pond ecosystem of Indian Sundarbans and to identify the real problems and find holistic solutions. This book can serve as a breakthrough in that track (Table 1.2).

Table 1.2 Chapter plan of the present book

Chapter No	Outline of the chapters
Chapter 2	This chapter detailed an overall scenario of the spatial distribution of the lentic water bodies of the Indian Sundarbans (in four blocks) and their ratio with the other prominent land-use classes
Chapter 3	This chapter comprehended the seasonal variation of several pond water quality parameters and assessed the surface water quality in the Indian part of Sundarbans. The study spanned over thirty-nine inhabited Mouzas from six ocean confronting Blocks (Sagar, Namkhana, Patharpratima, Kultali, Basanti, and Gosaba) of South 24 Parganas District (southwestern part of Sundarbans)
Chapter 4	Overall, this chapter gives a brief overview of the inland fish farming practice and the attitude of the local inhabitants towards this promising sector of earning revenues of Indian Sundarbans
Chapter 5	This chapter indicated that iron fertilization increases phytoplankton abundance and fish production in brackishwater ponds. In that way, controlled use of iron fertilizer may increase the income of fish farmers in Indian Sundarbans
Chapter 6	This chapter discussed the CO_2 dynamics emphasizing air–water CO_2 exchange from typical households and abandoned ponds of the Indian Sundarbans. The chapter also highlighted the role of these ponds in combatting climate change
Chapter 7	This chapter discussed the CH_4 dynamics emphasizing air–water CH_4 exchange from typical households and abandoned ponds of the Indian Sundarbans. The chapter also highlighted the role of these ponds in combatting climate change
Chapter 8	This chapter studied the air–water CO_2 and CH_4 fluxes from two typical aquaculture pond types in the Indian Sundarbans
Chapter 9	The present chapter describes the different types of pond ecosystem services of Indian Sundarbans
Chapter 10	This chapter discussed the socio-cultural and economic role of ponds in deltaic communities through a gender lens
Chapter 11	This chapter aims to develop a preliminary idea on certain selected heavy metal accumulation levels in the brackish water ponds of the different parts of Indian Sundarban
Chapter 12	The present chapter describes the optical properties of chromophoric dissolved organic matter (CDOM) in the pond ecosystems of Indian Sundarbans in different seasons and the relationship between CDOM and other physicochemical parameters of pond water
Chapter 13	Overall, this chapter describes the variations in nutrient dynamics vis-a-vis the partial pressure of CO_2 in water [pCO_2 (water)] from shallow aquaculture ponds of Indian Sundarbans—using a microcosm study

References

Ahmed N (2013) Linking prawn and shrimp farming towards a green economy in Bangladesh: Confronting climate change. Ocean Coast Manag 75:33–42

Ahmed M, Lorica MH (2002) Improving developing country food security through aquaculture development—Lessons from Asia. Food Policy 27(2):125–141

Ahmed M (1999) Policy issues deriving from the scope, determinants of growth, and changing structure of supply of fish and fishery products in developing countries. ICLARM conference proceedings (Philippines).

Allison MA, Khan SR, Goodbred SL, Kuehl SA (2003) Stratigraphic evolution of the late Holocene Ganges-Brahmaputra lower delta plain. Sed Geol 155:317–342. https://doi.org/10.1016/S0037-0738(02)00185-9

Belton B, Azad A (2012) The characteristics and status of pond aquaculture in Bangladesh. Aquaculture 358:196–204. https://doi.org/10.1016/j.aquaculture.2012.07.002

Belton B, Haque MM, Little DC (2012) Does size matter? Reassessing the relationship between aquaculture and poverty in Bangladesh. J Develop Stud 48(7):904–922

Benneyworth L, Gilligan J, Ayers JC, Goodbred S, George G, Carrico A, Karim MR, Akter F, Fry D, Donato K, Piya B (2016) Drinking water insecurity: Water quality and access in coastal south-western Bangladesh. Int J Environ Health Res 26(5–6):508–524

Bhattacharya BD, Nayak DC, Sarkar SK, Biswas SN, Rakshit D, Ahmed MdK (2015) Distribution of dissolved trace metals in coastal regions of Indian Sundarban mangrove wetland: a multivariate approach. J Cleaner Prod 96:233–243. https://doi.org/10.1016/j.jclepro.2014.04.030

Chand B, Trivedi R, Biswas A, Dubey S, Beg M (2012) Study on impact of saline water inundation on freshwater aquaculture in Sundarban using risk analysis tools. Explor Anim Med Res 2:170–178

Christensen SM, Tarp P, Hjortsø CN (2008) Mangrove forest management planning in coastal buffer and conservation zones, Vietnam: a multimethodological approach incorporating multiple stakeholders. Ocean Coast Manag 51:712–726. https://doi.org/10.1016/j.ocecoaman.2008.06.014

Colding J, Barthel S (2019) Exploring the social-ecological systems discourse 20 years later. Ecol Soc 24(1)

Davies B, Biggs J, Williams P, Whitfield M, Nicolet P, Sear D, Bray S, Maund S (2008) Comparative biodiversity of aquatic habitats in the European agricultural landscape. Agr Ecosyst Environ 125:1–8. https://doi.org/10.1016/j.agee.2007.10.006

Dayan AD, Paine AJ (2001) Mechanisms of chromium toxicity, carcinogenicity and allergenicity: review of the literature from 1985 to 2000. Hum Exp Toxicol 20:439–451. https://doi.org/10.1191/096032701682693062

Dubey SK, Chand BK, Trivedi RK, Mandal B, Rout SK (2016) Evaluation on the prevailing aquaculture practices in the Indian Sundarban delta: an insight analysis. J Food Agric Environ 14(2):133–141

Dubey SK, Trivedi RK, Chand BK, Mandal B, Rout SK (2017) Farmers' perceptions of climate change, impacts on freshwater aquaculture and adaptation strategies in climatic change hotspots: a case of the Indian Sundarban delta. Environ Develop 21:38–51. https://doi.org/10.1016/j.envdev.2016.12.002

Ericson J, Vorosmarty C, Dingman S, Ward L, Meybeck M (2006) Effective sea-level rise and deltas: causes of change and human dimension implications. Glob Planet Change 50:63–82. https://doi.org/10.1016/j.gloplacha.2005.07.004

FAO (2014) The state of world fisheries and aquaculture 2014. Rome

FAO (2020) Q&A: COVID-19 pandemic—impact on fisheries and aquaculture. http://www.fao.org/2019-ncov/q-and-a/impact-on-fisheries-and-aquaculture/en/

Gammage S, Swanburg K, Khandkar M, Islam MZ, Zobair M, Muzareba AM (2006) A gendered analysis of the shrimp sector in Bangladesh. Greater Access to Trade and Expansion. USAID, Bangladesh

Ganguly S (2020) India's coronavirus pandemic shines a light on the curse of caste. The Conversation. https://theconversation.com/indias-coronavirus-pandemic-shines-a-light-on-the-curse-of-caste-139550

Ghosal T, Kaviraj A, Das B (1997) Preliminary observation of heavy metal speciation in some brackish water ponds of Kakdwip and Tona of Sunderban, India. Proc Zool Soc Calcutta, India 50:146–152

Ghosh P (2015) Conservation and conflicts in the Sundarban biosphere reserve, India. Geogr Rev 105(4):429–440

Gopal B, Chauhan M (2006) Biodiversity and its conservation in the Sundarban mangrove ecosystem. Aquat Sci 68:338–354

Guhathakurta H, Kaviraj A (2004) Effects of salinity and mangrove detritus on desorption of metals from brackish water pond sediment and bioaccumulation in fish and shrimp. Acta Hydrochim Hydrobiol 32(6):411–418

Hambrey J (2017) The 2030 Agenda and the sustainable development goals: the challenge for aquaculture development and management. FAO Fisheries and Aquaculture Circular, C1141

Haque MZ, Saifuzzaman M (2003) Social and environmental effects of shrimp cultivation in Bangladesh: notes on study methods. Globalization, Environmental Crisis and Social Change in Bangladesh, Dhaka, Bangladesh, UPL

Hill MJ, Hassall C, Oertli B, Fahrig L, Robson BJ, Biggs J, Samways MJ, Usio N, Takamura N, Krishnaswarmy J, Wood PJ (2018) New policy directions for global pond conservation. Conserv Lett 11:e12447

Islam MS, Begum A, Khan SI, Sadique MA, Khan MNH, Albert MJ, Yunus M, Huq A, Well RC (2000) Microbiology of pond ecosystems in rural Bangladesh: its public health implications. Int J Environ Stud 58(1):33–46

Islam MA, Sakakibara H, Karim MR, Sekine M, Mahmud ZH (2011) Bacteriological assessment of drinking water supply options in coastal areas of Bangladesh. J Water Health 9(2):415–428

Islam MA, Sakakibara H, Karim MR, Sekine M (2013) Potable water scarcity: Options and issues in the coastal areas of Bangladesh. J Water Health 11(3):532–542

Islam MM, Hofstra N, Islam MA (2017) The impact of environmental variables on faecal indicator bacteria in the Betna river basin, Bangladesh. Environ Process 4(2):319–332

Jahan R, Khan S, Haque MM, Choi JK (2010) Study of harmful algal blooms in a eutrophic pond, Bangladesh. Environ Monit Assess 170:7–21

Johnston D, Lourey M, Van Tien D, Luu TT, Xuan TT (2002) Water quality and plankton densities in mixed shrimp-mangrove forestry farming systems in Vietnam. Aquac Res 33(10):785–798

Joshi D, Fawcett B, Mannan F (2011) Health, hygiene and appropriate sanitation: experiences and perceptions of the urban poor. Environ Urban 23(1):91–111

Kale E (2017) Problematic uses and practices of farm ponds in Maharashtra. Econ Polit Weekly 52(3):21

Khan AE, Ireson A, Kovats S, Mojumder SK, Khusru A, Rahman A, Vineis P (2011) Drinking water salinity and maternal health in coastal Bangladesh: implications of climate change. Environ Health Perspect 119(9):1328–1332

Kränzlin I (2000) Pond management in rural Bangladesh: problems and possibilities in the context of the water supply crisis. Nat Resour Forum 24:211–223

Lewis-Phillips J, Brooks S, Sayer CD, McCrea R, Siriwardena G, Axmacher JC (2019) Pond management enhances the local abundance and species richness of farmland bird communities. Agr Ecosyst Environ 273:130–140. https://doi.org/10.1016/j.agee.2018.12.015

Luttrell C (2006) Adapting to aquaculture in Vietnam: securing livelihoods in a context of change in two coastal communities. Environ Livelihoods Trop Coastal Zones Managing Agric Fishery Aquacult Conflicts 2:17–29

Manoj K, Padhy PK (2015) Environmental perspectives of pond ecosystems: global issues, services and Indian scenarios. Curr World Environ 10:848–867

McLane JR (1993) Land and local kingship in eighteenth century Bengal, South Asian Studies, pp. 53. Cambridge University Press, Cambridge

Mukherjee R, Halder D, Saha S, Shyamali R, Subhranshu C, Ramakrishnan R, Murhekar MV, Hutin YJ (2011) Five pond-centred outbreaks of Cholera in Villages of West Bengal, India: evidence for focused interventions. J Health Popul Nutr 29:421–428. https://doi.org/10.3329/jhpn.v29i5.8895

Nhan DK, Phong LT, Verdegem MJC, Duong LT, Bosma RH, Little DC (2007) Integrated freshwater aquaculture, crop and livestock production in the Mekong delta, Vietnam: determinants and the role of the pond. Agric Syst 94:445–458. https://doi.org/10.1016/j.agsy.2006.11.017

Oertli B, Biggs J, Céréghino R, Grillas P, Joly P, Lachavanne JB (2005) Conservation and monitoring of pond biodiversity: introduction. Aquat Conserv Marine Freshwater Ecosyst 15:535–540. https://doi.org/10.1002/aqc.752

Pucher J, Mayrhofer R, El-Matbouli M, Focken U (2015) Pond management strategies for small-scale aquaculture in northern Vietnam: fish production and economic performance. Aquacult Int 23:297–314. https://doi.org/10.1007/s10499-014-9816-0

Rabbani G, Rahman SH, Faulkner L (2013) Impacts of climatic hazards on the small wetland ecosystems (ponds): evidence from some selected areas of coastal Bangladesh. Sustainability 5(4):1510–1521

Rabbani M, Rahman A, Islam N (2010) Climate change and sea level rise: Issues and challenges for coastal communities in the Indian Ocean region. Coastal Zones and Climate Change 17–29

Ranängen H, Cöster M, Isaksson R, Garvare R (2018) From global goals and planetary boundaries to public governance—a framework for prioritizing organizational sustainability activities. Sustainability 10(8):2741. https://doi.org/10.3390/su10082741

Renaud FG, Syvitski JP, Sebesvari Z, Werners SE, Kremer H, Kuenzer C, Ramesh R, Jeuken A, Friedrich J (2013) Tipping from the Holocene to the Anthropocene: how threatened are major world deltas? Curr Opin Environ Sustain 5(6):644–654

Roy M, Nandi NC (2010) Effect of insolation and overhead canopy on macrozoobenthos of tropical freshwater ponds. Russ J Ecol 41(5):428–435. https://doi.org/10.1134/S1067413610050115

Roy P (2020) Coronavirus destroys Bangladesh's crab exports. Eco-Business. https://www.eco-business.com/news/coronavirus-destroys-bangladeshs-crab-exports/

Sathiadhas R, Najmudeen T (2004) Economic evaluation of mud crab farming under different production systems in India. Aquacult Econ Manage 8(1–2):99–110

Schriewer A, Odagiri M, Wuertz S, Misra PR, Panigrahi P, Clasen T, Jenkins MW (2015) Human and animal fecal contamination of community water sources, stored drinking water and hands in rural India measured with validated microbial source tracking assays. Am J Trop Med Hygiene 93(3):509–516

Sinha R, Das S, Ghosh T (2020) Pollution and its consequences at Ganga Sagar mass bathing in India. Environ Dev Sustain 22(2):1413–1430

Spalding M, Blasco F, Field C (1997) World mangrove atlas. International Society for Mangrove Ecosystems, Okinawa

Tho N, Merckx R, Ngoc UV (2014) Impacts of saline water irrigation and shrimp pond discharges on the surrounding waters of a coastal district in the Mekong delta of Vietnam. Environ Earth Sci 71:2015–2027

UN SDG (2019) Webpage: United Nations, sustainable development goals, goal 6. Ensure access to water and sanitation for all. Available at: https://www.un.org/sustainabledevelopment/water-and-sanitation/

Upadhyay B (2005) Women and natural resource management: illustrations from India and Nepal (vol 29). Blackwell Publishing, Ltd

Chapter 2
Spatial Distribution of Ponds in the Indian Sundarbans Biosphere Reserve: Special Emphasis on Size-Class

Tuhin Ghosh, Niloy Pramanik, Sourav Das, Abhra Chanda, and Anirban Mukhopadhyay

Abstract The Indian part of Sundarbans hosts a population of more than 4.4 million people and the land-use-land-cover dynamics of this region has witnessed significant changes in the recent past due to multifarious anthropogenic interventions. Though several pieces of research characterized the land-use-land-cover dynamics of this region, exclusive focus on the ponds of this region is yet to receive any such attention. We selected four administrative blocks of the Sundarban Biosphere Reserve (SBR), namely Sagar, Patharpratima, Gosaba, and Hingalganj, and analyzed the present scenario of the ponds along with their size classes and spatial distribution. The present analysis indicated that Sagar, Patharpratima, Gosaba, and Hingalganj comprise around 9000, 20,500, 11,500, and 7700 ponds. On the whole, the number of ponds varied proportionally with the total area of the blocks; however, the total surface area covered by the ponds did not show any significant relationship with the total number of ponds in each of the blocks. Most of the ponds belonged to the size class having an area between 100 m^2 and 200 m^2. Spatially, the majority of the ponds were situated within 500 m from the river boundary, plausibly indicating the abundance of aquaculture ponds near the river boundary. Besides, a gross scenario of the spatial distribution of the lentic water bodies and their ratio with the other prominent land-use classes were analyzed in this chapter. However, differentiation in terms of the type of ponds, like the household ponds, community ponds, and aquaculture ponds should be carried out in the future to develop a more comprehensive understanding of both spatial as well temporal dynamics of these crucial lentic ecosystems.

Keywords Pond · Spatial distribution · Size class · Land use · Land cover · Remote sensing. GIS

T. Ghosh · N. Pramanik · S. Das · A. Chanda (✉)
School of Oceanographic Studies, Jadavpur University, Kolkata, West Bengal 700032, India

A. Mukhopadhyay
Disaster Preparedness, Mitigation, and Management, Asian Institute of Technology, Klong Luang, Pathumthani 12120, Thailand

S. Das et al. (eds.), *Pond Ecosystems of the Indian Sundarbans*, Water Science and Technology Library 112, https://doi.org/10.1007/978-3-030-86786-7_2

2.1 Introduction

The Sundarban Biosphere Reserve (SBR) is perhaps a standalone example in the present date, where a population having very high density coexists with the world's largest single mangrove forest (De and Das 2021; Ghosh and Mistri 2021). This mangrove ecosystem is an abode of a highly biodiverse array of flora and fauna, which includes the iconic Royal Bengal Tiger (*Panthera tigris*), estuarine crocodile (*Crocodylus porosus*), Irrawaddy Dolphin (*Orcaela brevirostris*), and many more (Gopal and Chauhan 2006). The mangrove floral diversity of the Sundarbans is also renowned all over the world (Sreelekshmi et al. 2020). The coastal regions in many of the world's developing and underdeveloped countries are inhabited by largely impoverished poor people (Ambrosino et al. 2020; Debrot et al. 2020; Onyena and Sam 2020), and the Sundarbans is also no exception (Hajra and Ghosh 2018; Das and Hazra 2020; Lázár et al. 2020). A substantial part of the population residing in the SBR exclusively depends on the ecosystem services provided by the Sundarbans mangrove forest, like timber, fuelwood, protection from storms and surges, honey, fishes, crabs, prawns, shrimps, tourism, and what more (Dasgupta et al. 2021; Sannigrahi et al. 2019, 2020). However, the substantially high anthropogenic intervention and encroachment towards the mangrove forests have made this unique ecosystem vulnerable coupled with the threats that arise due to climate change (Sahana et al. 2019, 2021). Under such circumstances, it has become an urgent need of the hour to look for alternative options for earning livelihoods for the millions who reside on the edge of this mangrove forest so that this ecosystem can be saved from utter ruin.

For effective utilization of forest resources, the entire area demarcated as the SBR is segregated into three tiers, namely the core, the buffer, and the transition with varying degrees of restrictions and allowed activities (Ghosh and Ghosh 2019). The core is strictly restricted for human beings to enter (with occasional exceptions for research purposes), whereas some specified livelihood activities like fishing and tourism are allowed in the buffer (Ghosh 2015). The transition zone, which occupies approximately 5300 km^2 out of the 9630 km^2 of the total SBR, shelters a population as high as 4.4 million (according to Census 2011). By now it is quite likely that the population might have exceeded five million people as anticipated from the existing population growth rate. The land use and land cover of the SBR has witnessed substantial change in the last four to five decades, especially in the transition zone. At present, the dominant land use classes are the protected mangrove forests, agricultural plots, aquaculture ponds, peri-urban settlements, and roads (Dasgupta et al. 2019). Aquaculture fish farming has lately emerged as one of the most viable as well as a profitable option in the SBR to earn livelihoods (Marcinko et al. 2021); however, concerns exist that this option can be detrimental if it comes at the cost of destroying the important blue carbon habitats like that of mangrove forests (Hamilton 2013). As far as Sundarbans is concerned, a pond is an integral part of almost every household. It is a very common practice to excavate soils from residential plots to build kuchha houses and shades for cattle and allow rainwater to accumulate in the unearthed scar, which gives rise to ponds. The people of Sundarbans use this pond water for multifarious purposes including bathing, washing clothes and utensils, irrigation purposes during the dry season, cattle rearing, homestead gardening, and many more (Mandal et al. 2015). However, most of the products that come out of such household ponds (mostly fishes) are meant for personal consumption. The concept of commercializing the production of fish, shrimps, prawns, and crabs, though gained impetus in the last few decades, continues to be in its infancy. The land-use-land-cover change dynamics of the Sundarbans have received fair attention from the scientific community in recent times (Datta and Deb 2012; Sahana et al. 2016; Thakur et al. 2021); however, lentic ecosystems like ponds and lakes have never been the focal point of such pieces of research. Given the potential of these static water bodies to serve humankind, it is time we looked into the dynamics of this crucial ecosystem. This is why the present book has been framed and the present chapter has tried to furnish a brief overview of the spatial distribution and the size classes of these small water bodies in some of the selected administrative blocks of the SBR.

2.1.1 Significance of Ponds

In a world of a growing population, the demand for water and water body-derived products is never-ending. To meet this ever-increasing demand, ponds play a crucial role throughout the world (Céréghino et al. 2014). Downing et al. (2006) quantified that there can be almost 277 million ponds of size less than 1 hectare, and 24 million ponds/lakes of size 1 hectare to 10 hectares throughout the world. Pond ecosystems are unanimously recognized to support substantial biodiversity (Sousa et al. 2016; Simaika et al. 2016; Deacon et al. 2018) and provide multifarious ecosystem services to not only human beings but also other life forms (Moore and Hunt 2012; Ghermandi and Fitchman 2015; Stewart et al. 2017; Fu et al. 2018). de Macro et al. (2014) emphasized that there is scarcely any difference between the man-made artificial ponds and the natural ones in terms of their intrinsic ecology, which has increased the importance of these clusters of waterbodies. Moreover, several studies have inferred that the ponds have substantial potential to mitigate climate change and at the same time tackle several water management issues (Quinn et al. 2007; Steidl et al. 2008). Céréghino et al. (2014) mentioned that a 500 m^2 pond can sequester and store as much carbon as emitted by a car throughout an annual cycle. Downing et al. (2008) emphasized that owing to the very high numbers of these lentic bodies, they occupy a substantial portion of the land surface and if properly managed they are capable of sequestering equal amounts of carbon as the oceans do at any time. With the ongoing climate change, the global scientific community is compelled to seek refuge in many of the natural existing ecosystems to look for answers through which the evil of climate change could be effectively combatted. A school of scientists earnestly believes that ponds can play a crucial role in serving this purpose. Despite many exciting research findings, the ponds have received very little attention from policy managers and stakeholders throughout the world. In the present date, ponds continue to be one of the most neglected land-use classes. However, there is a ray of hope as international bodies like that of the European Pond Conservation Network and concepts like the Important Areas for Ponds have emerged. Many such initiatives are due to preserve and protect these priceless pieces of static water bodies, especially in tropical countries, where these can be effectively utilized to serve multifarious crisis issues.

2.1.2 Aquaculture Ponds: An Emerging Dimension

Aquafarming practices go a long way back in the historic timescale. These inland fishing practice has been a common sight in the banks of rivers and estuaries, along with the coastal margins, especially in the tropical Asian countries like Vietnam, China, and India (Duan et al. 2020a). China, at present, stands tall as the largest producer of aqua-farmed fishes (Liu et al. 2020; Yu and Han 2020), and India is believed to hold the second rank (Belton et al. 2017). This type of fishing practice requires a large surface area, water, and substantial quantities of fish feeds and nutrients (Wang et al. 2015). In the last few decades, the coastal landscape has witnessed rampant changes due to the emergence of aquaculture fishing practices (Ottinger et al. 2016). Especially, in the cases of shrimps and crabs, a substantial quantity of the global catch is derived through aquaculture fishing. The growing uncertainties in open marine fishing coupled with the ever-increase demand for fishes, shrimps, and crabs have intensified the growth of aquaculture farms (Clavelle et al. 2019). However, the indiscriminate expansion of aquaculture ponds has some far-reaching consequences leading to the destruction of several crucial coastal habitats like the mangroves and tidal flats (Murray et al. 2019). Significant aquatic pollution has also been recognized in the coastal regions arising from the effluents of these aquaculture ponds, which leads to eutrophication and red tides (Neofitou et al. 2019). Nonetheless, these aquatic bodies have an important land-use class in the present era and have become susceptible to ongoing climate change also (Islam et al. 2019; Jayasinghe et al. 2019). Compared to the significance and the spatial and temporal dynamics of this land-use class, endeavors of mapping these aquatic bodies are scarce.

2.1.3 Mapping of Ponds

Given the significance of the aquaculture ponds, characterization of the spatiotemporal dynamics of these aquatic bodies has gained impetus since the last decade (Duan et al. 2020b). Understanding the spatial dynamics and distribution of ponds where fish farming is practiced is elemental in delineating proper management strategies to ensure a sustainable return from these ponds (FAO 2016). Remotely sensed images have come up as an efficient tool to study the distribution of such lentic water bodies like many other land-use and land-cover classes (Lacaux et al. 2007;

Yao et al. 2016; Ren et al. 2019). Visual interpretation along with supervised and unsupervised classification, and object-oriented classification techniques are the principal protocols followed for the extraction of water bodies like ponds (Tschudi et al. 2008; Virdis 2014; Al Sayah et al. 2020). The land-use and land-cover changes from mangrove patches, agricultural land, and tidal flats to aquaculture ponds have been the focal point of several studies (Daly et al. 2017; Veettil et al. 2019; Xu et al. 2020). However, since the size of the ponds in many of the coastal regions is small, high-resolution images are required to study their distribution and dynamics (Prasad et al. 2019). Since high-resolution images are costly and at the same time not available for retrospective time intervals in the past, characterizing the temporal dynamics in the past three decades remains a challenging endeavor (Duan et al. 2020a, b). Moreover, though ponds play a crucial role in terrestrial ecology, most of the focus has remained exclusively on the aquaculture ponds. This chapter, thus, aimed to contribute to assessing the spatial distribution of all types of ponds in selected administrative blocks within the SBR.

2.1.4 Crucial Findings on SBR from Land-Use-Land-Cover Related Studies

Several studies have been carried out in the last decade trying to characterize the land-use-land-cover dynamics of the Indian SBR with varying objectives. Thakur et al. (2021) took into account eleven land-use classes namely, mangrove forest, mangrove swamp, mudflats, rivers, sand beach, open scrubs, settlements, plantations, aquaculture, water-logged areas, and agricultural plots. They observed that between the years 2000 and 2017, both mangrove forests and swamps have reduced in the area accompanied by open scrubs and plantations. Whereas the other land-use classes, especially, agricultural and aquaculture land have increased substantially. They inferred from their study that the reduction in green cover in the two decades indicates significant stress on this ecosystem as a whole. An exclusive study on the mangrove forest cover indicated that the mangrove cover of this region witnessed a 3.76% reduction between the years 1990 and 2019 (Halder et al. 2021). In a case study in Satjelia Island, Thakur et. al. (2019) also reported the loss of natural vegetation for the use of cooking the school mid-day meals. Hajra et al. (2017) studied the land-use-land-cover dynamics in three islands situated in the western part of the SBR namely Sagar, Mousuni, and Ghoramara, between the years 1990 and 2015. They observed that the brunt of sea-level rise has led to considerable erosion in these islands. The population growth of more than 2% per year has led to a reduction in agricultural lands associated with a concomitant rise in settlement areas, ultimately leading to lesser land availability per capita. They further observed that the nature and degree of the anthropogenic activities played a deciding role in the character of land-use changes in these islands. Mondal et al. (2019) analyzing the change dynamics in the same Sagar Island between the years 1975 and 2015 reflected similar trends. As per their observations, mangroves to croplands, wetlands to aquaculture ponds, and croplands to human settlements have been the fundamental transitions in this sea-facing island. Debnath (2018) has seen a similar increase in settlement area and decrease in agricultural land in one of the northern islands of SBR, namely Gosaba. He observed that despite being situated up north, this island was heavily affected by the cyclone Aila in the year 2009, and since then it has witnessed a tremendous increase in population pressure. Taking the SBR as a whole a drastic reduction of mangrove cover is reported by Ranjan et al. (2017). DasGupta et al. (2019); however, emphasized the loss of tidal flats and increase in riverine areas coupled with a moderate reduction in mangrove area. They mentioned the rise in aquaculture ponds in the SBR and they anticipated that under a strong demand scenario the rate of this rise can increase manifold. Thus, several studies exist which tried to throw light on the various land-use-land-cover classes of the Indian SBR, but exclusive attention on the ponds is yet to be paid.

2.2 Materials and Methods

2.2.1 Study Area Selection

The SBR encompasses two districts of the West Bengal state, namely 24 Parganas North and 24 Parganas South. Thirteen administrative blocks of North 24 Parganas and six administrative blocks of South 24 Parganas comprise the entire SBR. We have selected four administrative blocks within the SBR, namely Hingalganj, Gosaba, Patharpratima, and Sagar (Fig. 2.1). Out of these four blocks, Hingalgang is the northernmost block that receives ample freshwater through the Ichamati, Gosaba lies immediately next down south.

Patharpratima and Sagar are situated in the southern end. Sagar receives substantial freshwater from the main flow of Hooghly; however, due to the proximity of the sea, the salinity in the adjoining waters of these two blocks remains substantially high throughout the year (Nath et al. 2021). Hingalganj is a community development block comprising a population close to 1,80,000 and 44% of the households live below the poverty line making it one of the most impoverished regions of the SBR (Dutta 2019; District Census Handbook 2011). As of 2011, almost 1210 ha of this block was used for aquaculture purposes approximately engaging 7% of the block's population in this occupation (District Statistical Handbook 2010–11a). This block is directly connected to the mainland. Gosaba block lying next to Hingalganj accommodates almost 2,50,000 people and encompasses an area of almost 300 km^2 (District Statistical Handbook 2010–11b). Besides tourism and honey collection, rice farming, and rice-shrimp mixed farming is popular in this block (Ghosh et al. 2019; Ghosh and Mistri 2020). This block is not directly connected to the mainland yet. Patharpratima, directly connected to the mainland is one of the largest and at the same time climate-sensitive blocks of SBR (Sahana et al. 2021). Almost 5,800 ha of

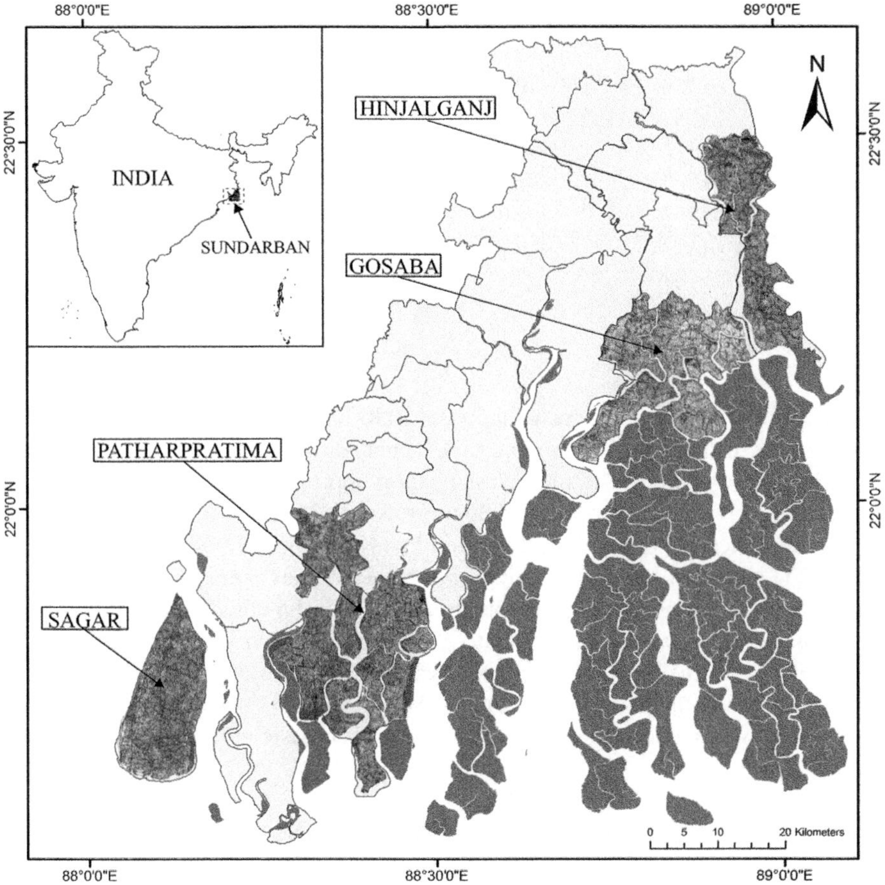

Fig. 2.1 The study area map showing the location of the four selected administrative blocks for the present study in Indian Sundarbans

aquaculture ponds exist in this pond that acts as a means of earning livelihood for more than 41,000 people (District Statistical Handbook 2010–11b). Lastly, Sagar, which is not connected to the mainland directly is one of the most developed blocks of this island (Haldar and Bhattacharya 2020). This island is a popular tourist destination, especially, for the Hindu pilgrims who come to visit the Kapil Muni Ashram in numbers throughout the year and on an auspicious occasion where millions come to take a holy dip in the oceanic waters (Sinha et al. 2020).

Table 2.1 The percentage cover of the most prominent land-use classes in the selected four administrative blocks of the SBR

	Agricultural Land	Mangrove Area	Settlement with Vegetation	Water Body	River
Sagar	46.7	7.6	36.9	7.7	1.1
Patharpratima	49.0	10.5	30.4	9.8	0.2
Gosaba	62.1	4.1	21.8	10.8	1.2
Hingalganj	50.9	5.5	24.4	17.1	2.1

2.2.2 Methodology

The Sentinel-2 satellite images of the year 2020 have been used (Table 2.1). Two scenes cover the total study area. The LULC analysis is based on the object-oriented classification approach. The final classification approach is applied by combining the machine learning algorithms of Support Vector Machine and Random Forest. For the identification of pond indices, Modified Normalized Difference Water Index (MNDWI) has been used. Analyzing open water bodies precisely is a significant and rudimentary application of remote sensing. Several methods for mapping water bodies have been established to identify water bodies from multispectral satellite images. The spectral-based water index, like (MDNWI) (Xu 2006), which is designed from the green and Shortwave-Infrared (SWIR) bands, is one of the most proven methods for water body study (Du et al. 2016). The basic equation of the index is given below:

$$\text{MNDWI} = (\text{Green} - \text{SWIR})/(\text{Green} + \text{SWIR}) \tag{2.1}$$

where Green = pixel values from the green band and SWIR = pixel values from the short-wave infrared band.

2.3 Results and Discussion

2.3.1 Present Scenario of Land Use Land Cover in the Selected Administrative Blocks

Table 2.1 reflects the area covered by the five major land-use classes in this region, namely agricultural land, mangrove cover, settlement with vegetation, water bodies, and rivers. Overall, agricultural lands comprise the highest percentage of land use in the respective blocks followed by settlement with vegetation (Figs. 2.2, 2.3, 2.4, and 2.5). Mangroves, water bodies, and river networks together encompass less than 20% of the respective block's total area. Amongst the four blocks, Hingalganj has the

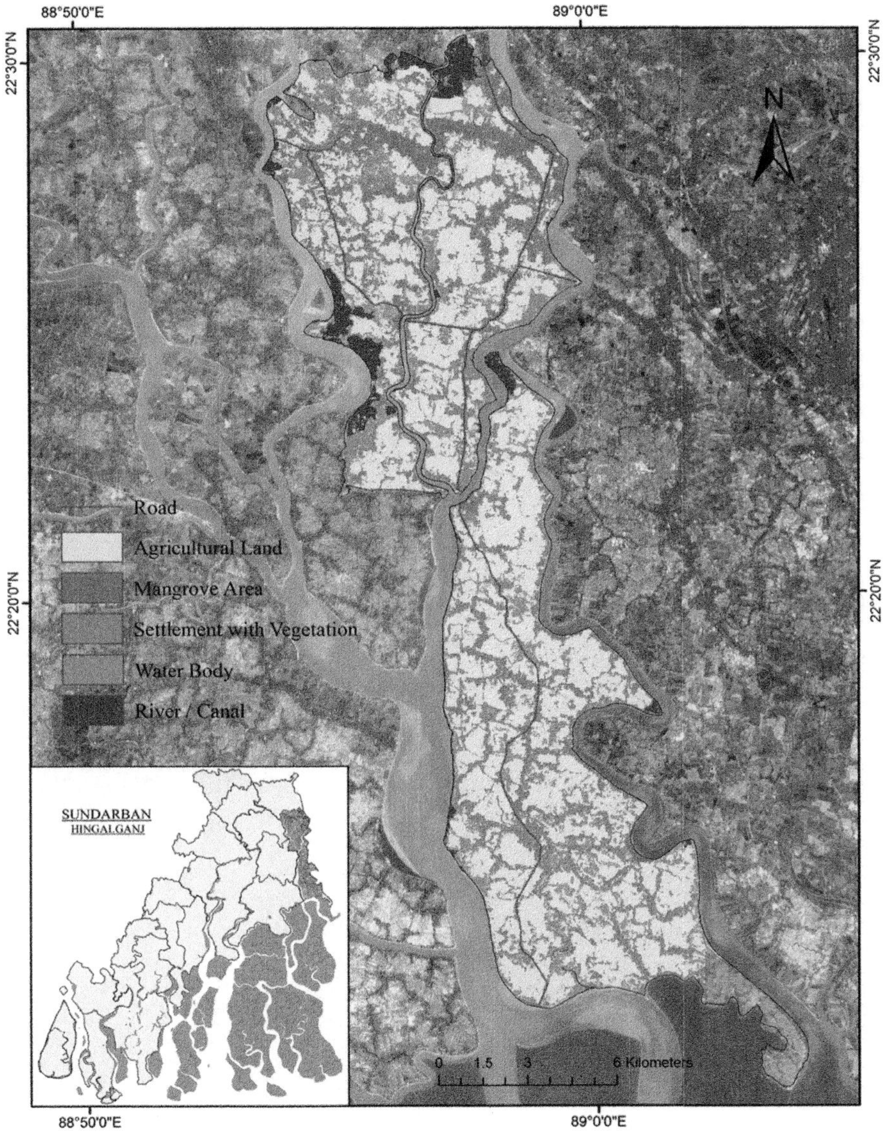

Fig. 2.2 The land-use-land-cover map of the Hingalganj block in the SBR

highest percentage of water bodies (17.1%) followed by Gosaba (10.8%), Patharpratima (9.8%), and Sagar (7.7%). This observation indicates that the more we move north away from the sea, the area covered by inland water bodies has increased. It is worth mentioning that the inland water bodies do not exclusively indicate the ponds, rather a total of ponds, canals, wetlands with stagnant water, etc. We analyzed the ratio between the class 'settlement with vegetation' and 'water bodies'. This ratio

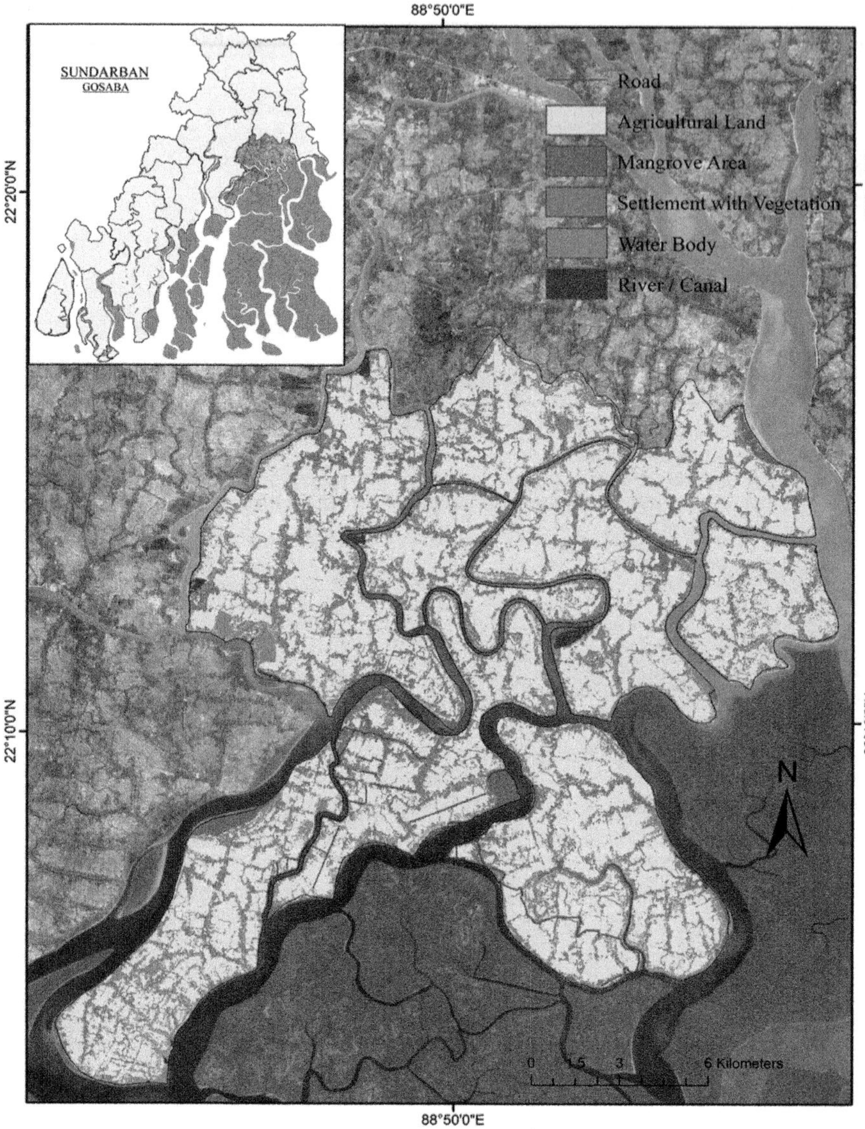

Fig. 2.3 The land-use-land-cover map of the Gosaba block in the SBR

was found highest in Sagar (4.8) followed by Patharpratima (3.1), Gosaba (2.0), and Hingalganj (1.4). This observation indicates that surface area covered by inland water bodies in comparison to the settlement area are fewer in the seaward block like Sagar and gradually increased as we moved north. Similarly, we analyzed the ratio between the class 'agricultural land' and 'water bodies'. This ratio followed a trend: Hingalganj (3.0) < Patharpratima (5.0) < Gosaba (5.7) < Sagar (6.1). This

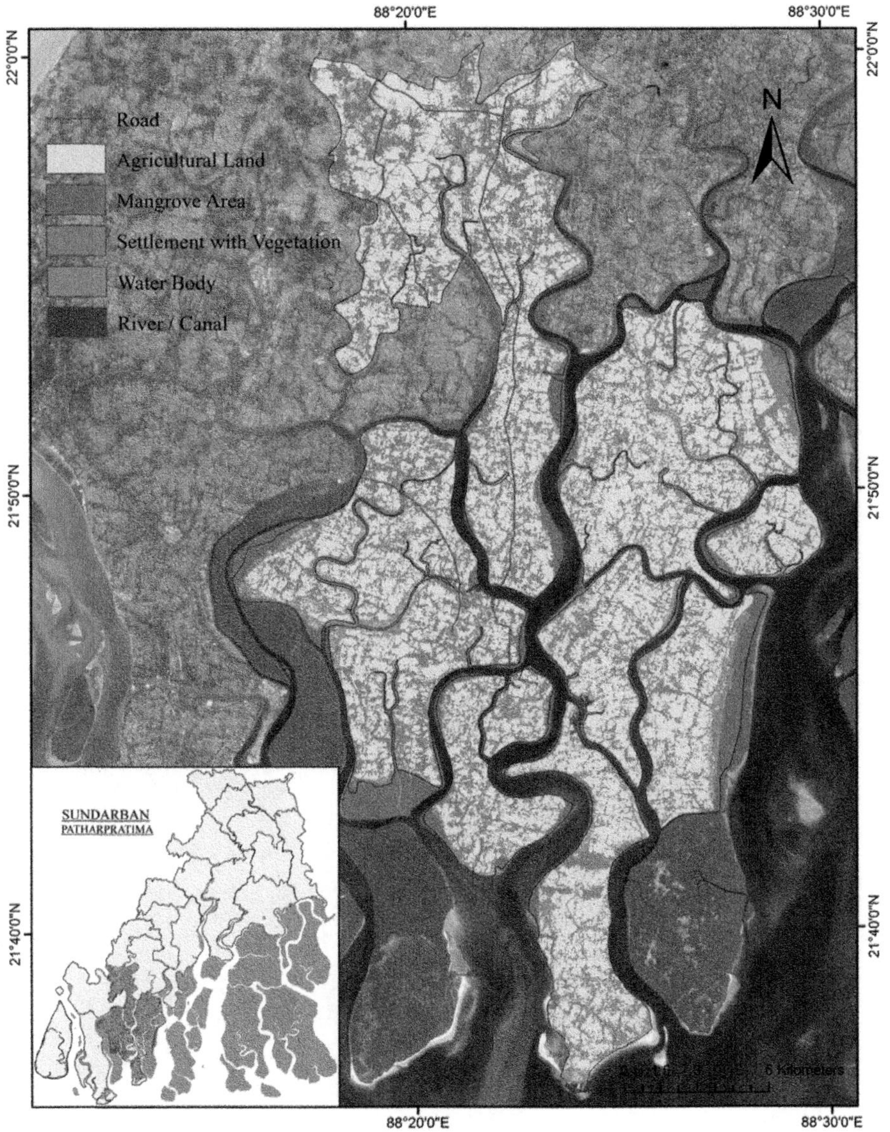

Fig. 2.4 The land-use-land-cover map of the Patharpratima block in the SBR

observation also shows that surface water body which can be utilized for agriculture is in good amount in Hingalganj and the least in Sagar. However, drawing such inferences would be an oversimplification of the scenario as the surface area itself cannot give us information on parameters like the salinity of the pond waters, their depth, and hence volume, etc. Moreover, the inland water bodies in this chapter took into

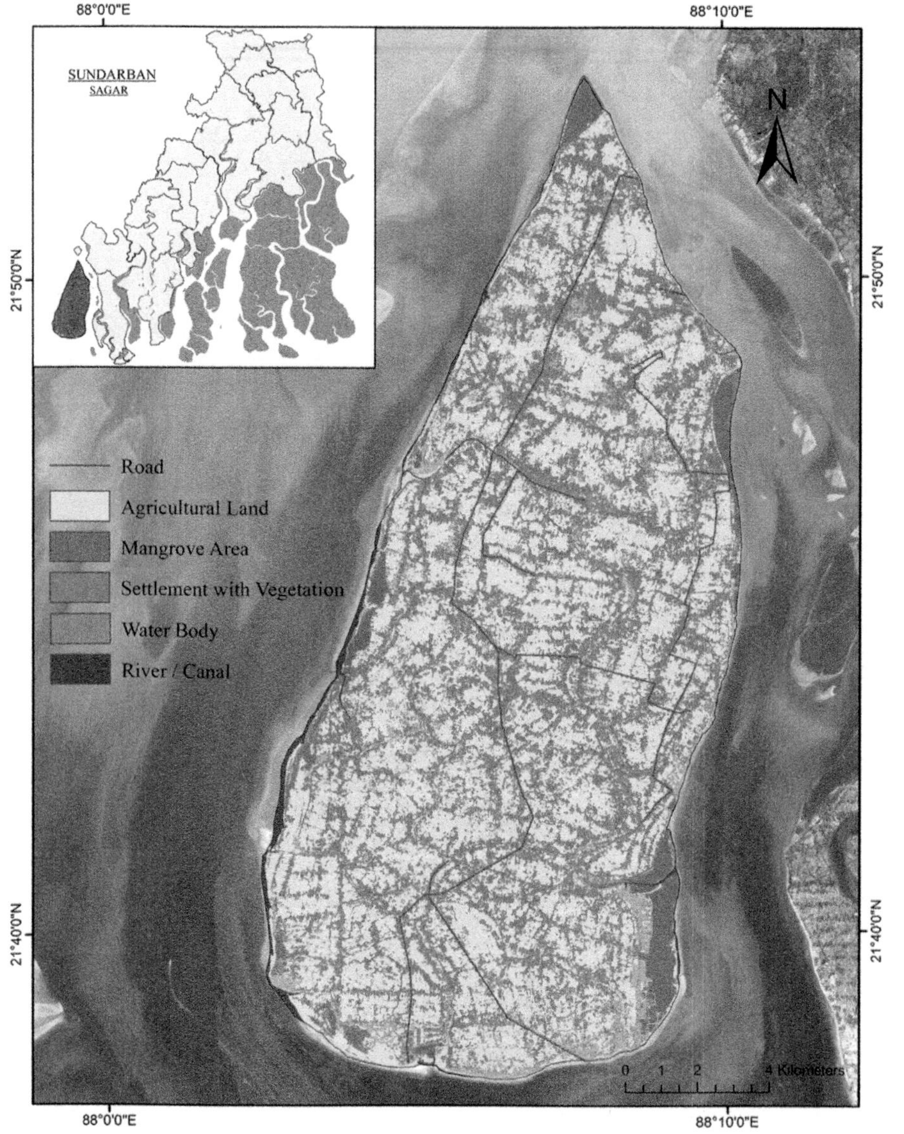

Fig. 2.5 The land-use-land-cover map of the Sagar block in the SBR

account the household ponds as well as the aquaculture ponds. The differentiation between the two classes would require further focused study.

2.3.2 *The Descriptive Statistics Related to Ponds in the Study Area*

Table 2.2 gives us a glance at the size classes of the ponds in the selected administrative blocks of the SBR. The percentage area covered by the ponds compared to the total block area of the respective blocks followed the order: Hingalganj (9.2%) > Gosaba (4.6%) > Patharpratima (4.4%) > Sagar (3.4%). This observation is in parity with the earlier observation made from the land-use classes that as we move north, the number of ponds has increased. We considered four size classes in this analysis: 100–200 m^2, 200–300 m^2, 300–400 m^2, and >400 m^2. Patharpratima had the highest number of ponds (30,000) followed by Gosaba (19,000), Sagar (16,000), and Hingalganj (14,000). This observation indicates that there is no relationship between the area covered and the number of ponds amongst the blocks. Hingalganj despite having the largest percentage area covered by ponds recorded the lowest number of total ponds amongst the four blocks. Among the different size classes, ponds of area varying between 100 and 200 m^2 are substantially higher in number than the other size classes in all the respective blocks. The number of ponds (of the area between 100 and 200 m^2) per square km in the respective blocks varied between 37 (Gosaba and Hingalganj) and 46 (Sagar). The number of ponds of the other size classes did not show much variation (5 to 6 for ponds having an area between 200 and 300 m^2; 3 to 5 for ponds having an area between 300 and 400 m^2; and 11 to 13 for ponds having an area >400 m^2). This observation leads us to infer that the number of ponds is higher in Sagar Island; however, the size of the ponds are overall lower; whereas, the number of ponds is lower in Hingalganj; however, the size of the ponds, on the whole, is substantially higher, which led to the difference in total pond area cover between the two blocks.

Table 2.2 The size class of the ponds and their distribution in the selected four administrative blocks of the SBR

Block Name	Sagar	Pathar-pratima	Gosaba	Hingal-ganj
Total Area of blocks (km^2)	238.0	494.1	335.9	232.8
Total area covered by ponds (km^2)	8.0	21.9	15.4	21.3
Number of ponds (100–200 m^2)	10,964	19,263	12,266	8723
Number of ponds (200–300 m^2)	1507	2814	1793	1264
Number of ponds (300–400 m^2)	1153	2000	1137	940
Number of ponds (>400 m^2)	2527	6187	3837	3064
Number of ponds within 500 m from river boundary	3713	5228	4080	3513
Number of ponds within 1000 m from river boundary	7255	9736	7493	6273
Population (Census 2011)	212,037	331,823	246,598	175,545

2.3.3 *Visual Interpretation of the Spatial Distribution of Ponds*

Table 2.2 along with Figs. 2.6, 2.7, 2.8, and 2.9 indicate that most of the ponds exist within 500 m from the river boundary in all the blocks, and the numbers decline in

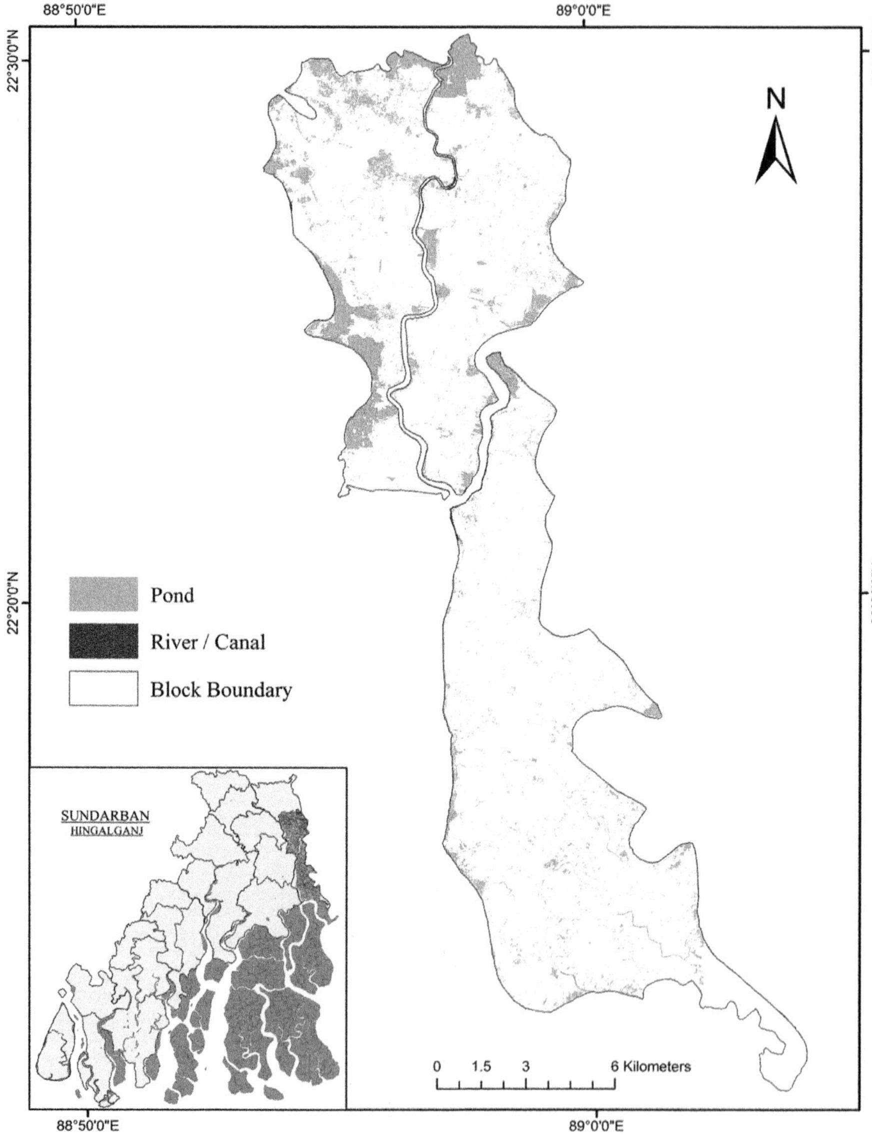

Fig. 2.6 The map showing the spatial distribution of ponds in the Hingalganj block of SBR

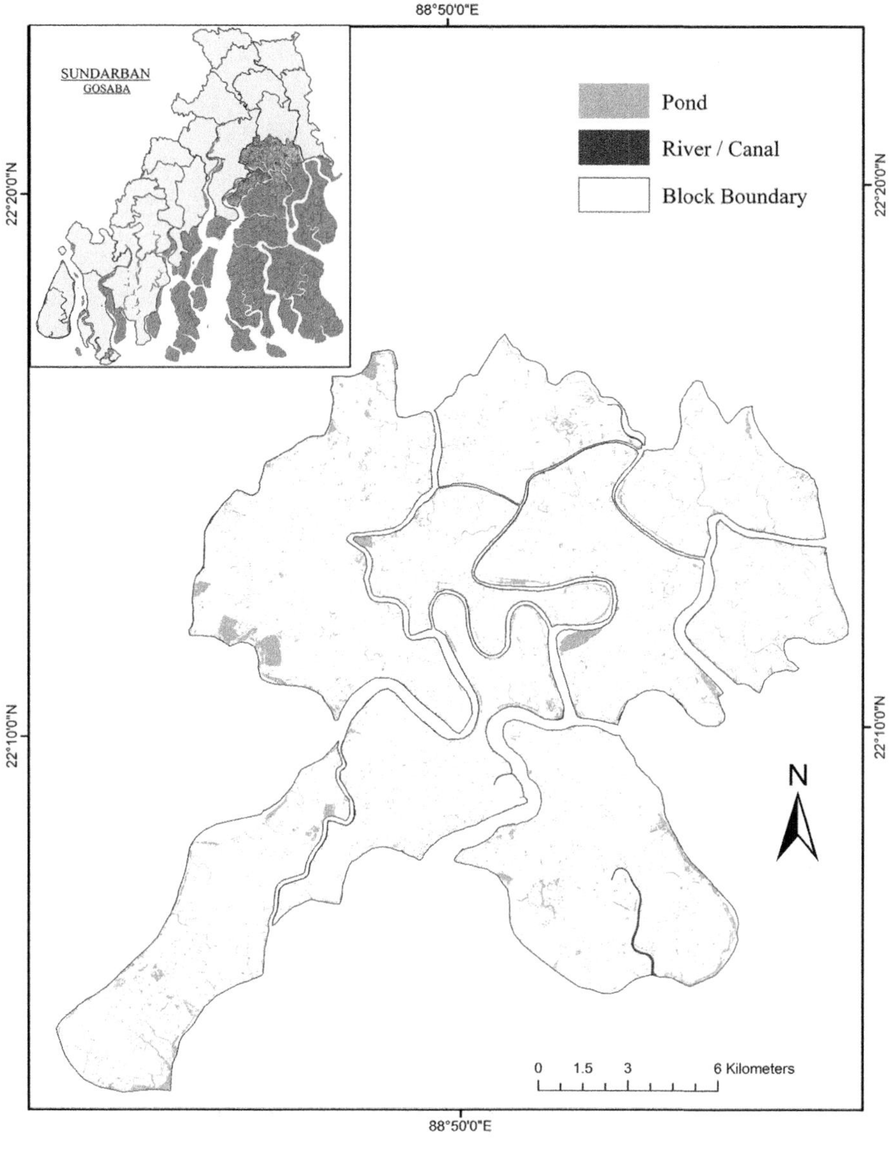

Fig. 2.7 The map showing the spatial distribution of ponds in the Gosaba block of SBR

the 500–1000 m distances from the river boundary. Analyzing the data, we observed that almost 45% of the ponds on both Sagar and Hingalganj were within 1 km from the river boundary. In the case of Gosaba and Patharpartima, the percentage decrease to 40 and 32, respectively. However, it is worth mentioning that a substantial part of Patharpratima is land-locked in the north. Spatially, the northern part of Hingalganj has some dense clusters of ponds. Most of these are aquaculture ponds and are of

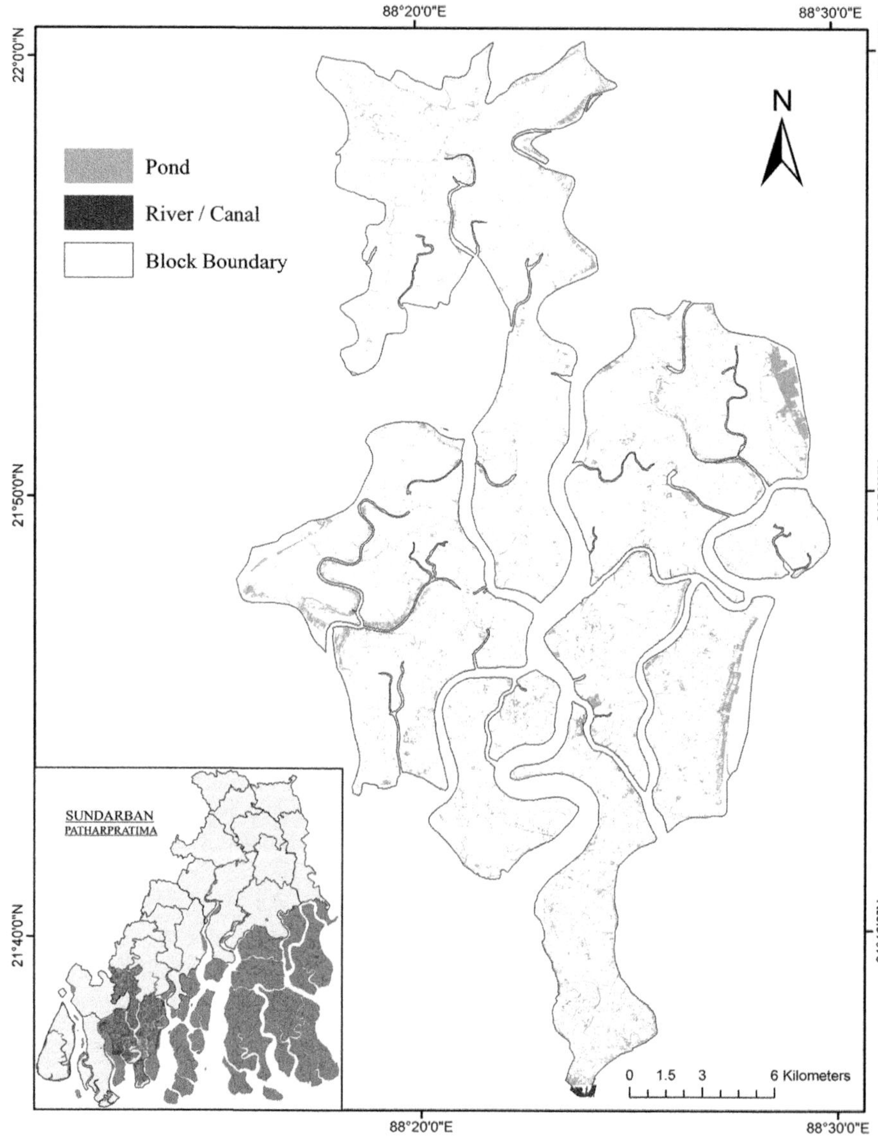

Fig. 2.8 The map showing the spatial distribution of ponds in the Patharpratima block of SBR

substantially large areas. The Gosaba block has some big clusters of ponds in the west and the central portions. Besides these two clusters, more or less the rest of the ponds has a homogenous distribution. Most of the ponds in clusters are visible near the river boundary in the case of the Patharpratima block. Comparatively, the northern part of the block, which is close to the mainland had a lower number of ponds than that observed in the south. Similarly, the sea-facing end of the Sagar Island had more

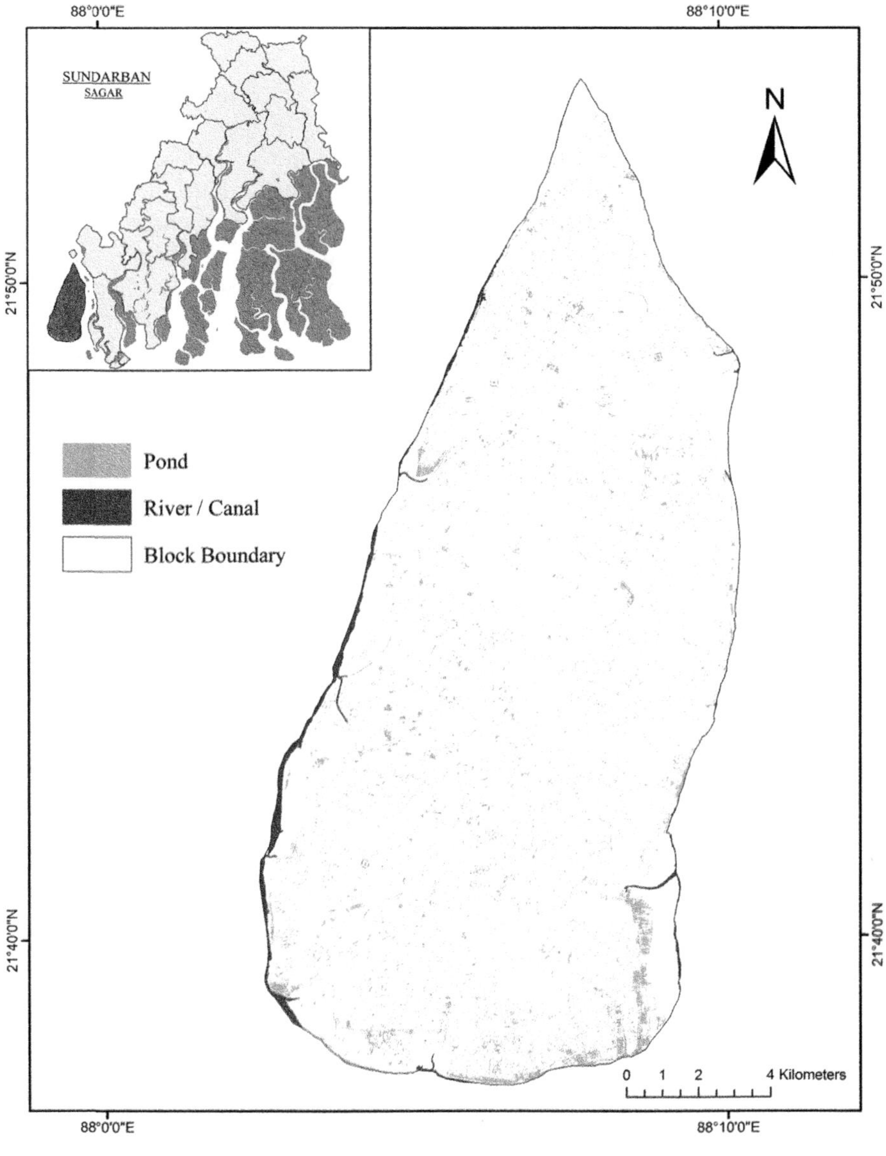

Fig. 2.9 The map showing the spatial distribution of ponds in the Sagar block of SBR

ponds than that observed near the tip of the island. Some thick patches indicating mostly aquaculture ponds are located close to the river-boundary. Apart from that, the ponds are homogeneously distributed throughout the island.

2.3.4 *Future Scope of Work*

In this study, we analyzed the size class and the spatial distribution of the ponds in four selected blocks of SBR. The same exercise should be carried out for all the nineteen blocks of the SBR to draw a holistic scenario about the entire biosphere reserve. Mapping such small water bodies requires high-resolution satellite images and most of these images have become available in the last decade. Hence, characterizing the long-term temporal variation of this land-use class is a difficult endeavor; however, the short-term variability within a calendar year needs to be assessed to see the seasonal variability. Many of these ponds get dried up during the dry season and get replenished during the monsoon. Studying the seasonal variability can enable us to understand the number of ponds that can hold water even during the dry season. This study did not differentiate the type of ponds based on their usage. A detailed classification of the household ponds, community ponds, and aquaculture ponds would fetch much more meaningful outcomes that could be useful for policy managers.

2.4 Conclusion

Overall, this study indicated that ponds cover almost 3% to 9% of the block areas in the SBR. Typically, the ponds with an area lying between 100 and 200 m^2 are in plenty and the ponds with an area >400 m^2 come next in terms of abundance. There was no proper relationship between the percentage area covered and the total number of ponds in a block. The numbers and percentage area covered varied significantly among the blocks. Overall, the number of ponds was higher in the sea-facing blocks like Sagar and Patharpratima; however, the total surface area was lower in the same. Whereas, in the northern blocks like Hingalganj and Gosaba, a completely reverse scenario prevailed. Spatially, most of the ponds in the respective blocks were close to the river boundary indicating aquaculture practices.

References

Al Sayah MJ, Nedjai R, Abdallah C, Khouri M (2020) On the use of remote sensing to map the proliferation of aquaculture ponds and to investigate their effect on local climate, perspectives from the Claise watershed, France. Environ Monit Assess 192(5):1–17. https://doi.org/10.1007/s10661-020-08250-0

Ambrosino C, Hufton B, Nyawade BO, Osimbo H, Owiti P (2020) Integrating climate adaptation, poverty reduction, and environmental conservation in Kwale County, Kenya. In: Filho WL et al (eds) African handbook climate change adaptation. Springer BV, pp 1–19. https://doi.org/10.1007/978-3-030-42091-8_118-1

Belton B, Padiyar A, Ravibabu G, Rao KG (2017) Boom and Bust in Andhra Pradesh: development and transformation in India's domestic aquaculture value chain. Aquaculture 470:196–206. https://doi.org/10.1016/j.aquaculture.2016.12.019

Census (2011) Census data. Government of India
Céréghino R, Boix D, Cauchie HM, Martens K, Oertli B (2014) The ecological role of ponds in a changing world. Hydrobiologia 723(1):1–6. https://doi.org/10.1007/s10750-013-1719-y
Clavelle T, Lester SE, Gentry R, Froehlich HE (2019) Interactions and management for the future of marine aquaculture and capture fisheries. Fish Fish 20(2):368–388. https://doi.org/10.1111/faf.12351
Daly P, Halim A, Hundlani D, Ho E, Mahdi S (2017) Rehabilitating coastal agriculture and aquaculture after inundation events: Spatial analysis of livelihood recovery in post-tsunami Aceh, Indonesia. Ocean Coast Manag 142:218–232. https://doi.org/10.1016/j.ocecoaman.2017.03.027
Das S, Hazra S (2020) Trapped or resettled: coastal communities in the Sundarbans Delta, India. Forced Migr Rev 64:15–17
DasGupta R, Hashimoto S, Okuro T, Basu M (2019) Scenario-based land change modelling in the Indian Sundarban delta: an exploratory analysis of plausible alternative regional futures. Sustain Sci 14(1):221–240. https://doi.org/10.1007/s11625-018-0642-6
Dasgupta R, Hashimoto S, Basu M, Okuro T, Johnson BA, Kumar P, Dhyani S (2021) Spatial characterization of non-material values across multiple coastal production landscapes in the Indian Sundarban delta. Sustain Sci. https://doi.org/10.1007/s11625-020-00899-3
Datta D, Deb S (2012) Analysis of coastal land use/land cover changes in the Indian Sunderbans using remotely sensed data. Geospat Info Sci 15(4):241–250. https://doi.org/10.1080/10095020.2012.714104
De Marco P, Nogueira DS, Correa CC et al (2014) Patterns in the organization of Cerrado pond biodiversity in Brazilian pasture landscapes. Hydrobiologia 723:87–101. https://doi.org/10.1007/s10750-013-1695-2
De D, Das CS (2021) Measuring livelihood sustainability by PCA in Indian Sundarban. Environ Dev Sustain. https://doi.org/10.1007/s10668-021-01451-8
Deacon C, Samways MJ, Pryke JS (2018) Artificial reservoirs complement natural ponds to improve pondscape resilience in conservation corridors in a biodiversity hotspot. PLoS ONE 13(9):e0204148. https://doi.org/10.1371/journal.pone.0204148
Debnath A (2018) Land use and land cover change detection of Gosaba Island of the Indian Sundarban Region by using multitemporal satellite image. Int J Hum Soc Sci 7(1):209–217
Debrot AO, Veldhuizen A, Van Den Burg SW, Klapwijk CJ, Islam M, Alam M et al (2020) Non-timber forest product livelihood-focused interventions in support of mangrove restoration: a call to action. Forests 11(11):1224. https://doi.org/10.3390/f11111224
District Census Handbook (2011a) North 24 Parganas, West Bengal. Census of India. Directorate of Census Operations, West Bengal
District Statistical Handbook (2010–11a) North 24 Parganas, West Bengal. Bureau of Applied Economics & Statistics, Department of Statistics & Programme Implementation, West Bengal
District Statistical Handbook (2010–11b) South 24 Parganas, West Bengal. Bureau of Applied Economics & Statistics, Department of Statistics & Programme Implementation, West Bengal
Downing JA, Prairie YT, Cole JJ, Duarte CM, Tranvik LJ, Striegl RG, McDowell WH, Kortelainen P, Caraco NF, Melack JM, Middelburg JJ (2006) The global abundance and size distribution of lakes, ponds, and impoundments. Limnol Oceanogr 51:2388–2397. https://doi.org/10.4319/lo.2006.51.5.2388
Downing JA, Cole JJ, Middelburg JJ, Striegl RG, Duarte CM, Kortelainen P, Prairie YT, Laube KA (2008) Sediment organic carbon burial in agriculturally eutrophic impoundments over the last century. Glob Biogeochem Cyc 22(1). https://doi.org/10.1029/2006GB002854
Du Y, Zhang Y, Ling F, Wang Q, Li W, Li X (2016) Water bodies' mapping from Sentinel-2 imagery with modified normalized difference water index at 10-m spatial resolution produced by sharpening the SWIR band. Remote Sens 8(4):354. https://doi.org/10.3390/rs8040354
Duan Y, Li X, Zhang L, Chen D, Ji H (2020a) Mapping national-scale aquaculture ponds based on the Google Earth Engine in the Chinese coastal zone. Aquaculture 520:734666. https://doi.org/10.1016/j.aquaculture.2019.734666

Duan Y, Li X, Zhang L, Liu W, Chen D, Ji H (2020b) Detecting spatiotemporal changes of large-scale aquaculture ponds regions over 1988–2018 in Jiangsu Province, China using Google Earth Engine. Ocean Coast Manag 188:105144. https://doi.org/10.1016/j.ocecoaman.2020.105144

Dutta U (2019) Socioeconomic condition of shrimp collectors in Sundarbans region, special reference-Hingalganj Block. Int J Res Soc Sci 8(2):662–670

FAO (2016) The state of world fisheries and aquaculture 2016. Contributing to food security and nutrition for all. FAO, Rome, Italy, p 200

Fu B, Xu P, Wang Y, Yan K, Chaudhary S (2018) Assessment of the ecosystem services provided by ponds in hilly areas. Sci Total Environ 642:979–987. https://doi.org/10.1016/j.scitotenv.2018.06.138

Ghermandi A, Fichtman E (2015) Cultural ecosystem services of multifunctional constructed treatment wetlands and waste stabilization ponds: time to enter the mainstream? Ecol Eng 84:615–623. https://doi.org/10.1016/j.ecoleng.2015.09.067

Ghosh P (2015) Conservation and conflicts in the Sundarban biosphere reserve, India. Geogr Rev 105(4):429–440. https://doi.org/10.1111/j.1931-0846.2015.12101.x

Ghosh P, Ghosh A (2019) Is ecotourism a panacea? Political ecology perspectives from the Sundarban Biosphere Reserve, India. GeoJournal 84(2):345–366. https://doi.org/10.1007/s10708-018-9862-7

Ghosh S, Mistri B (2020) Drainage-induced waterlogging problem and its impact on the farming system: a study in Gosaba Island, Sundarban, India. Spat Info Res 28(6):709–721. https://doi.org/10.1007/s41324-020-00328-8

Ghosh S, Mistri B (2021) Assessing coastal vulnerability to environmental hazards of Indian Sundarban delta using multi-criteria decision-making approaches. Ocean Coast Manag 209:105641. https://doi.org/10.1016/j.ocecoaman.2021.105641

Ghosh A, Nanda MK, Sarkar D, Sarkar S, Brahmachari K, Ray K (2019) Application of multi-dated sentinel-2 imageries to assess the cropping system in Gosaba Island of Indian Sundarbans. J Indian Soc Coast Agric Res 37(2):32–44

Gopal B, Chauhan M (2006) Biodiversity and its conservation in the Sundarban Mangrove Ecosystem. Aquat Sci 68:338–354. https://doi.org/10.1007/s00027-006-0868-8

Hajra R, Ghosh A, Ghosh T (2017) Comparative assessment of morphological and landuse/landcover change pattern of Sagar, Ghoramara, and Mousani Island of Indian Sundarban Delta Through Remote Sensing. In: Hazra S, Mukhopadhyay A, Ghosh A, Mitra D, Dadhwal V (eds) Environment and earth observation. Springer Remote Sensing/Photogrammetry. Springer, Cham. https://doi.org/10.1007/978-3-319-46010-9_11

Hajra R, Ghosh T (2018) Agricultural productivity, household poverty and migration in the Indian Sundarban Delta. Elem Sci Anthr 6:3. https://doi.org/10.1525/elementa.196

Haldar A, Bhattacharya A (2020) Sustainable energy development and participatory management scenario in the Sundarban: a case study in the Sagar Island, West Bengal, India. In: Sahdev S, Singh R, Kumar M (eds) Geoecology of landscape dynamics. Advances in geographical and environmental sciences. Springer, Singapore. https://doi.org/10.1007/978-981-15-2097-6_20

Halder S, Samanta K, Das S, Pathak D (2021) Monitoring the inter-decade spatial-temporal dynamics of the Sundarban mangrove forest of India from 1990 to 2019. Reg Stud Mar Sci 44:101718. https://doi.org/10.1016/j.rsma.2021.101718

Hamilton S (2013) Assessing the role of commercial aquaculture in displacing mangrove forests. Bull Mar Sci 89(2):585–601. https://doi.org/10.5343/bms.2012.1069

Islam MM, Barman A, Kundu GK, Kabir MA, Paul B (2019) Vulnerability of inland and coastal aquaculture to climate change: evidence from a developing country. Aquac Fish 4(5):183–189. https://doi.org/10.1016/j.aaf.2019.02.007

Jayasinghe JMPK, Gamage DGND, Jayasinghe JMHA (2019) Combating climate change impacts for shrimp aquaculture through adaptations: Sri Lankan perspective. In: Sarkar A, Sensarma S, vanLoon G (eds) Sustainable solutions for food security. Springer, Cham, pp 287–309. https://doi.org/10.1007/978-3-319-77878-5_15

Lacaux JP, Tourre YM, Vignolles C, Ndione JA, Lafaye M (2007) Classification of ponds from high-spatial resolution remote sensing: Application to Rift Valley Fever epidemics in Senegal. Remote Sens Environ 106(1):66–74. https://doi.org/10.1016/j.rse.2006.07.012

Lázár AN, Adams H, Adger WN, Nicholls RJ (2020) Modelling household well-being and poverty trajectories: an application to coastal Bangladesh. PLoS ONE 15(9):e0238621. https://doi.org/10.1371/journal.pone.0238621

Liu B, Wang J, Xu M, Wang Z, Zhao L (2020) Evaluation of the comprehensive benefit of various marine exploitation activities in China. Mar Pol 116:103924. https://doi.org/10.1016/j.marpol.2020.103924

Mandal S, Burman D, Sarangi SK, Bandyopadhyay BK, Maji B (2015) Homestead production systems in Sundarbans region of West Bengal, India–Current status and opportunities. CGIAR-CPWF

Marcinko CL, Nicholls RJ, Daw TM, Hazra S, Hutton CW, Hill CT et al (2021) The development of a framework for the integrated assessment of SDG trade-offs in the Sundarban biosphere reserve. Water 13(4):528. https://doi.org/10.3390/w13040528

Mondal I, Thakur S, Ghosh P, De TK, Bandyopadhyay J (2019) Land use/land cover modeling of Sagar Island, India using remote sensing and GIS techniques. In: Abraham A, Dutta P, Mandal J, Bhattacharya A, Dutta S (eds) Emerging technologies in data mining and information security. Advances in intelligent systems and computing, vol 755. Springer, Singapore. https://doi.org/10.1007/978-981-13-1951-8_69

Moore TL, Hunt WF (2012) Ecosystem service provision by stormwater wetlands and ponds–a means for evaluation? Water Res 46(20):6811–6823. https://doi.org/10.1016/j.watres.2011.11.026

Murray NJ, Phinn SR, DeWitt M, Ferrari R, Johnston R, Lyons MB et al (2019) The global distribution and trajectory of tidal flats. Nature 565(7738):222–225. https://doi.org/10.1038/s41586-018-0805-8

Nath A, Samanta S, Banerjee S, Danda AA, Hazra S (2021) Threat of arsenic contamination, salinity and water pollution in agricultural practices of Sundarban Delta, India, and mitigation strategies. SN Appl Sci 3(5):1–15. https://doi.org/10.1007/s42452-021-04544-1

Neofitou N, Papadimitriou K, Domenikiotis C, Tziantziou L, Panagiotaki P (2019) GIS in environmental monitoring and assessment of fish farming impacts on nutrients of Pagasitikos Gulf, Eastern Mediterranean. Aquaculture 501:62–75. https://doi.org/10.1016/j.aquaculture.2018.11.005

Onyena AP, Sam K (2020) A review of the threat of oil exploitation to mangrove ecosystem: insights from Niger Delta, Nigeria. Glob Ecol Conserv 22:e00961. https://doi.org/10.1016/j.gecco.2020.e00961

Ottinger M, Clauss K, Kuenzer C (2016) Aquaculture: Relevance, distribution, impacts and spatial assessments–a review. Ocean Coastal Manag 119:244–266. https://doi.org/10.1016/j.ocecoaman.2015.10.015

Prasad KA, Ottinger M, Wei C, Leinenkugel P (2019) Assessment of coastal aquaculture for India from Sentinel-1 SAR time series. Remote Sens 11(3):357. https://doi.org/10.3390/rs11030357

Quinn PF, Hewett JM, Jonzyk J, Glenis V (2007) The PROACTIVE approach to farm integrated runoff management (FIRM) plans. Flood storage on Farms. Newcastle University. http://www.ncl.ac.uk/iq/download/PROACTIVEFloods.doc

Ranjan AK, Vallisree S, Verma SK, Murmu L, Kumar PBS (2017) Spatio-temporal variation in the Indian part of Sundarban Delta over the years 1990–2016 using Geospatial Technology. Int J Geomat Geosci 7(3):275–292

Ren C, Wang Z, Zhang Y, Zhang B, Chen L, Xi Y et al (2019) Rapid expansion of coastal aquaculture ponds in China from Landsat observations during 1984–2016. Int J Appl Earth Obs Geoinfo 82:101902. https://doi.org/10.1016/j.jag.2019.101902

Sahana M, Ahmed R, Sajjad H (2016) Analyzing land surface temperature distribution in response to land use/land cover change using split window algorithm and spectral radiance model in

Sundarban Biosphere Reserve, India. Model Earth Syst Environ 2(2):81. https://doi.org/10.1007/s40808-016-0135-5
Sahana M, Hong H, Ahmed R, Patel PP, Bhakat P, Sajjad H (2019) Assessing coastal island vulnerability in the Sundarban Biosphere Reserve, India, using geospatial technology. Environ Earth Sci 78(10):1–22. https://doi.org/10.1007/s12665-019-8293-1
Sahana M, Rehman S, Paul AK, Sajjad H (2021) Assessing socio-economic vulnerability to climate change-induced disasters: evidence from Sundarban Biosphere Reserve, India. Geol Ecol Landsc 5(1):40–52. https://doi.org/10.1080/24749508.2019.1700670
Sannigrahi S, Chakraborti S, Joshi PK, Keesstra S, Sen S, Paul SK et al (2019) Ecosystem service value assessment of a natural reserve region for strengthening protection and conservation. J Environ Manag 244:208–227. https://doi.org/10.1016/j.jenvman.2019.04.095
Sannigrahi S, Zhang Q, Pilla F, Joshi PK, Basu B, Keesstra S et al (2020) Responses of ecosystem services to natural and anthropogenic forcings: a spatial regression based assessment in the world's largest mangrove ecosystem. Sci Total Environ 715:137004. https://doi.org/10.1016/j.scitotenv.2020.137004
Simaika JP, Samways MJ, Frenzel PP (2016) Artificial ponds increase local dragonfly diversity in a global biodiversity hotspot. Biodivers Conserv 25(10):1921–1935. https://doi.org/10.1007/s10531-016-1168-9
Sinha R, Das S, Ghosh T (2020) Pollution and its consequences at Ganga Sagar mass bathing in India. Environ Dev Sustain 22(2):1413–1430. https://doi.org/10.1007/s10668-018-0255-3
Sousa E, Quintino V, Palhas J, Rodrigues AM, Teixeira J (2016) Can environmental education actions change public attitudes? An example using the pond habitat and associated biodiversity. PLoS ONE 11(5):e0154440. https://doi.org/10.1371/journal.pone.0154440
Sreelekshmi S, Nandan SB, Kaimal SV, Radhakrishnan CK, Suresh VR (2020) Mangrove species diversity, stand structure and zonation pattern in relation to environmental factors—a case study at Sundarban delta, east coast of India. Reg Stud Mar Sci 35:101111. https://doi.org/10.1016/j.rsma.2020.101111
Steidl J, Kalettka T, Ehlert V, Quast J, Augustin J (2008) Mitigation of pressures on water bodies by nutrient retention from agricultural drainage effluents using purification ponds. In: Proceedings of the 10th international drainage workshop, vol 16. Helsinki University of Technology, pp 187–194
Stewart RI, Andersson GK, Brönmark C, Klatt BK, Hansson LA, Zülsdorff V, Smith HG (2017) Ecosystem services across the aquatic-terrestrial boundary: linking ponds to pollination. Basic Appl Ecol 18:13–20. https://doi.org/10.1016/j.baae.2016.09.006
Thakur S, Ghosh A, Hazra S, Ghosh T (2019) Biomass burning and natural vegetation loss as a consequence to government policy: a case study of Satjelia Island, Indian Sundarban Delta. Int J Res Anal Rev 6(1):338–345
Thakur S, Maity D, Mondal I, Basumatary G, Ghosh PB, Das P, De TK (2021) Assessment of changes in land use, land cover, and land surface temperature in the mangrove forest of Sundarbans, northeast coast of India. Environ Dev Sustain 23(2):1917–1943. https://doi.org/10.1007/s10668-020-00656-7
Tschudi MA, Maslanik JA, Perovich DK (2008) Derivation of melt pond coverage on Arctic sea ice using MODIS observations. Remote Sens Environ 112(5):2605–2614. https://doi.org/10.1016/j.rse.2007.12.009
Veettil BK, Quang NX, Trang NTT (2019) Changes in mangrove vegetation, aquaculture and paddy cultivation in the Mekong Delta: a study from Ben Tre Province, southern Vietnam. Estuar Coast Shelf Sci 226:106273. https://doi.org/10.1016/j.ecss.2019.106273
Virdis SGP (2014) An object-based image analysis approach for aquaculture ponds precise mapping and monitoring: a case study of Tam Giang-Cau Hai Lagoon, Vietnam. Environ Monit Assess 186(1):117–133. https://doi.org/10.1007/s10661-013-3360-7
Wang Q, Cheng L, Liu J, Li Z, Xie S, De Silva SS (2015) Freshwater aquaculture in PR China: trends and prospects. Rev Aquac 7(4):283–302. https://doi.org/10.1111/raq.12086

Xu H (2006) Modification of normalised difference water index (NDWI) to enhance open water features in remotely sensed imagery. Int J Remote Sens 27(14):3025–3033. https://doi.org/10.1080/01431160600589179

Xu M, Cui B, Lan S, Li D, Wang Y, Jiang B (2020) Exploring dynamic change of the tidal flat aquaculture area in the Shandong Peninsula (China) using multitemporal landsat imagery (1990–2015). J Coast Res 99(SI):197–202. https://doi.org/10.2112/SI99-028.1

Yao YC, Ren CY, Wang ZM, Wang C, Deng PY (2016) Monitoring of salt ponds and aquaculture ponds in the coastal zone of China in 1985 and 2010. Wetl Sci 14(6):874–882

Yu J, Han Q (2020) Food security of mariculture in China: evolution, future potential, and policy. Mar Pol 115:103892. https://doi.org/10.1016/j.marpol.2020.103892

Chapter 3
Assessment of Pond Water Quality and Its Impact on Local Livelihood in the Indian Sundarbans

Sourav Das

Abstract The present study is an attempt to comprehend the seasonal variation of several water quality parameters and assess the surface water quality in the Indian part of Sundarbans. 39 inhabited *Mouzas* (smaller administrative blocks) from six ocean confronting Blocks (Sagar, Namkhana, Patharpratima, Kultali, Basanti, and Gosaba) of South 24 Parganas District and southwestern part of Sundarbans have been chosen for the current study. Water quality analysis has been carried out by measuring eight parameters such as pH, Salinity, Total Dissolved Solids (TDS), Electrical Conductivity (EC), Dissolved Oxygen (DO), Temperature, Total Hardness, and Turbidity. Several statistical analyses have been performed to extract meaningful inferences. Mann–Whitney Test has been conducted to find out Pre and Post Monsoon variations in selected water quality parameters. Wilcoxon's Signed Rank Test has been employed to identify the intra-Block variation and Mann Whitney (pairwise) Test and Kruskal Wallis Test have been carried out to examine the Block wise and inter-Block-wise variation of each water quality parameter. A correlation matrix following Pearson's correlation has been done to analyze the interrelations among water quality parameters and their level of significance. Primary household surveys (611 households in Six Blocks) have been conducted to develop a comprehensive understanding of water-related risks faced by the coastal habitat and communities. The result of the study indicates a grave situation of water quality and security consequent upon persistent climatic hazards, embankment breaching, coastal flooding, and salinization. The findings of the study seem crucial in framing future adaptation policies related to surface water in the estuarine ecosystem to safeguard the vulnerable communities in the current juncture of climate change.

Keywords Pond water quality · Water security · Indian Sundarbans · Climate change · Salinization · Coastal communities

---

S. Das (✉)
School of Oceanographic Studies, Jadavpur University, Kolkata, India

S. Das et al. (eds.), *Pond Ecosystems of the Indian Sundarbans*, Water Science and Technology Library 112, https://doi.org/10.1007/978-3-030-86786-7_3

3.1 Introduction

Indian Sundarbans is situated at the fringe between the apex of the Bay of Bengal and the southernmost landmass of West Bengal, India. The Indian Sundarbans is the largest contiguous tract of mangroves covering an area of 4264 km^2 (Sarkar and Bhattacharya 2010; Sánchez-Triana et al. 2016). The term Sundarbans came from "*Sunderbunds*" mentioned by Walter Hamilton of the East India Gazetteer, meaning the forests of Heritiera fomes, locally called "*Sundri*" (Mitra et al. 2009). The Indian Sundarbans was declared a national park in 1984, a UNESCO World Heritage Site in 1987, a Biosphere Reserve in 1989, and recently has achieved the status of Ramsar Site in 2019. It is an archipelago of 102 islands out of which 54 are inhabited (Mukherjee and Siddique 2018). The region is a transitional zone between freshwaters coming from the Hooghly River and the saline water from the Bay of Bengal (Akhter et al. 2018). The estuarine ecosystem of the Sundarbans is mainly dependent on the balance between fresh and saline water (Rahman et al. 2013). But the balance has been disrupted by the impacts of climate change enforcing persistent sea-level rise; higher storm surges, tidal ingression; intensification of depressions into a cyclone, increase in temperature, rainfall, and the intensity of storm waves leading to embankment breaching and flooding (Nicholls 2003; Hanson et al. 2011; Whitehead et al. 2018). These are ensuing saline water incursion in the area (Dasgupta et al. 2014; Mukherjee et al. 2019; Mukherjee and Siddique 2019a) and raising the level of salinity in this active deltaic zone of Ganga and Brahmaputra. The rise of salinity in such marginal coastal areas is detrimental to fishing activity as salinity retards the growth and profusion of aquatic species (Jahan 2018). It also affects crop yield (Rahman et al. 2019) and the productivity of forest species (Sadik et al. 2018). *Heritiera fomes* (*Sundri*) and *Nypa fruticans* (*Golpata*) are declining rapidly (Das and Siddiqi 1985; Gopal and Chauhan 2006) in the Sundarbans and the prime cause for top-dying of the species (Chaffey et al. 1985; Zaman et al. 2013) is salinization of water bodies. Recent studies have also attested that increased salinity in potable water leads to gestational hypertension of pregnant mothers that causes very high blood pressure, risk of postnatal death, and even infant mortality (Khan et al. 2011, 2014; Vineis et al. 2011; Scheelbeek et al. 2016; Siddique 2018; Shammi et al. 2019). Thus augmented water salinity as a probable corollary of climate change promotes issues of environmental justice, and if it remains unrestrained, the crisis of freshwater availability will be realized with severe human health implications (Vineis et al. 2011).

The research outcomes of water quality assessment and its impact on local livelihood in the Sundarbans reveal that water-related hazards have a critical impact on the livelihood of coastal communities. Monitoring and assessment of river water quality in low-lying active delta for management and improvement of water resources have been worked out by Mitra (2019) and Das and Mukherjee (2019). Das and Bandyopadhyay (2019) have researched the effects of water pollution on the ecology of fish life. Mondal et al. (2016) compared river water quality in pre-monsoon, monsoon, and post-monsoon seasons in Icchamati River which has proved increased water pollution caused by the change in water quality parameters. Salinity difference between the western and central Indian Sundarbans over more than two decades and their impact on mangrove forest have been identified by Zaman et al. (2013). Saline water-related health hazards such as diarrhea (Saha et al. 2019), skin disease (Kanjilal et al. 2010; Mukherjee and Siddique 2019a, b), gestational hypertension has been discussed in the studies conducted by Dasgupta et al. (2015) and Scheelbeek et al. (2016) in coastal deltas. Management of water-related risk from changing climate through policy framing has been put forward by Sánchez-Triana et al. (2014) and Dasgupta et al. (2015), and Mitra (2019) have searched out threats on household water supplies by saline water intrusion from sea level rise and embankment breaching. But, none of these works, however, have covered the entire southern stretch of the Indian Sundarbans region for analyzing pond water quality, especially, in the vulnerable *Mouzas* (i.e., the smaller administrative blocks).

Water is a precious natural resource for livelihood security and sustainable development (Mishra 2003; Hossain et al. 2016). Thus proper planning and management of water resources are extremely important in a fragile landscape like Sundarbans for ensuring environmental sustainability, which is one of the Millennium Development Goals of UNDP (WHO and UNICEF 2017). At the same time, understanding the threats on water resources related to climate change is a key requirement for building an adaptation strategy in the low-lying active delta of Sundarbans. Realizing the need for better understanding, the current paper elucidates seasonal variation of pond water quality by analyzing the parameters such as pH, Salinity, Total Dissolved Solids (TDS), Electrical Conductivity (EC), Dissolved Oxygen (DO), Temperature, Total Hardness, and Turbidity. It also tries to analyze problems related to water quality and security faced by the coastal people through a perception survey and field observation of villagers by questionnaire. The current study fills the gap in the literature by conducting pond water quality analysis in *Mouzas* suffering currently from climate change-related effects most importantly sea level rise, higher storm and tidal surges; embankment breaching, and saline water intrusion. It also employs primary data on household information for future policy framing related to water. The study has been structured under the following sections such as introduction, study area, and survey design, results and discussion, policy implications, and conclusion.

3.2 Study Area and Sampling Plan

The study area includes six blocks located in the southern part of South 24 Parganas district, spread in the south-western part of Indian Sundarbans—Sagar, Namkhana, Patharpratima, Kultali, Basanti, and Gosaba. According to the District Disaster Management Plan (2012, 2015, 2019–2020) of West Bengal, these Blocks fall under the severe category of multi-hazard-prone areas. Climate change intensifies the occurrence of natural hazards like coastal erosion, flood, higher storm surges, violent cyclones, etc. in these ocean-confronting Blocks of the Sundarbans. The area is encircled by the Hooghly River in the west and Hariabhanga River in the east, and the Bay of Bengal lies in the southern part. The other most important rivers in the study area include Gabtala, Baratala, Saptamukhi, Hetania Doania, Thakuran, Matla, Bidyadhari, Gomor, Garal, Bangaduani and Gosaba. Indian Sundarbans occupies the south-western part of the lower Ganga–Brahmaputra delta, which started forming in the Jurassic period of the Mesozoic era and the sedimentation took place in the quaternary period. The study area falls under the active delta that was accreted during 5–2.5 Ka BP (thousand years before present) and here tides play a crucial role in the development of landforms (Bandyopadhyay 2020). The average maximum and minimum temperature rise to 29.29 °C and 22.31 °C, respectively. The mean monthly precipitation is 151.05 mm and the average annual rainfall is 1812.66 mm (IMD 2018; Mukherjee and Siddique 2019a) (Fig. 3.1).

Pond water samples have been collected from 39 *Mouzas* located in the six Blocks (Table 3.1). These *Mouzas* have been selected as the area under review based on the severity of erosion examined through a review of several pieces of literature, Annual Flood Report (2014), field observation, and discussion with experts and respondents. Water samples (1.5 m below the surface layer) were collected randomly from 81 ponds located in these 39 *Mouzas* for the analysis of physicochemical properties during the Pre-Monsoon (March–May) and Post-Monsoon (June–Sept) period in the year 2016–2017 (Fig. 3.2). The parameters have been selected by a rigorous review

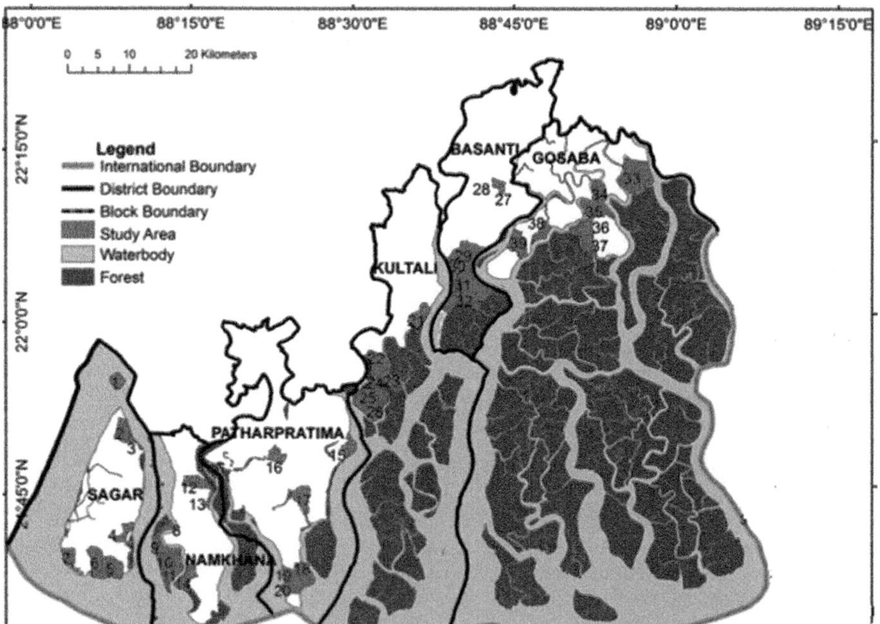

Fig. 3.1 The study area map comprising the six blocks (namely Sagar, Namkhana, Patharpratima, Kultali, Basanti, and Gosaba) and thirty-nine *Mouzas* sampled in Indian Sundarbans

Table 3.1 Blocks and mouzas (please refer Fig. 3.1 also for mouzas number) of the present study area

Blocks	Mouzas (number)
Sagar	Ghoramara (1), Kachubaria (2), Muri Ganga (3), Bankimnagar (4), Sibpur (5), Dhablat (6) and Beguyakhali (7)
Namkhana	Mousuni (8), Bagdanga (9), Kusumtala (10), Baliara (11), Narayanpur (12), Iswaripur (13) and Patibania (14)
Patharpratima	Purba Sripatinagar (15), Ramganga (16), Gangapur (17), Sitarampur (18), Buraburir Tat (19) and Gobardhanpur (20)
Kultali	Deulbari Debipur (21), Bhubaneshwari (22), Maipit (23), Baikunthapur (24), Binodpur (25) and Kisorimohanpur (26)
Basanti	Chandrakona (27), Radhaballavpur (28), Birinchibari (29), Parbatipur (30), Laskarpur (31) and Lot 126 (32)
Gosaba	Kumirmari (33), Hetalbari (34), Satjelia (35), Dayapur (36), Hamilton Abad (37), Sonagram (38) and Bali (39)

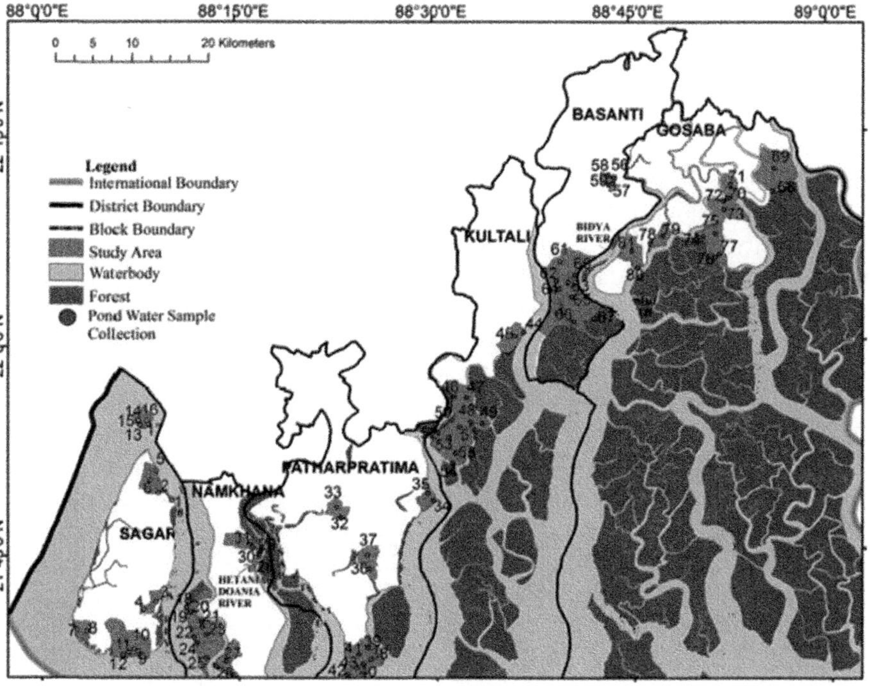

Fig. 3.2 The pond locations in the six blocks of water sample collection in Indian Sundarbans

of the literature (Mitra et al. 2009; Dasgupta et al. 2014; Zaman et al. 2015; Mondal et al. 2016). One water sample is collected from the extreme marginal location where vulnerable embankments are located (Questionnaire survey, expert judgment, and literature review helped reveal such locations) and the other is located >100 m away from the margin in each of the *Mouzas*. Three collected water samples were mixed for making one composite sample for a particular pond. Samples were collected in 100 ml plastic bottles, which were washed with dilute HCl followed by distilled water (1:1) and taken to the chemical laboratory of the School of Oceanographic Studies, Jadavpur University, Kolkata, India for further analysis. Statistical analysis has been carried out using the Statistical Package for the Social Sciences (SPSS) and Paleontological Statistics (PAST) software. The graphs are generated from MS Excel and maps are prepared using ArcGIS 9.1 software.

A statistically significant sample size of 611 households from 39 *Mouzas* has been selected from six Blocks (Sagar—101, Namkhana—105, Patharpratima—100, Kultali—102, Basanti—103, and Gosaba—100 Households) for the primary household survey. The survey took place from January 2016 to December 2016.

3.3 Results and Discussion

Management of water resources must ensure that the water quality parameters of surface waters should be below pollution threshold levels (Gupta et al. 2017). Poor water quality does not only affect humans or aquatic life but also the surrounding ecosystem as well (Mondal et al. 2016). Two Samples-Mann Whitney U test of significance has been applied in the study to work out the pre and post-monsoon variation of water quality parameters (Table 3.2). The result shows that pH and DO of the ponds have increased from pre-monsoon to post-monsoon period, while Salinity, TDS, Conductivity, Total Hardness, and Turbidity have decreased during the same period. This is mostly due to the addition of freshwater from monsoonal rainfall. Significant pre-monsoon and post-monsoon variations in certain water quality parameters in few Blocks have been reported from the study area. One water sample from each *Mouza* has been collected from extreme marginal locations where vulnerable embankments are located, and the other sample has been collected away from such intrusion cases. One-Sample Wilcoxon's Signed Ranked Test (significantly correlated at 1% level of significance in pre-monsoon and post-monsoon) has confirmed that there is significant variation in each water quality parameter in each of the Blocks. Mann Whitney Pairwise Test has revealed that there is significant variation of DO, Water Temperature, and Turbidity among all the Blocks. Kruskal Wallis Test has shown a significant variation of each parameter between each of the Blocks (Table 3.3).

Table 3.2 Pre and post monsoon variation of water quality parameters

Block	pH	Salinity	Conductivity	TDS	DO	Water temp	Total hardness	Turbidity
Sagar	0.013**	0.18	0.12	0.13	0.008***	0.0001***	0.25	0.10*
Namkhana	0.24	0.17	0.32	0.14	0.21	0.0001***	0.25	0.41
Patharpratima	0.18	0.49	0.93	0.40	0.24	0.0001***	0.17	0.31
Kultali	0.03**	0.07*	0.18	0.23	0.06*	0.0001***	0.005***	0.09*
Basanti	0.23	0.06*	0.03**	0.10*	0.20	0.0001***	0.12	0.29
Gosaba	0.015**	0.44	0.20	0.28	0.03**	0.03**	0.03**	0.40

Two sample Mann Whitney U test of significance where *** indicate significantly correlated at 1% level of significance, ** indicate significantly correlated at 5% level of significance, * indicate significantly correlated at 10% level of significance

Table 3.3 Block wise and inter block variation of each water quality parameters

	Block	Inter block
pH	0.27	Namkhana and Kultali (0.013**)
Salinity	0.19	Sagar and Kultali (0.02**), Kultali and Basanti (0.013**)
Conductivity	0.30	–
TDS	0.20	Sagar and Kultali (0.02**), Kultali and Basanti (0.013**)
DO	0.056*	Sagar and Gosaba (0.002***), Kultali and Gosaba (0.004***)
Water temperature	0.00***	Sagar and Patharpratima (0.00***), Sagar and Kultali (0.00**), Sagar and Basanti (0.03**), Sagar and Gosaba (0.005***), Patharpratima and Kultali (0.002***), Patharpratima and Basanti (0.002***), Patharpratima and Gosaba (0.04**), Kultali and Basanti (0.00***), Basanti and Gosaba (0.03**)
Total hardness	0.18	Basanti and Gosaba (0.001***)
Turbidity	0.00***	Sagar with all the other Blocks (0.00***), Patharpratima and Kultali (0.003***), Patharpratima and Basanti (0.024**), Patharpratima and Gosaba (0.009***)

Mann Whitney pair wise Test of significance for Block wise variation and Kruskal Wallis Test of equal median for inter Block variation where *** indicate significantly correlated at 1% level of significance, ** indicate significantly correlated at 5% level of significance, * indicate significantly correlated at 10% level of significance

3.3.1 Water Quality Status of the Pond of Indian Sundarban

Figures 3.3 and 3.4 show the water quality parameters for 81 stations in the Sundarbans along with the Bureau of Indian Standards (BIS) permissible limit.

pH is a measure of the concentration of hydrogen ions (Mondal et al. 2016). The pH standards proposed by WHO (World Health Organization) (2006), ICMR (Indian Council of Medical Research) (1975), CPCB (Central Pollution Control Board) (2020), and BIS suggest that the range should recline between 6.5 and 8.5 because pH lower than 6.5 discontinues the production of vitamins and minerals in the human body while greater than 8.5 pH causes water salinization, human eye irritation and skin disorder (Leo and Dekkar 2000; Adarsh and Mahantesh 2006; Gupta et al. 2017). In most of the ponds of the study area, pH value is within the standard limit while few of them indicate extreme acidity, especially in the pre-monsoon period. Ghoramara (Pond No. 16), Dhablat *Mouza* (Pond No. 180 11), Bagdanga (20), Baliara (24), Narayanpur (30, 31), Purba Sripatinagar (34), Gangapur (36), Gobardhanpur (42), Baikunthapur (52), Birinchibari (60), Parbatipur (62), Kumirmari (68), and Hamilton Abad (76) *Mouza* have recorded pH below 6.

Recorded Salinity, EC, and TDS values of the study area are high in pre-monsoon compared to post-monsoon season. High temperature and evaporation lower the water level during summer which makes sodium ions present in the water more

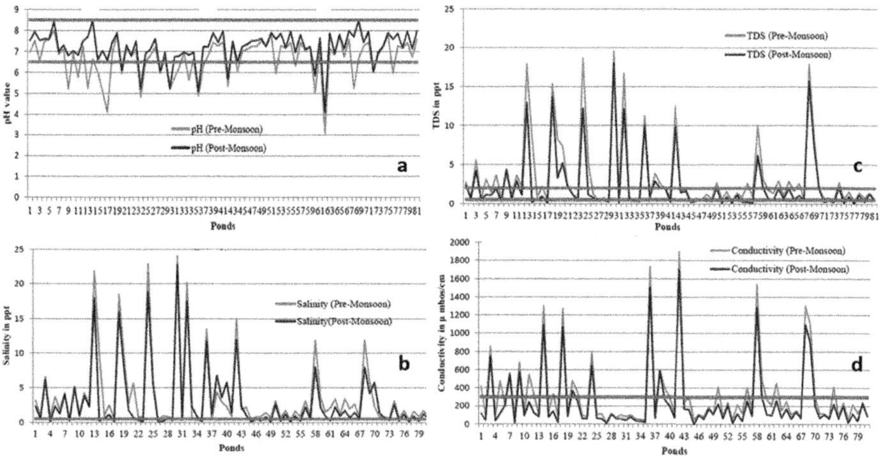

Fig. 3.3 Pond water quality (**a** pH, **b** Salinity, **c** TDS and **d** Conductivity) in the study area during pre-monsoon and post-monsoon season along with the Bureau of Indian Standards (BIS) permissible limit (red color line)

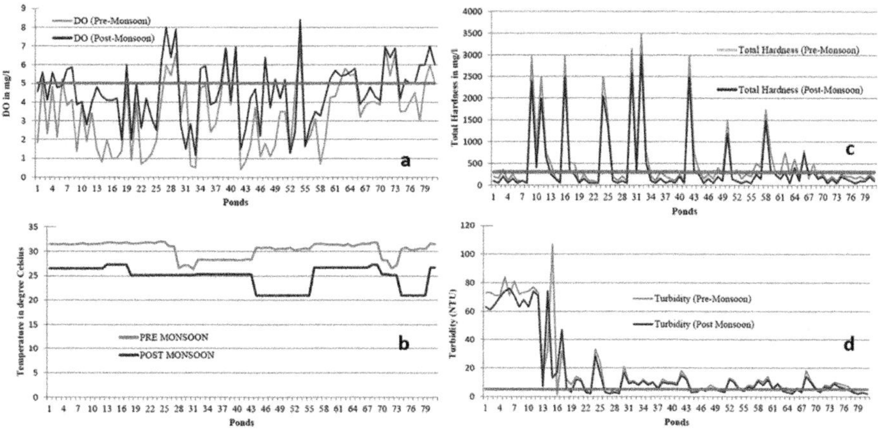

Fig. 3.4 Pond water quality (**a** DO, **b** Water temp, **c** Total Hardness and **d** Turbidity) in the study area during pre-monsoon and post-monsoon seasons along with the Bureau of Indian Standards (BIS) permissible limit (red color line)

concentrated, contributing to higher salinity and conductivity (LCRA 2014). Electrical conductivity indicates the total concentration of ionized constituents of water (Huq and Alam 2005) and it is a surrogate measure of salinity (Corwin and Yemoto 2017). Salinity levels are extremely high above 10 ppt in Ghoramara (13), Mousuni (18), Baliara (24), Narayanpur (30), Ramganga (32), Gangapur (36), Gobardhanpur (42) during pre-monsoon and post-monsoon seasons, and in Radhaballavpur (58), Kumirmari (68) only during pre-monsoon season. After cyclone *Aila*, the average water salinity in Sundarbans was altered from 13.64 ± 6.24 ppt to 17.08 ± 8.03 ppt with an ascend of 25.2% in concentration (Mitra et al. 2011; Dubey et al. 2016). Studies on salinity acclimation tolerance in fishes show that high salinity is lethal for freshwater fishes, though different fish have a different tolerance level of salinity and temperature (Dubey et al. 2016). Ghoramara (14), Mousuni (18), Gangapur (36), Gobardhanpur (42), Radhaballavpur 58 and Kumirmari(68) recorded EC value >1000 μ mhos/cm. TDS determines solid materials (calcium, magnesium, sodium, potassium, carbonates, bicarbonates, chlorides, sulfate, phosphate, silica, iron) dissolved in the water (Ahamed et al. 2015) and its higher concentration is detrimental to the human central nervous system, causing irritability and dizziness of face and paralysis of lips and tongue. It also harms the physiological processes of fishes and aquatic plants (Chang 2005). The Upper Permissible level of TDS in water is 2 ppt according to BIS (2012). Higher TDS values above 10ppt have been found in those ponds that recorded high salinity values.

Dissolved oxygen is required for living organisms to maintain their biological process. As per the standards proposed by CPCB and BIS, the permissible limit of DO ranges between 4 and 6 mg/L. DO level is extremely low in 29 ponds across the study area during the pre-monsoon period. The lowest DO has been found in Bagdanga (20), Kusumtala (22, 23) of Namkhana, Gobardhanpur (42, 43) of Patharpratima Block, and Radhaballavpur (58) of Basanti Block.

Temperature enhances the growth and death rate of aquatic organisms irrespective of DO level; it affects plants' photosynthesis, the solubility of oxygen in water (Mondal et al. 2016), and generates pollution (Xu et al. 2012; World Bank 2019). Temperature is a major factor in assessing water quality as it affects several other parameters and can modify the physical and chemical properties of water (Wilde 2006). Pre-monsoon temperature in the study area varies from 26.5 to 31.5 °C while post-monsoon temperature varies from 21 to 27.3 °C.

Total hardness reflects the soap-neutralizing power of water and is a measure of dissolved Ca^{2+} and Mg^{2+} substance expressed as $CaCO_3$ (Li et al. 2013). Sawyer and McCarthy (1967) classified water based on total hardness into four categories: soft (75 mg/L), moderately hard (75–150 mg/L), hard (150–300 mg/L), and very hard (300 mg/L) (Ahamed et al. 2015). The study area, located in a coastal zone possesses hard surface water. Except for two ponds (48, 50) in Kultali Block, all the other ponds reported total hardness below 350 mg/L, in Gosaba Block, except Kumirmari (68) all of the ponds recorded hardness below BIS limit. Except for few ponds in other Blocks, most notified hardness below or slightly above the BIS limit. The ponds that show exceptionally high hardness are pond no. 9, 11, and 16 in Sagar; 24, 25, 30 in Namkhana; 32, 42 in Patharpratima; 50 in Kultali, and 58 in Basanti.

Turbidity is generated from plankton, microscopic organisms in water as well as from suspended and inorganic materials (Efendi et al. 2015), which deter light penetration in the water column and thus limit photosynthesis in the bottom layer of water bodies. Higher turbidity leads to stratification of temperature and DO (Kale 2016; Halim et al. 2018; Anyanwu and Umeham 2020). Turbidity values are quite high in Sagar Block (7–74 NTU) while most of the stations of the study area reported turbidity above BIS permissible limit (5–20 NTU). Few stations of Kultali, Basanti, and Gosaba Block recorded lower turbidity.

3.3.2 Correlation Among Water Quality Parameters

Tables 3.4 and 3.5 show the correlation coefficient and their level of significance between water quality parameters in the pre-monsoon and post-monsoon seasons. Pearson's correlation has been applied in the study to understand the linkage among the parameters considered for the current research. Many of these parameters are significantly correlated with one another, and some are considerably higher than the recommended values. The tables clearly show that pH and Salinity are significantly negatively correlated (Namkhana, Patharpratima in pre and post-monsoon; Gosaba Block in pre-monsoon), as salinity causes a drop in pH values. It is thought that Salinity, TDS, and EC are related to each other, but they are different too. While Salinity is the concentration of salt, TDS denotes the amount of dissolved solids that do not necessarily contain salt. These dissolved salts and inorganic materials induce electrical current (increase conductivity) in water. A significant correlation among Salinity, TDS, and Conductivity has been found in this study. There is a significant negative correlation between DO and water temperature in most of the Blocks because the solubility of oxygen decreases with the increase of temperature. A significantly high positive correlation has been found between TDS and Total Hardness in this study. Water hardness represents the number of ions that have lost two electrons dissolved in the water and is, therefore, related to the TDS. Tables 3.4 and 3.5 also show that there is a correlation between turbidity, salinity, and TDS in most of the Blocks. Turbidity is a measure of water clarity while Salinity, TDS, and in some cases Conductivity affects water clarity. Thus Turbidity is related to these parameters. All of these water quality constituents are directly or indirectly linked with various ecosystem services.

3.3.3 Water-Related Challenges Faced by the Coastal Communities

Water-borne disease in the study primarily includes gastroenteric disease, skin disease, fever, and malaria. 23.57% of the surveyed households on average have reported such diseases. Baliara *Mouza* in Namkhana Block recorded the highest percentage (33.05%) of households suffering from such cases (Fig. 3.5). These diseases rise immediately after any natural hazards (cyclone, flood inundation) either due to consumption of contaminated water or by use of polluted water for daily household activities like cooking, washing, bathing (Kanjilal et al. 2010). Survey also mentioned that women suffer the most from skin diseases as they use saline

Table 3.4 Correlation coefficient and level of significance between pond water quality parameters (pre-monsoon season) in six blocks of Indian Sundarban

Block	Parameter	pH	Salinity	Conductivity	TDS	DO	Water temp	Total hardness	Turbidity
Sagar	pH	1							
	Salinity	−0.29	1						
	Conductivity	−0.011	0.29	1					
	TDS	−0.29	0.99***	0.30	1				
	DO	0.76***	−0.43*	−0.46*	−0.43*	1			
	Water temp	−0.53**	−0.33	−0.65	0.32	−0.44*	1		
	Total hardness	−0.55**	0.02	0.16	0.03	−0.44*	−0.085	1	
	Turbidity	0.54**	0.47**	−0.08	−0.47**	0.46**	−0.6***	−0.31	1
Namkhana	pH	1							
	Salinity	−0.46*	1						
	Conductivity	−0.19	0.59**	1					
	TDS	−0.52*	0.98***	0.62**	1				
	DO	0.24	−0.32	−0.48*	−0.39	1			
	Water temp	0.38	0.09	0.36	0.12	−0.62**	1		
	Total hardness	−0.73***	0.81***	0.19	0.80***	−0.27	−0.12	1	
	Turbidity	−0.59**	0.57**	0.43	0.56**	−0.31	0.22	0.58**	1
Patharpratima	pH	1							
	Salinity	−0.57**	1						
	Conductivity	−0.66**	0.60**	1					
	TDS	−0.58**	0.99***	0.60**	1				
	DO	0.36	−0.52*	−0.30	−0.51*	1			

(continued)

Table 3.4 (continued)

Block	Parameter	pH	Salinity	Conductivity	TDS	DO	Water temp	Total hardness	Turbidity
	Water temp	−0.53*	0.50*	0.58**	0.51*	−0.58**	1		
	Total hardness	−0.34	0.82***	0.31	0.83***	−0.67**	0.54*	1	
	Turbidity	−0.32	0.47	0.60**	0.48	−0.33	0.60**	0.55*	1
Kultali	pH	1							
	Salinity	−0.44	1						
	Conductivity	−0.46	0.99***	1					
	TDS	−0.45	0.99***	0.99***	1				
	DO	0.25	0.007	0.01	0.01	1			
	Water temp	0.25	−0.20	−0.25	−0.24	0.026	1		
	Total hardness	−0.028	0.82***	0.81***	0.82***	0.04	−0.21	1	
	Turbidity	−0.44	0.013	0.040	0.035	−0.10	−0.81***	−0.098	1
Basanti	pH	1							
	Salinity	−0.03	1						
	Conductivity	−0.10	0.97***	1					
	TDS	−0.05	0.99***	0.97***	1				
	DO	−0.18	−0.63**	−0.67**	−0.63**	1			
	Water temp	0.33	−0.024	0.05	−0.007	−0.53*	1		
	Total hardness	−0.13	0.94***	0.91***	0.95***	−0.67**	0.046	1	
	Turbidity	−0.50*	0.48*	0.60**	0.52**	−0.47	0.056	0.51*	1
Gosaba	pH	1							
	Salinity	−0.67***	1						

(continued)

Table 3.4 (continued)

Block	Parameter	pH	Salinity	Conductivity	TDS	DO	Water temp	Total hardness	Turbidity
	Conductivity	−0.61**	0.96***	1					
	TDS	−0.67***	0.99***	0.95***	1				
	DO	−0.028	−0.31	−0.34	−0.27	1			
	Water temp	0.002	0.42	0.45*	0.41	−0.47*	1		
	Total hardness	−0.63**	0.81***	0.66***	0.79***	−0.41	0.36	1	
	Turbidity	−0.65***	0.81***	0.78***	0.81***	−0.37	0.16	0.69***	1

Correlation statistics by Pearson's measure and p value analysis where *** indicate significantly correlated at 1% level of significance, ** indicate significantly correlated at 5% level of significance, * indicate significantly correlated at 10% level of significance

Table 3.5 Correlation coefficient and level of significance between pond water quality parameters (post-monsoon season) in six blocks of Indian Sundarban

Block	Parameter	pH	Salinity	Conductivity	TDS	DO	Water temp	Total hardness	Turbidity
Sagar	pH	1							
	Salinity	0.04	1						
	Conductivity	0.29	0.07	1					
	TDS	0.05	0.99***	0.04	1				
	DO	0.35	−0.07	−0.18	−0.09	1			
	Water Temp	−0.11	0.17	0.06	0.17	−0.19	1		
	Total Hardness	0.43*	0.06	0.09	0.10	0.60***	0.07	1	
	Turbidity	0.32	−0.41*	0.33	−0.43*	0.08	−0.75***	−0.13	1
Namkhana	pH	1							
	Salinity	−0.52*	1						
	Conductivity	−0.12	0.53**	1					
	TDS	−0.55**	0.95***	0.60**	1				
	DO	0.36	−0.39	−0.40	−0.44	1			
	Water Temp	0.01	0.01	0.01	0.01	0.01	1		
	Total Hardness	−0.74***	0.81***	0.12	0.73***	−0.39	0.01	1	
	Turbidity	−0.56**	0.51*	0.42	0.43	−0.18	0.01	0.56**	1
Patharpratima	pH	1							
	Salinity	−0.58**	1						
	Conductivity	−0.80***	0.57*	1					
	TDS	−0.71***	0.96***	0.66***	1				
	DO	0.47	−0.43	−0.33	−0.45	1			

(continued)

Table 3.5 (continued)

Block	Parameter	pH	Salinity	Conductivity	TDS	DO	Water temp	Total hardness	Turbidity
	Water Temp	−0.01	−0.01	−0.01	−0.01	−0.01	1		
	Total Hardness	−0.38	0.79***	0.32	0.79***	0.56*	−0.01	1	
	Turbidity	−0.44	0.46	0.61**	0.50*	−0.37	−0.01	0.52*	1
Kultali	pH	1							
	Salinity	−0.40	1						
	Conductivity	−0.41	0.83***	1					
	TDS	−0.67**	0.92***	0.79***	1				
	DO	−0.30	−0.001	−0.07	0.18	1			
	Water Temp	0.00	0.00	0.00	0.00	0.00	1		
	Total Hardness	0.26	0.88***	0.61**	0.77***	0.010	0.00	1	
	Turbidity	0.33	0.06	0.21	−0.10	0.49*	0.00	−0.18	1
Basanti	pH	1							
	Salinity	−0.17	1						
	Conductivity	0.11	0.98***	1					
	TDS	−0.19	0.95***	0.94***	1				
	DO	−0.17	−0.45	−0.46	−0.23	1			
	Water Temp	0.01	0.01	0.01	0.00	−0.01	1		
	Total Hardness	0.08	0.90***	0.90***	0.89***	−0.47	0.00	1	
	Turbidity	−0.59**	0.54*	0.55*	0.48*	−0.31	0.00	0.45	1
Gosaba	pH	1							
	Salinity	0.17	1						

(continued)

Table 3.5 (continued)

Block	Parameter	pH	Salinity	Conductivity	TDS	DO	Water temp	Total hardness	Turbidity
	Conductivity	0.28	0.82***	1					
	TDS	0.22	0.85***	0.95***	1				
	DO	−0.35	−0.58**	−0.44	−0.40	1			
	Water Temp	0.01	0.49*	0.54**	0.48*	0.17	1		
	Total Hardness	−0.11	0.77***	0.68***	0.72***	−0.45*	0.35	1	
	Turbidity	0.01	0.69***	0.79***	0.80***	−0.56**	0.23	0.64**	1

Correlation statistics by Pearson's measure and p value analysis where *** indicate significantly correlated at 1% level of significance, ** indicate significantly correlated at 5% level of significance, * indicate significantly correlated at 10% level of significance

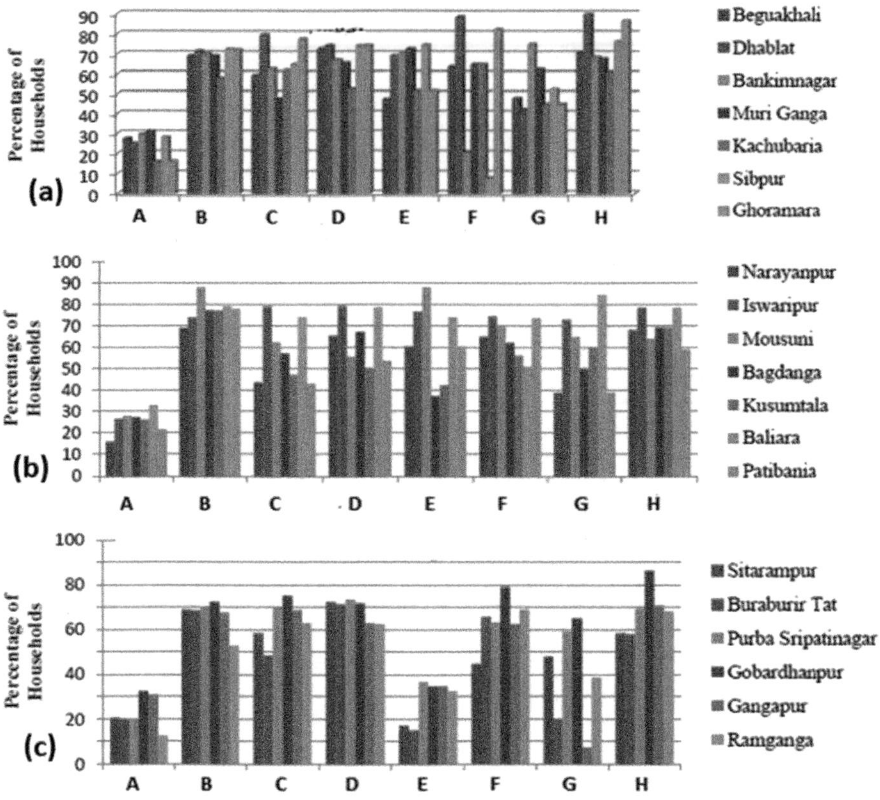

Fig. 3.5 Water-related hazards faced by coastal communities of **a** Sagar, **b** Namkhana and **c** Patharpratima blocks (A: Suffer from water-related diseases, B: Decreasing regeneration of green leafy vegetables, C: Decreasing agricultural production, D: Decreasing fish production, E: Walk more than two kilometers for household water, F: Use of unsafe water, G: Reported water consumption related argument and H: Climatic fluctuation)

pond water for cooking, bathing, washing, and *meen* farmers (prawn seed farmers) spend 4–6 h in saline river water for catching shrimp (Figs. 3.5 and 3.6). They also suffer from reproductive tract infection by use of saline water. Research studies have already stated that during post *Aila* period, there has been a constant decline in crop productivity owing to saline water ingression in agricultural land and increased evaporation in the Sundarbans. These may pose a serious threat to the future food security of the region (Chakraborty and Satpati 2019; Haldar and Debnath 2014). Survey results have also evidently indicated that 68.04% of households on an average reported decreasing availability of certain green leafy vegetables and almost 56.39% of households stated a post-*Aila* decline in overall agricultural production. 79.4% of households of Dhablat and Iswaripur *Mouzas* have reported such decline in crop production, which is the highest amongst all. A decrease in agricultural production forced people to shift their occupation, thus some turned into migrant workers

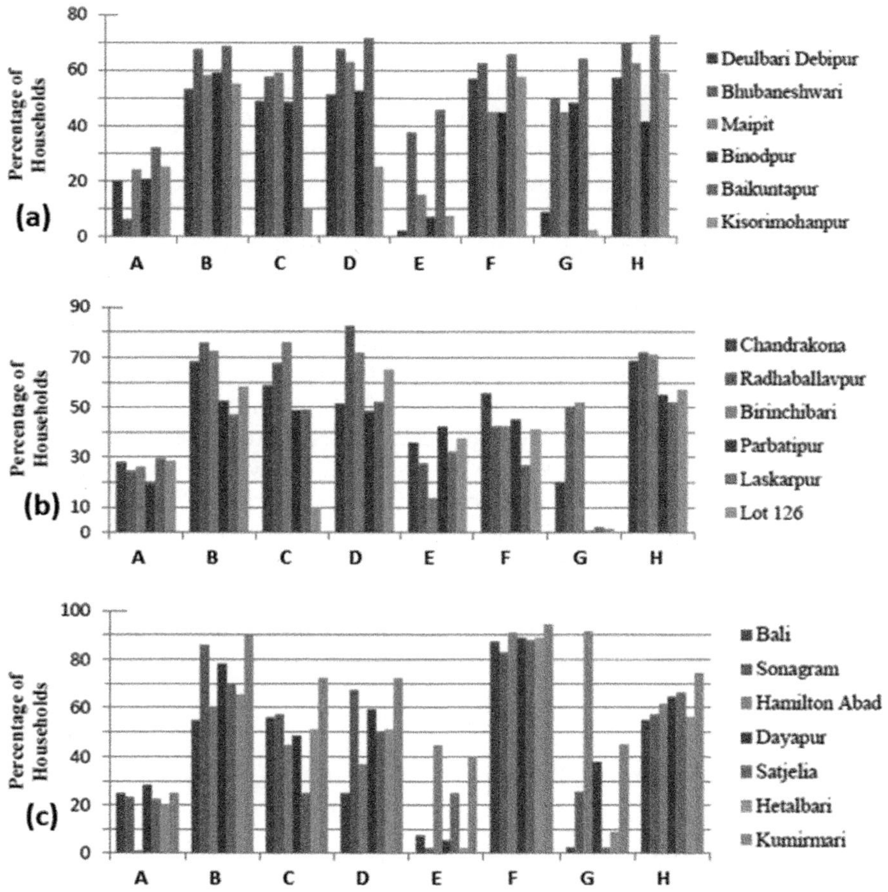

Fig. 3.6 Water-related hazards faced by coastal communities of **a** Kultali, **b** Basanti and **c** Gosaba blocks (A: Suffer from water-related diseases, B: Decreasing regeneration of green leafy vegetables, C: Decreasing agricultural production, D: Decreasing fish production, E: Walk more than two kilometers for household water, F: Use of unsafe water, G: Reported water consumption related argument and H: Climatic fluctuation)

while many of them started saline water fishing. As the sweet water bodies turned saline, reports on the decrease of size and amount of fishes became common here. At least 61.81% of households mentioned a decrease in fish production owing to saline water intrusion in the ponds located in marginal sites (Figs. 3.5 and 3.6). Higher Salinity, TDS, Total Hardness and lower DO content affects aquatic life vehemently. Many ponds of the study area recorded poor quality of water in terms of the aforesaid parameters. Sometimes they pump out saline water to desalinize the pond. Contaminated water does not only affect the swimming animals but also the rustic habited communities. On average, 61.93% of households use pond water for drinking, cooking, bathing, and washing. According to Census 2011, pond water is not considered for drinking in Ghoramara, Muri Ganga, Bagdanga, Kusumtala, Gangapur, Bali, Satjelia, Dayapur and all the *mouzas* of Kultali. Here the availability of safe drinking water relies on tube well, tap water but during flood inundation or saltwater ingression, tube wells remain submerged. Water access is also another problem in this area along with the quality. 37.14% of household heads stated that they have to walk more than 2 km to reach the water source for meeting household requirements. There are three tube wells in Iswaripur *Mouza* and more than 100 households are dependent fully on tube well. Thus conflicts related to water exist very much in the area that has been confirmed by the survey (39.89% household). During the household surveys, several banners or wall writings were observed near the tube wells with appeals "use tube well water only for drinking". Thus despite the poor quality of water, people are compelled to use pond water for washing or bathing. Households boil the saline water for cooking to make it germ-free. About 61.93% of households use unsafe water daily. Coastal erosion and embankment breaching are maximum in the *Mouzas* located near the Bay of Bengal; they are the worst sufferers with problems of saline water such as Ghoramara, Dhablat, Sibpur, Baliara, Gobardhanpur, Baikunthapur, and Kumirmari. Iswaripur *Mouza* is encircled by the Hatania-Doania River on one side and Saptamukhi on the other and thus suffers from severe erosion. 66.33% of households perceive that rising temperature, rainfall anomaly, increasing frequency of would flood have direct or indirect consequences on food security of the region in near future.

3.4 Policy Recommendation

Public Health Engineering Department, Govt. of West Bengal (PHED) is entrusted to provide safe and quality drinking water supply service to the communities and manage technical services during the period of natural hazard/disaster. PHED also dig tube well, extract groundwater and provide schemes of rainwater harvesting. Rajiv Gandhi National Drinking Water Mission is also implemented here (World Bank 2014). Sundarban Social Development Centre (SSDC), an NGO, has been organizing several programs with local youth and women to deal with water-related problems. They have facilitated more tube wells and introduced "*jalabandhus*" (friends of water) who are the trained individuals recruited to maintain tube wells. They also organize campaigns for awareness on sanitation, water contamination, etc. (SSDC 2020). During the surveys, it has been found that when tube wells installed by *Panchayat* (local governing body) become defunct, villagers contribute money to recuperate it. Even whenever an embankment gets breached, Govt. does not make a timely decision in re-erection of embankments. Villagers with their joint efforts somehow manage to rebuild breached parts of the embankments.

Maintenance of vulnerable embankment and tube wells during a flood and apt reconstruction of embankments after breaching are the actions appropriate to resolve water-related problems. As per Census 2011, the percentage of unirrigated land is very high in the study area (Fig. 3.7). Thus proper maintenance of sweet surface water

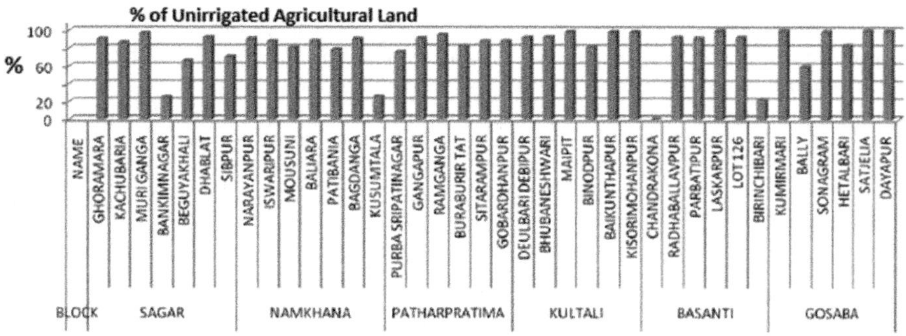

Fig. 3.7 Un-irrigated areas of Sagar, Namkhana, Patharpratima, Kultali, Basanti, and Gosaba (*Data Source* District Census Handbook, South 24 Parganas District, 2011)

can help to water those unirrigated agricultural lands. The major cereal crop cultivated in the area is *Aman* (Rice) which depends on monsoon rainfall for water input. If irrigation and proper protection from salinization are undertaken, then more than two or three crops can be cultivated in the same field. Agriculture and fishing activity are the two climate-sensitive subsistence sectors affected badly by the changing climate. Thus the Union and the State Govt. must work hand in hand for building adaptation capacity. Environmental monitoring and grass-root level implementation of Governmental schemes must be checked by Officials from time to time. Water quality must be tested frequently by collecting samples and checking their condition in authentic laboratories.

3.5 Conclusion

The objective of the research is satisfied as it is shown that the water security and water quality of the study region have been threatened by persistent sea-level rise, coastal flooding, climatic hazards, embankment breaching, and salinization. Extremely low pH, DO and extremely high Salinity, Conductivity, TDS, Turbidity, and Total Hardness of the water of the ponds located in the vulnerable sites are an indubitable problem to the coastal society. Poor water quality is related to several health hazards discussed in the present study. Proper environmental monitoring is thus a key requirement to address the problem. Block Development Office and non-profit Non-Government Organizations (NGOs) must arrange workshops for awareness generation on health, sanitation, rainwater harvesting, water purification measures, etc. to ensure water quality and security. The present study can be beneficial in framing future adaptation policies related to water in the estuarine ecosystem to safeguard the vulnerable communities in the current era of climate change.

Acknowledgements The author acknowledges all the local inhabitants of Indian Sundarbans who collected several water samples and sent to the laboratory. The author is also grateful to all the people who helped acquiring information on the functioning of these ponds.

References

Adarsh S, Mahantesh B (2006) Personal communication. Resident Doctors of Medicine. Hanagal Shree Kumareshwara Hospital, Bagalkot, Karnataka, India

Ahamed AJ, Loganathan K, Jayakumar R (2015) Hydrochemical characteristics and quality assessment of groundwater in Amaravathi river basin of Karur district, Tamil Nadu, South India. Sustain Water Resour Manag. https://doi.org/10.1007/s40899-015-0026-3

Akhter M, Reza MdS, Jamil NAHM, Uddin Nazim Md (2018) Assessment water quality and seasonal variations based on aquatic biodiversity of Sundarbans mangrove forest, Bangladesh. IOSR J Biotechnol Biochem 4(1):06–15

Annual Flood Report (2014) Irrigation and waterways department. Government of West Bengal, India

Anyanwu ED, Umeham SN (2020) Identification of waterbody status in Nigeria using predictive index assessment tools: a case study of Eme River, Umuahia, Nigeria. Int J Energy Water Resour. https://doi.org/10.1007/s42108-020-00066-5

Bandyopadhyay S (2020) Sundarban: a review of evolution & geomorphology. In: World Bank. http://documents.worldbank.org/curated/en/119121562735959426/pdf/Sundarban-A-Review-of-Evolution347and-Geomorphology.pdf. Accessed 10 April 2021

BIS (Bureau of Indian Standards) 10500 (2012) Specification for drinking water. Indian Standards Institution, New Delhi, India

Census (2011) South 24 Parganas District, West Bengal, Government of India

Central Pollution Control Board (CPCB) report (2000) Water quality status of Yamuna River. Assessment and Development of River Basin

Chaffey DR, Miller FR, Sandom JH (1985) A forest inventory of the Sundarban, Bangladesh. Overseas Development Administration Project Report 140, England, 196

Chakraborty A, Satpati LN (2019) Vulnerability, resilience and quality of life: a micro-level study of Ghoramara Island in the Sundarban region of West Bengal, India. In: Sinha BRK (ed) Multidimensional approach to quality of life issues. Springer Nature, Singapore, pp 241–252

Chang H (2005) Spatial and temporal variations of water quality in the Han River and its tributaries, Seoul, Korea, 1993–2002. Water Air Soil Pollut 161:267–284

Corwin D, Yemoto K (2017) Salinity: electrical conductivity and total dissolved solid. Methods Soil Anal. https://doi.org/10.2136/msa2015.0039

Das P, Bandyopadhyay J (2019) The present status of the coastal ecological estuarine water bodies and its water quality analysis through geospatial technology and assessment of fishery resources in Sundarban, West Bengal, India. Int J Sci Res Rev 8(1):992–1006

Das K, Mukherjee A (2019) Depth-dependent groundwater response to coastal hydrodynamics in the tropical, Ganges river mega-delta front (the Sundarbans): impact of hydraulic connectivity on drinking water vulnerability. J Hydrol. https://doi.org/10.1016/j.jhydrol.2019.05.053

Das S, Siddiqui NA (1985) Effect of time and method of storage on the survivability of uprooted Keora (Sonneratia apetala) seedlings, Bano Bigyan Patrika

Dasgupta S, Kamal FA, Khan ZH, Choudhury S, Nishat A (2014) River salinity and climate change evidence from Coastal Bangladesh. The World Bank, Development Research Group, Environment, and Energy Team

Dasgupta S, Huq M, Wheeler D (2015) Drinking water salinity and infant mortality in Coastal Bangladesh. Development Research Group, Environment and Energy Team, February 2015, Policy Research Working Paper

District Disaster Management Plan (2012, 2015, 2019–20) Government of West Bengal, India
Dubey SK, Trivedi RK, Chand BK, Mandal B, Rout SK (2016) Farmers' perceptions of climate change, impacts on freshwater aquaculture and adaptation strategies in climatic change hotspots: a case of the Indian Sundarban delta. Environ Dev. https://doi.org/10.1016/j.envdev.2016.12.002
Efendi H, Romanto B, Wardiatno Y (2015) Water quality status of Ciambulawung River Banten Province based on pollution index and NSF-WQI. Procedia Environ Sci 24:228–237
Gopal B, Chauhan M (2006) Biodiversity and its conservation in the Sundarban Mangrove Ecosystem. Aquat Sci. https://doi.org/10.1007/s00027-006-0868-8
Gupta N, Pandey P, Hussain J (2017) Effect of physicochemical and biological parameters on the quality of river water of Narmada, Madhya Pradesh, India. Water Sci. https://doi.org/10.1016/j.wsj.2017.03.002
Haldar A, Debnath A (2014) Assessment of climate-induced soil salinity conditions of Gosaba Island, West Bengal and its influence on local livelihood. In: Singh M et al (eds) Climate change and biodiversity: proceedings of IGU Rohtak conference, Volume 1, Advances in geographical and environmental sciences. Springer Nature, Japan, pp 27–44. https://doi.org/10.1007/978-4-431-54838-6_3
Halim A, Sharmin S, Rahman H, Haque M, Rahman S, Islam S (2018) Assessment of water quality parameters in baorenvironment, Bangladesh: a review. Int J Fish Aquat Stud 6(2):259–263
Hanson S, Nicholls R, Ranger N, Hallegatte S, Corfee-Morlot J, Herweijer C, Chateau J (2011) A global ranking of port cities with high exposure to climate extremes. Clim Change 104:89–111
Hossain RMd, Shahriyar S, Md RA, Sabab FR, Nawshin N, Forhad N (2016) Suitability of pond sand filter as an alternative safe drinking water technology at the Sundarban Region, Bangladesh. J Biodivers Environ Sci 8(2):283–292
Huq SMI, Alam MD (2005) A handbook on analysis of soil, plant, and water. University of Dhaka, Bangladesh, BACERDU
ICMR-Council of Medical Research (1975) Manual of standards of quality for drinking water supplies, Indian Special Report No. 44
IMD-Indian Meteorological Department (2018) Government of India, 1901–2018
Jahan A (2018) The effect of salinity in the flora and fauna of the Sundarbans and the impacts on local livelihood. Department of Earth Sciences, Uppsala University. www.geo.uu.se
Kale VS (2016) Consequence of temperature, pH, turbidity, and dissolved oxygen water quality parameters. Int Adv Res J Sci Eng Technol 3(8):186–190
Kanjilal B, Mazumdar PG, Mukherjee M, Mondal S, Barman D, Singh S, Mandal A (2010) Health care in the Sundarbans (India): challenges and plan for a better future. Institute of Health Management Research
Khan AE, Ireson A, Kovats S, Mojumder SK, Khusru A, Rahman A, Vineis P (2011) Drinking water salinity and maternal health in Coastal Bangladesh: implications of climate change. Environ Health Perspect. https://doi.org/10.1289/ehp.1002804
Khan AE, Scheelbeek PFD, Shilpi AB, Chan Q, Mojumder SK et al (2014) Salinity in drinking water and the risk of (pre)eclampsia and gestational hypertension in Coastal Bangladesh: a case-control study. PLoS ONE. https://doi.org/10.1371/journal.pone.0108715
LCRA (2014) Water quality indicators. In: Colorado river watch network. http://www.lcra.org/water/quality/colorado-river-watch-network/Pages/water-qualityindicators.aspx. Accessed 13 July 2020
Leo ML, Dekkar M (2000) Hand book of water analysis. Marcel Dekker, New York, pp 1–25, 115–117, 143, 175, 223–226, 261, 273, 767
Li P, Wu J, Qian H (2013) Assessment of groundwater quality for irrigation purposes and identification of hydrogeochemical evolution mechanisms in Pengyang County, China. Environ Earth Sci 69:2211–2225
Mishra VN (2003) Water—a threatened global resource. In: Proceedings international conference on water and environment (WE-2003), Bhopal, India, pp 11–16
Mitra A (2019) Climate change and pollution. In: Mitra A (ed) Estuarine pollution in the lower Gangetic Delta. Springer Nature, Switzerland AG, pp 207–227

Mitra A, Gangopadhyay A, Dube A, Schmidt ACK, Banerjee K (2009) Observed changes in water mass properties in the Indian Sundarbans (northwestern Bay of Bengal) during 1980–2007. Curr Sci 97(10):25

Mitra A, Halder P, Banerjee K (2011) Changes of selected hydrological parameters in Hooghly estuary in response to severe tropical cyclone (Aila). Indian J Geomar Sci 40:32–36

Mondal I, Bandyopadhyay J, Paul AK (2016) Water quality modeling for seasonal fluctuation of Ichamati river, Model Earth Syst Environ. https://doi.org/10.1007/s40808-016-0153-3

Mukherjee N, Siddique G (2018) Climate change and vulnerability assessment in Mousuni Island: South 24 Parganas District. Spat Inf Res. https://doi.org/10.1007/s41324-018-0168-0

Mukherjee N, Siddique G (2019a) Assessment of climatic variability risks with application of livelihood vulnerability indices. Environ Dev Sustain. https://doi.org/10.1007/s10668-019-00415-3

Mukherjee N, Siddique G (2019b) Gendered vulnerability of climate change: experiences of women in the Sundarbans. In: Chattopadhyay et al (eds) Gender and development: aspects of social and economic change. New Delhi Publishers, New Delhi, pp 15–29

Mukherjee N, Siddique G, Basak A, Roy A, Mandal MH (2019) Climate change and livelihood vulnerability of the local population on Sagar Island, India. https://doi.org/10.1007/s11769-019-1042-2

Nicholls RJ (2003) An expert assessment of storm surge "Hotspots." Final Report (Draft Version) to Center for Hazards and Risk Research, Lamont-Dohert Observatory, Columbia University

Rahman MMd, Rahman MT, Md RS, Ahmad JU, Shakera B, Md HA (2013) Water quality of the world's largest mangrove forest. Can Chem Trans. https://doi.org/10.13179/canchemtrans.2013.01.02.0018

Rahman MM, Penny G, Zaman MH, Kryston A, Salehin M, Nahar Q, Islam MS, Bolster D, Tank JL, Müller MF (2019) Salinization in large river deltas: drivers, impacts and socio-hydrological feedbacks. Water Secur. https://doi.org/10.1016/j.wasec.2019.100024

Sadik SMd, Shaw R, Rahman RMd, Nakagawa H, Kawaike K (2018) Event-consequence chain of climate change-induced salinity intrusion in Sundarbans mangrove socioecological system, Bangladesh. In: Science and Technology in Disaster Risk Reduction in Asia. https://doi.org/10.1016/B978-0-12-812711-7.00006-7

Sánchez-Triana E, Paul T, Ortolano L, Ruitenbeek J (2014) Building resilience for sustainable development for the West Bengal Sundarbans—strategy report (Report no. 88061-In). World Bank, Washington, DC. http://documents.worldbank.org/curated/en/2014/01/20162806/building-resilience-sustainable-development-sundarbans-strategy-report. Accessed 10 April 2020

Sánchez-Triana E, Ortolano L, Paul T (2016) Managing water-related risks in 481 the West Bengal Sundarbans: policy alternatives and institutions. Int J Water Resour Dev. https://doi.org/10.1080/07900627.2016.1202099

Saha S, Halder M, Mookerjee S, Palit A (2019) Seasonal influence, enteropathogenic microbial load and diarrhoeal enigma in the Gangetic Delta, India: present scenario and health implications. J Infect Public Health. https://doi.org/10.1016/j.jiph.2019.01.066

Sarkar SK, Bhattacharya BD (2010) Water quality analysis of the Coastal Regions of Sundarban Mangrove Wetland, India using multivariate statistical techniques. In: Sarkar SK (ed) Environmental management. Sciyo, Croatia

Sawyer GN, McCarthy DL (1967) Chemistry of sanitary engineers, 2nd edn. McGraw Hill, New York, p 518

Scheelbeek PFD, Khan AE, Mojumder S, Elliott P, Vinei P (2016) Drinking water sodium and elevated blood pressure of healthy pregnant women in salinity affected coastal areas. Hypertension. https://doi.org/10.1161/HYPERTENSIONAHA.116.07743

Shammi M, Rahman MMd, Bondad SE, Bodrud-Doza Md (2019) Impacts of salinity intrusion in community health: a review of experiences on drinking water sodium from coastal areas of Bangladesh. Healthcare. https://doi.org/10.3390/healthcare7010050

Siddique A (2018) Increase in drinking water salinity in Bangladesh heightens risk of pregnancy complications. In Earth Journalism. https://earthjournalism.net/stories/increase-in-drinking-water-salinity-in-Bangladesh-heightens-risk-of-pregnancy-complications. Accessed 19 April 2021

Sundarban Social Development Centre (SSDC) (2020) Our focus area. http://ssdcindia.org.in/. Accessed 08 April 2020

Whitehead PG, Jin L, Macadam I, Janes T, Sarkar S, Rodda HJE, Sinha R, Nicholls RJ (2018) Modelling impacts of climate change and socio-economic change on the Ganga, Brahmaputra, Meghna, Hooghly, and Mahanadi river systems in India and Bangladesh. Sci Total Environ. https://doi.org/10.1016/j.scitotenv.2018.04.362

WHO and UNICEF (2017) Progress on drinking water, sanitation and hygiene: 2017 update and SDG baselines. World Health Organization and the United Nations Children's Fund, Geneva, p108

Wilde F (2006) Temperature 6.1. In: USGS field manual. http://water.usgs.gov/owq/FieldManual/Chapter6/6.1_ver2.pdf. Accessed 8 Jan 2020

World Bank (2014) Building resilience for sustainable development of the Sundarbans. South Asia Region, Sustainable Development Department, Environment & Water Resources Management Unit, The World Bank, Report No. 88061

World Bank (2019) Inventory of freshwater resources in the Sundarbans landscape, Limited to the part of Landscape Locate. http://documents.worldbank.org/curated/en/328371562736987775/text/Inventory-of-Freshwater-Resources-in-the-Sundarbans-Landscape-Limited-to-the-part-of-LandscapeLocated-in-India.txt. Accessed 21 April 2021

World Health Organization (WHO) (2006) Guidelines for drinking water quality. Recommendations (1), 3rd ed, pp 491–493

Vineis P, Chan Q, Khan A (2011) Climate change impacts on water salinity and health. J Epidemol Glob Health. https://doi.org/10.1016/j.jegh.2011.09.001

Xu L, Li H, Liang X, Yao Y, Zhou L, Cui X (2012) Water quality parameters response to temperature change in small shallow lakes. Phys Chem Earth. https://doi.org/10.1016/j.pce.2010.11.005

Zaman S, Bhattacharyya SB, Pramanick P, Raha AK, Chakraborty S, Mitra A (2013) Rising water salinity: a threat to mangroves of Indian Sundarbans. Water insecurity: a social dilemma community. In: Abedin MA et al (eds) Environment and disaster risk management. Emerald Group Publishing Limited, pp 167–183

Zaman S, Agarwal S, Mitra A, Amin G, Prmanick P, Mitra A (2015) Impact of Aila on the dissolved oxygen level in the Indian Sundarbans Region. J Energy Environ Carb Credits 5(3):1–4

Chapter 4
Fishing and Aquaculture Practice in the Ponds of the Indian Sundarbans

Abhra Chanda and Sourav Das

Abstract Though agriculture continues to be the predominant livelihood option for the millions residing within the Indian Sundarban Biosphere Reserve (SBR), inland fishing has lately emerged as a promising and viable option to alleviate the financial status of the grossly impoverished population of this region. Ponds and rearing of fishes in those ponds are integral components of the daily life of the people of Sundarbans. However, fishes are grown in most of the household ponds for self-consumption, and extensively without much investment. Aquaculture operations, on the other hand, require substantial scientific intervention to commercialize the fish yield and get a handsome return against the time and labor input. The present chapter collated the knowledge acquired so far in this domain, from the perspective of SBR. The fish diversity and the aquaculture management practices are detailed in a nutshell. The salient features and the typical characteristics of the ponds of the SBR are also discussed. This chapter also identifies the basic threats and challenges that the aquaculture farms of this region are facing at present. Overall, this chapter gives a brief overview of the inland fish farming practice and the attitude of the local inhabitants towards this promising sector of earning revenues.

Keywords Household ponds · Aquaculture ponds · Freshwater aquaculture · Brackishwater aquaculture · Fish diversity · Pond management practices · Threats to fish farming

4.1 Introduction

The Sundarban Biosphere Reserve (SBR), which is renowned for being the largest single tract of contiguous mangrove forest, is also home to more than four million people in the Indian counterpart (Mitra et al. 2021). The population density in the transition zone of the SBR is substantially high, and agriculture is the principal livelihood option for most of the people who live in this region (Ghosh and Mistri

A. Chanda (✉) · S. Das
School of Oceanographic Studies, Jadavpur University, 188 Raja S. C. Mullick Road, Kolkata 700032, India

S. Das et al. (eds.), *Pond Ecosystems of the Indian Sundarbans*, Water Science and Technology Library 112, https://doi.org/10.1007/978-3-030-86786-7_4

2020). Besides agriculture, apiculture, the collection of forest resources like fuelwood and other non-timber forest produce (NTFPs) comprise another means of livelihood (De and Das 2021). However, another livelihood option that has been traditionally practiced by many in the SBR since time immemorial is fishing and fish rearing (Kumaran et al. 2020). The adjacency of the open oceanic regions of the Bay of Bengal made capture fisheries a viable source of income for many (Dutta et al. 2021). However, offshore fishing requires substantial capital investment and it is not a feasible option for the impoverished and marginalized section of the society. Though offshore fishing engages hundreds and thousands as laborers, it is not a suitable option for the millions to earn a secure livelihood. Moreover, the recent past has witnessed uncertain turnovers from offshore fishing mostly due to practicing unsustainable means of fishing (Suresh et al. 2021). However, fishing in open estuarine waters with small country boats by an individual or a small group of people is a very common sight in the SBR. According to Das and Manadal (2016), in the forest fringe areas of the SBR, more than 20% of the population practices this type of fishing. These people mainly go for various types of finfish, shellfish, prawn seed, and prawn seed collection. A substantial number of women in this fringe area are professional fishers. Women often set sail with small boats and fish not only in the open estuaries but also enters the forest creeks (Sundaray et al. 2019). Women with no boats are often seen walking in the intertidal zones and they collect small fish and prawn seeds with bag nets or bare hands. However, these practices have considerable risk as occasional human-animal conflict and casualties due to crocodile and tiger attacks are reported from this region (Paul 2020).

Inland fishing is traditionally practiced in the local household ponds throughout the rural setup of India and the Indian Sundarbans is no exception. In the SBR, ponds are an integral part of almost all households. The soil dug out from a portion of the land is used to make the Kucha houses or shelter for cattle and thus, it serves a two-way purpose by creating a means of water availability and raw construction material (Mandal et al. 2015). Rainwater is usually stored in these ponds. The monsoon rain fills these ponds during the rainy season from June to September. The stored water is used for multiple purposes throughout the year and especially during the dry months. Bathing and washing utensils are some of the common household activities that utilize pond water. Irrigation in agricultural plots or household agriculture also finds its use. Almost all the household ponds are also used to rear fishes of various types. The local people introduce fingerlings of various species and rely on natural photosynthetic processes for the fishes to grow. This type of fish rearing is only conducted for subsistence and local consumption (Mandal et al. 2015). However, in the recent past commercial aquaculture practice to grow fishes have gained significant popularity in this region (Kumaran et al. 2020). Fish is an extremely common and, at the same time, a popular cuisine among the Bengali community (Das Chaudhuri 2019). Thus, the demand for fishes in this region is never-ending. Given the uncertainty witnessed in the field of capture fisheries and the risk of fishing in forested estuaries, aquaculture emerged as one of the potent options within the SBR (Sundaray et al. 2019; Kumaran et al. 2020). The introduction of aquaculture, especially amongst the marginalized section of Asia and Africa, was meant to act as a relief from the burden of poverty and a means of earning a decent livelihood (Ahmed and Lorica 2002; Mwaijande and Lugendo 2015). The people of Sundarbn have also lately realized the potential of aquaculture practice in providing an alternate livelihood and self-reliance (Kumaran et al. 2020).

The scope of aquaculture in India has a promising future, and already at present, India ranks second after China (the largest aquaculture industry in the world) accounting for 6% of the global total aquaculture produce (FAO 2016). Owing to a diverse landscape and geomorphological setting, India offers significant potential for carrying out both freshwater and brackish water aquaculture. Both these types of aquaculture are prevalent in the SBR. Followed by the state of Andhra Pradesh, West Bengal ranks second in terms of aquaculture production within the country and it involves an estimated 3.2 million people in this sector. Within West Bengal, SBR plays the most crucial role in aquaculture production. Singh et al. (2010) estimated that almost two million people from the SBR are directly or indirectly engaged with fishing activities. These observations indicate that the SBR is witnessing a paradigm shift from agriculture to aquaculture and such radical change can play a crucial role in alleviating socio-economic conditions of a large section of this coastal community. However, it is also feared at the same time that the effect of unsustainable and indiscriminate conversion of land to aquaculture can lead to severe consequences in this ecosystem (Halder et al. 2021).

4.2 Salient Features of Aquaculture Ponds in SBR

Various types of aquaculture practice are prevalent in the SBR, which includes freshwater monoculture, brackishwater monoculture, fresh and brackishwater composite culture or mixed culture, rice-shrimp mixed farming, etc. The aquaculture operations are mostly handled by men and their income lies around Rs. 50,000 per annum (Dubey et al. 2016). Inland aquaculture is practiced in the SBR since time immemorial and most of the present practitioners inherited the farms from their ancestors. However, in the recent past, a substantial rate of conversion of agricultural lands to aquaculture farms is witnessed (DasGupta et al. 2019). A comparatively better income and lesser risk than traditional agriculture (due to erratic rainfall and climate variability) are the principal reasons behind this large-scale conversion to aquaculture (Sarkhel 2015). They mostly rely on their traditional knowledge when it comes to fish farming (Abraham et al. 2010). Though efforts from the government as well as non-government sectors to disseminate the knowledge of the modern scientific advancements in this domain to the fish farmers exist, in reality, it is scarcely implemented due to lack of coordination (Dubey et al. 2016). It is also worth mentioning that the people of SBR seldom carry out aquaculture exclusively throughout the year. Most of these people practice fish farming as an additional source of income, keeping other livelihood options intact (Dubey et al. 2016).

According to Dubey et al. (2016) farming of finfish and shellfish, both are popular in the SBR. Almost three out of four ponds in the SBR practice freshwater aquaculture. One of four ponds practices both freshwater and brackishwater aquaculture in SBR. In the present date, polyculture is mostly preferred by the fish farmers of SBR (Biswas et al. 2019). Most of the aquaculture ponds are perennial ($\approx$70%), while the rest are seasonal ($\approx$30%). Inlets and outlets are maintained in almost half of the ponds whereas it is absent in the other half. In the case of freshwater aquacultures, mainly rainwater is utilized to practice fish farming; however, several ponds use a mixture of rainwater and groundwater. Brackishwater aquacultures are mostly practiced in the island peripheries, where the estuarine water of varying salinity is allowed to enter the ponds (Shyne Anand et al. 2018). Very few of the ponds in SBR have additional nursery ponds or grow-out ponds (Dubey et al. 2016). However, at present, the number of ponds with the grow-out facility is increasing. In the present date, the majority of the fish farmers purchase juvenile fish stock from hatcheries (Dubey et al. 2016).

4.3 Pond Management and Farming Practices

A suite of pond management strategies is adopted by local people of SBR for their household ponds, specifically for the aquaculture ponds. Dewatering of ponds is one such common strategy often implemented for household as well as aquaculture ponds. The fresh initiation of the cycle of an aquaculture production often begins with dewatering the entire pond, followed by sun-drying (Biswas et al. 2019). This effort enables the pond managers to carve out the pond shape properly before allowing the waters to refill the ponds and remove any unwanted weeds and shrubs. The anaerobic soil bottoms get rejuvenated to some extent through this practice. This practice is occasionally carried out in household ponds as well. The pond owners often mentioned that sometimes they feel the entire water needs to be drained out. They rely on their traditional knowledge and experience in deciding the time of dewatering. A foul odor and blackish-green color are some of the indicators that compel the dewatering of household ponds. Removal of bottom sediment from the ponds is also a common practice in the SBR (Dubey et al. 2016). This practice is mostly seen in the aquaculture ponds are scarcely in the household ponds. The aquaculture ponds accumulate substantial organic loads and detritus material in the pond bottom (Kalous et al. 2012). Bottom feeders like some prawns can collect their food resources from such deposited materials (Moraes-Valenti and Valenti 2010; Franchini et al. 2020); however, an excess of such materials in the pond bottom often lead to undesirable consequences. Bottom sediment removal is often practiced in the presence of water but mostly takes place when the ponds are dewatered. The presence of excess organic detritus makes the pond bottom extremely anoxic (Musyoka 2016), which in turn favors some anaerobic bacterial communities to reduce sulfate to hydrogen sulfide and this leads to foul odor and pungent gas ebullition (Antony and Philip 2006). Accumulation of hydrogen sulfide levels beyond a certain threshold affects the respiratory process of several fishes and indirectly incurs various diseases (Jasmin et al. 2020). De-weeding is another technique that is quite prevalent in the SBR. Removal of unnecessary weeds from the ponds leads to better functioning of the ponds (Rahman et al. 2011). Besides, the above-mentioned management practices, lime treatment, and compost manure application are two of the most common practices observed in both the household ponds and aquaculture ponds. Lime treatment is carried out for multifarious purposes. Maintaining the pH of the water column is the prime objective of lime treatment (Chanda et al. 2019). Calcium is known to play a crucial role in the bone development of fishes (Fontagné et al. 2009) which comes into the aquaculture system through lime treatment. Lime treatment is done in two ways in the ponds of SBR. Sometimes, when the ponds are de-watered solid lime (powdered) is applied to the bottom soil and allowed to sun-dry. Besides, a lime slurry is sprayed throughout the pond when the pond remains filled with water. Again in this perspective, the fishermen use their traditional knowledge to decide the time and magnitude of lime required for their ponds to perform well in terms of fish production. Occasionally potash alum is mixed with lime and applied to the water body. The alum is known to play a disease preventive role (Anderson 1992), as well as, it acts as a coagulating agent and facilitates the deposition of unwanted suspended particles, which in turn, enhances the clarity of the water column (Igwegbe and Onukwuli 2019). Even in the present date, a substantial number of fish farmers rely upon manure to fertilize their ponds, which includes vermicompost, farmyard, and many other types, like cow dung, and oilseed cakes (Ghosh et al. 2019). According to Dubey et al. (2016), 30% of the fish farmers practice single stocking, i.e., they stock the fish once in a cycle and harvest altogether at once. However, the majority of the fish farmers practice multiple stockings, i.e., they stock the fish at regular intervals and periodically harvest depending on the size and shape attained by the fishes or species (in the case of polyculture). Fish seeds can be broadly categorized into three types, namely the fries, the hatchlings, and the fingerlings. Dubey et al. (2016) observed that fingerlings are the most popular followed by the fries, and the hatchlings are the least used. A sizable number of farmers also prefer a mixture of all three types. The fish stocking density varies largely depending on the fishes being cultured; however, the stocking combination has some prominent preferences. The Indian carps and Tilapia are the most preferred couple followed by the minor carps mixture. A substantial number of fish farmers do not provide any external fish and rely on the natural photosynthetic process for the fishes to grow; whereas a large number of people use natural items like rice bran and oil cakes as fish feed (Dubey et al. 2016). Those with a higher tenacity to commercialize the fish produce purchase fish feeds available in the market and apply them in the ponds to amplify the fish growth. Feeding intervals vary from daily to weekly depending on the type of fish being farmed and the particular farmer's traditional knowledge.

4.4 Freshwater and Brackishwater Species

The Sundarbans host a wide variety of fishes in the estuaries, creeks, and near-shore waters of the Bay of Bengal. Dubey et al. (2015) recorded 62 freshwater fish species, out of which 8% belonged to the endangered and critically endangered category. The admixture of freshwater and saline water in and around the intricate network of waterways of Sundarbans provides a suitable habitat for an array of both freshwater and marine fishes (Gopal and Chauhan 2006). The adjacency of the vast mangrove stretch provides a suitable shelter and acts as a nursing ground for these fishes (Dutta et al. 2016). Several anadromous fishes that spend most of their life cycle in the marine water proceed towards the estuarine reaches during spawning (Giri et al. 2020), whereas several freshwater fishes and prawns that usually inhabit the freshwaters access the estuaries during spawning (Sundaray et al. 2019). Thus, this unique eco-region acts as an abode for all types of fishes, which enhances the fish diversity of this region. This high ichthyofaunal abundance in the estuaries is reflected in the inland fishing scenario as well (Fig. 4.1). The Indian major carps (*Catla catla*, *Labeo rohita*, and *Cirrhinus mrigala*) still dominate freshwater aquaculture; however, several minor carps (*Labeo bata*, *Labeo calbasu*, etc.) are also cultivated throughout the SBR. Tilapia (*Oreochromis niloticus*, *Oreochromis mossambicus*) is still a very popular choice among many fish farmers (Fig. 4.2). Compared to freshwater finfishes, options for freshwater shellfishes are very few, out of which, *Macrobrachium rosenbergii* is farmed by many, especially during the rainy season (Sundaray et al. 2019) (Fig. 4.3). Brackishwater finfishes like *Mystus gulio* and *Lates calcarifer* are occasionally farmed (Fig. 4.4); however, among the brackishwater

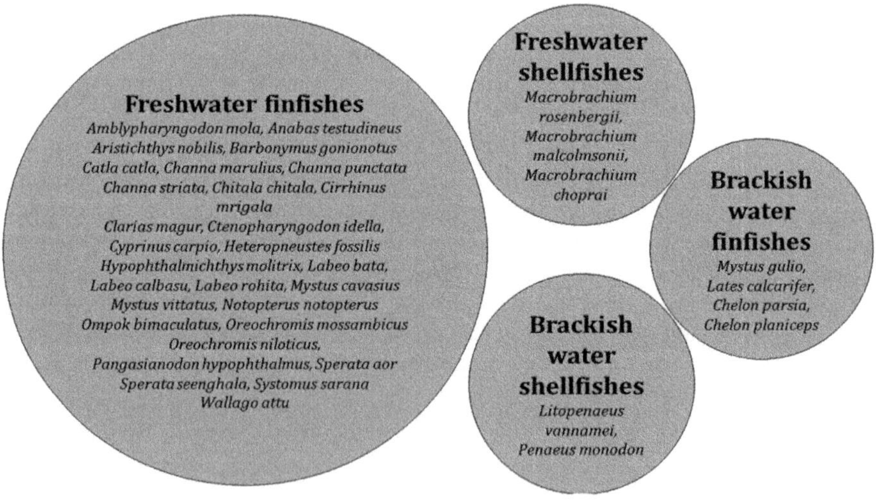

Fig. 4.1 The dominant fish species farmed in the aquacultures of SBR

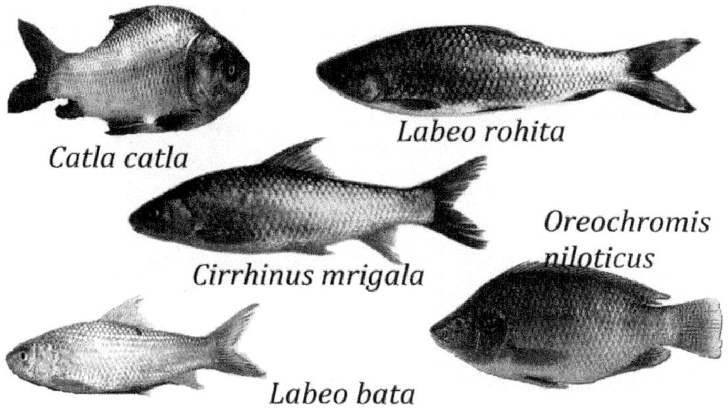

Fig. 4.2 The images of few dominant freshwater finfishes which are grown through aquaculture farming in the SBR

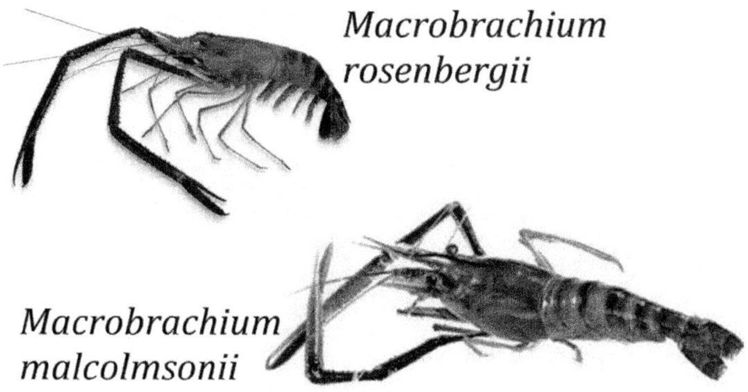

Fig. 4.3 The images of the dominant freshwater shellfishes which are grown through aquaculture farming in the SBR

shellfishes, the exotic *Litopenaeus vannamei* is at present outplaying the indigenous *Penaeus monodon* (Fig. 4.5).

4.5 Challenges and Threats to Aquaculture Fishing

4.5.1 Serious Lack of Scientific Intervention

Sundaray et al. (2019) emphasized that there exists a serious lack of coordination among the fish farmers remotely based in the different islands of Sundarban. A significant number of fish farmers have no idea about the modern scientific breakthroughs

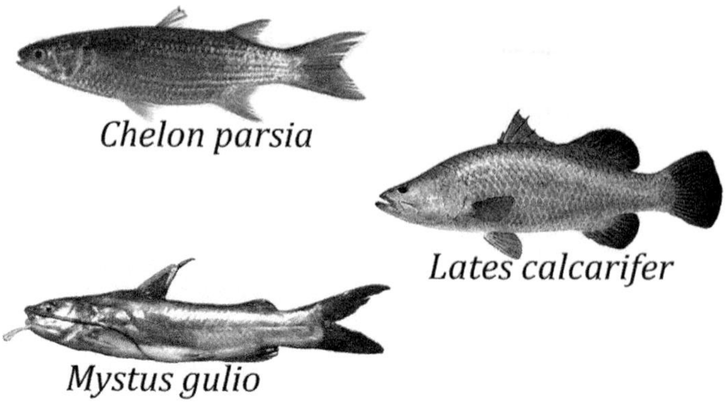

Fig. 4.4 The images of few dominant brackishwater finfishes which are grown through aquaculture farming in the SBR

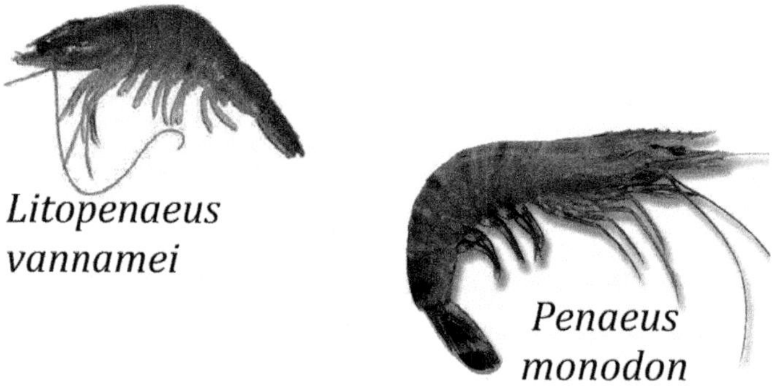

Fig. 4.5 The images of the dominant brackishwater shellfishes which are grown through aquaculture farming in the SBR

in the domain of aqua-farming. Their traditional knowledge is not up to the mark to enhance the yield of the fish to that extent, which can be commercially profitable. Inadequate pond management strategies, absence of fish feed and fertilizers bar the fish farmers of SBR from getting the expected return, even when this unique eco-region has a huge potential in terms of aquaculture farming. Sundaray et al. (2019) also stressed that the fish farmers of this region are not all aware of the fish nursing and rearing, and purchase the fish seeds from external markets. This lack of knowledge or proper intent incurs heavy financial losses to the fish farmers or reduces the profit margin.

4.5.2 Cyclonic Disasters

The Sundarbans is prone to the frequent occurrence of severe cyclonic storms. In the last three decades the intensity, as well as the frequency of these tropical cyclones, have increased manifold (Mandal and Hosaka 2020). The high wind velocity not only leads to the destruction of life and property, but massive storm-induced surges also ravage the rural areas (Sahana and Sajjad 2019). Breaching of embankments and saline water intrusion into the agricultural fields and ponds are some of the most inevitable events that follow after a cyclone passes through the SBR (Pramanik et al. 2021). Saline water floods most of the freshwater ponds and imposes a salinity shock to the present ichthyofaunal diversity leads to mass mortality of fishes (Sundaray et al. 2019). Moreover, extreme rainfall events lead to the overtopping of the ponds and most of the fishes leave the pond enclosures, which lead to substantial losses to fish farmers (Paul and Chatterjee 2019). Several fish farmers also claim that blown away materials like broken branches of trees and leaves due to the storm-passe leads to deterioration of the water quality of the ponds. These observations are very common in the case of household ponds, which are often associated with trees planted nearby.

4.5.3 Fish Diseases

Dubey et al. (2016) noted that many of the farmers of the SBR strongly believe that the fishes of this region do not suffer from any diseases and the same was advocated by Chand et al. (2012) due to the presence of salt in these waters, which kept several diseases at bay. However, with the changing climate and the recurrent climate extreme events like cyclone, droughts, and increase in the ambient temperature, several fish diseases have become prominent in this region (Sundaray et al. 2019). Epizootic ulcerative syndrome, fin and tail rot, malnutrition, dropsy, and parasitic outbreaks like Lernaeasis, Argulosis, and Myxoboliasis, along with fluke diseases like Gyrodatylosis and Dactylogyrosis, are some of the most common diseases among the finfishes of SBR (Dubey et al. 2016, 2017; Sundaray et al. 2019). The crustaceans on the other hand mostly suffer from white spots, fungal infection in shells, blackening of gills, and softening of shells (Dubey et al. 2016; Sundaray et al. 2019). These diseases were found to onset any time of the year; however, the beginning of the post-monsoon season till the winter end, is the most vulnerable time of the year for disease outbreak (Khan and Lilley 2002; MacRae et al. 2002). The monsoonal runoff-induced proliferation of industrial chemicals and pollutants in the ponds coupled with lowered immunity at this time of the year is held accountable for the susceptibility of diseases in the fishes (Bly et al. 1997).

4.5.4 *Intrinsic Soil Character*

The ponds of this SBR are earthen and the water stored within these enclosures remain in direct contact with the unearthed region. Mondal (2003) observed that the SBR comprises two types of soils, the Aridisols (which thrives in arid to semi-arid environments influenced by salinity) and the Alfisols (alluvium which experiences a steady humid environment). However, both of these types often encompass soils rich in acid-sulfate at varying depths. Though the ponds of SBR are mostly shallow, during unearthing these acid soil layers get exposed and come in direct contact with water (Sundaray et al. 2019). Direct exposure to such soils drastically reduces the pH of the entire water column. The local fishermen though recognize the problem but often due to a lack of proper lime treatment and other necessary measures that need to be implemented, the fishes die in numbers. Sundaray et al. (2019) observed that such effects of acid soils diminish over three to five years, and this accounts for substantial loss to the fish farmers.

4.6 Conclusion

Collating the very few existing pieces of literature on inland fishing in the Indian Sundarban Biosphere Reserve, it can be inferred that fish rearing and farming is integral to the lives of the local inhabitants of this region, however, in terms of commercialization of the aquaculture practice, it is still in infancy. Most of the households practice traditional extensive fishing in their ponds, and the fishes grown are meant for self-consumption. The aquaculture sector has seen tremendous growth in the last decade; however, technological intervention is yet to spread across the nooks and corners of this ecosystem. In the last decade, the Central Institute of Freshwater Aquaculture (CIFA) founded by the Indian Council of Agricultural Research (ICAR), New Delhi has taken several measures to properly educate the fish farmers regarding the scientific advancements of this domain; however, still, it is a long way to go to fetch the expected results. Many scholars believe that the multifarious challenges that the aquaculture and inland fishing sector of SBR experience in the present date can be handled by implementing proper management strategies, and that this sector can in the truest sense, alleviate the impoverished conditions of the majority of the population within the SBR.

References

Abraham TJ, Sil SK, Vineetha P (2010) A comparative study of the aquaculture practices adopted by fish farmers in Andhra Pradesh and West Bengal. India J Fish 57:41–48

Ahmed M, Lorica MH (2002) Improving developing country food security through aquaculture development lessons from Asia. Food Pol 27(2):125–141. https://doi.org/10.1016/S0306-9192(02)00007-6

Anderson DP (1992) Immunostimulants, adjuvants, and vaccine carriers in fish: applications to aquaculture. Ann Rev Fish Dis 2:281–307. https://doi.org/10.1016/0959-8030(92)90067-8

Antony SP, Philip R (2006) Bioremediation in shrimp culture systems. Naga World Fish Center Q 29(3 & 4):62–66

Biswas G, Kumar P, Kailasam M, Ghoshal TK, Bera A, Vijayan KK (2019) Application of integrated multi trophic aquaculture (IMTA) concept in brackishwater ecosystem: the first exploratory trial in the Sundarban, India. J Coast Res 86(SI):49–55. https://doi.org/10.2112/SI86-007.1

Bly JE, Quiniou SM, Clem LW (1997) Environmental effects on fish immune mechanisms. Dev Biol Stand 90:33–43

Chand BK, Trivedi RK, Dubey SK, Beg MM (2012) Aquaculture in changing climate of Sundarbans. Survey report on climate change vulnerabilities, aquaculture practices & coping measures in Sagar and Basanti Blocks of Indian Sundarban, West Bengal University of Animal & Fishery Sciences, Kolkata, India. Available at India environment portal: http://re.indiaenvironmentportal.org.in

Chanda A, Das S, Bhattacharyya S, Das I, Giri S, Mukhopadhyay A et al (2019) CO_2 fluxes from aquaculture ponds of a tropical wetland: potential of multiple lime treatment in reduction of CO_2 emission. Sci Total Environ 655:1321–1333. https://doi.org/10.1016/j.scitotenv.2018.11.332

Das Chaudhuri R (2019) Menus matter: examining class and Bengali cuisine culture through restaurant menus in Kolkata. Contemp South Asia 27(3):422–435. https://doi.org/10.1080/09584935.2019.1648383

Das CS, Mandal RN (2016) Coastal people and mangroves ecosystem resources vis-à-vis management strategies in Indian Sundarban. Ocean Coast Manag 134:1–10. https://doi.org/10.1016/j.ocecoaman.2016.09.025

DasGupta R, Hashimoto S, Okuro T, Basu M (2019) Scenario-based land change modelling in the Indian Sundarban delta: an exploratory analysis of plausible alternative regional futures. Sustain Sci 14(1):221–240. https://doi.org/10.1007/s11625-018-0642-6

De D, Das CS (2021) Measuring livelihood sustainability by PCA in Indian Sundarban. Environ Dev Sustain. https://doi.org/10.1007/s10668-021-01451-8

Dubey SK, Chand BK, Trivedi RK, Mandal B, Rout SK (2016) Evaluation on the prevailing aquaculture practices in the Indian Sundarban delta: an insight analysis. J Food Agric Environ 14(2):133–141

Dubey SK, Trivedi RK, Chand BK, Mandal B, Rout SK (2017) Farmers' perceptions of climate change, impacts on freshwater aquaculture and adaptation strategies in climatic change hotspots: a case of the Indian Sundarban delta. Environ Dev 21:38–51. https://doi.org/10.1016/j.envdev.2016.12.002

Dubey SK, Trivedi RK, Chand BK (2015) Indigenous freshwater piscine resources of Indian sundarban biosphere reserve: status and prospects. World J Fish Mar Sci 7(1):21–28. https://doi.org/10.5829/idosi.wjfms.2015.7.1.92103

Dutta S, Chakraborty K, Hazra S (2016) The status of the marine fisheries of West Bengal coast of the northern Bay of Bengal and its management options: a review. Proc Zool Soc 69(1):1–8. https://doi.org/10.1007/s12595-015-0138-7

Dutta S, Orlov A, Hazra S (2021) Population biology and exploitation status of four commercially important marine fishes of the northern Bay of Bengal, India. Iranian J Fish Sci 20(1):62–83. https://doi.org/10.22092/IJFS.2021.123477

FAO (2016) Food and Agriculture Organization of the United Nations. The state of world fisheries and aquaculture (2016) In: Contributing to food security and nutrition for all. FAO, Rome, p 200

Fontagné S, Silva N, Bazin D, Ramos A, Aguirre P, Surget A et al (2009) Effects of dietary phosphorus and calcium level on growth and skeletal development in rainbow trout (*Oncorhynchus mykiss*) fry. Aquaculture 297(1–4):141–150. https://doi.org/10.1016/j.aquaculture.2009.09.022

Franchini AC, Costa GA, Pereira SA, Valenti WC, Moraes-Valenti P (2020) Improving production and diet assimilation in fish-prawn integrated aquaculture, using iliophagus species. Aquaculture 521:735048. https://doi.org/10.1016/j.aquaculture.2020.735048

Ghosh S, Mistri B (2020) Coastal agriculture and its challenges: a case study in Gosaba Island, Sundarban, India. Sp Cult India 8(2):140–154. https://doi.org/10.20896/saci.vi0.747

Ghosh S, Sahu NC, Rahaman FH, Das KS (2019) Periphyton based climate smart aquaculture for the farmers of Indian Rural Sunderban Areas. Indian Res J Ext Educ 19(1):60–72

Giri S, Purkayastha S, Hazra S, Chanda A, Das I, Das S (2020) Prediction of monthly Hilsa (*Tenualosa ilisha*) catch in the Northern Bay of Bengal using Bayesian structural time series model. Reg Stud Mar Sci 39:101456. https://doi.org/10.1016/j.rsma.2020.101456

Gopal B, Chauhan M (2006) Biodiversity and its conservation in the Sundarban Mangrove Ecosystem. Aquat Sci 68(3):338–354. https://doi.org/10.1007/s00027-006-0868-8

Halder S, Samanta K, Das S, Pathak D (2021) Monitoring the inter-decade spatial–temporal dynamics of the Sundarban mangrove forest of India from 1990 to 2019. Reg Stud Mar Sci 44:101718. https://doi.org/10.1016/j.rsma.2021.101718

Igwegbe CA, Onukwuli OD (2019) Removal of total dissolved solids (TDS) from aquaculture wastewater by coagulation-flocculation process using Sesamum indicum extract: effect of operating parameters and coagulation-flocculation kinetics. Pharma Chem J 6(4):32–45

Jasmin MY, Syukri F, Kamarudin MS, Karim M (2020) Potential of bioremediation in treating aquaculture sludge. Aquaculture 519:734905. https://doi.org/10.1016/j.aquaculture.2019.734905

Kalous L, Bui AT, Petrtýl M, Bohlen J, Chaloupková P (2012) The South American freshwater fish *Prochilodus lineatus* (Actinopterygii: characiformes: Prochilodontidae): new species in Vietnamese aquaculture. Aquac Res 43(6):955–958. https://doi.org/10.1111/j.1365-2109.2011.02895.x

Khan MH, Lilley JH (2002) Risk factors and socio-economic impacts associated with epizootic ulcerative syndrome (EUS) in Bangladesh. In: Arthur JR, Phillips MJ, Subasinghe RP, Reantaso MB, Macrae IH (eds) Primary aquatic animal health care in rural, small-scale aquaculture development. FAO Fisheries Technical Paper No. 406, FAO, Rome, pp 27–39

Kumaran M, Ghoshal TK, De D, Biswas G, Raja RA, Anand PS et al (2020) Aquaculture-based production systems for the livelihood security of coastal farm families in the risk-prone agroecosystem of India: an appraisal. Aquac Int 28(2):805–814. https://doi.org/10.1007/s10499-019-00495-y

MacRae IH, Chapman G, Nabi SMN, Dhar GC (2002) A survey of health issues in carp/ Macrobrachium culture in rice fields in Bangladesh. In: Arthur JR, Phillips MJ, Subasinghe RP, Reantaso MB, Macrae IH (eds) Primary aquatic animal health care in rural, small-scale aquaculture development. FAO Fisheries Technical Paper No. 406, FAO, Rome, pp 27–39

Mandal MSH, Hosaka T (2020) Assessing cyclone disturbances (1988–2016) in the Sundarbans mangrove forests using Landsat and Google Earth Engine. Nat Hazard 102(1):133–150. https://doi.org/10.1007/s11069-020-03914-z

Mandal S, Burman D, Sarangi SK, Bandyopadhyay BK, Maji B (2015) Homestead production systems in Sundarbans region of West Bengal, India–Current status and opportunities. CGIAR-CPWF

Mitra S, Roy S, Paul SK (2021) Rehabilitative planning strategy for transformative changes of vulnerable communities in Indian Sundarban. Social Transformations in India, Myanmar, and Thailand, vol I. Palgrave Macmillan, Singapore, pp 259–277

Mondal AK (2003) The Sundarbans of India. Indus Publishing Company, New Delhi, pp 26–27

Moraes-Valenti P, Valenti WC (2010) Culture of the Amazon River prawn *Macrobrachium amazonicum*. In: New MB, Valenti WC, Tidwell JH et al (eds) Freshwater prawns: biology and farming. Wiley-Blackwell, USA, pp 485–501

Musyoka SN (2016) Concept of microbial bioremediation in aquaculture wastes: review. Int J Adv Sci Tech Res 5:(6). http://repository.seku.ac.ke/handle/123456789/2696

Mwaijande FA, Lugendo P (2015) Fish-farming value chain analysis: policy implications for transformations and robust growth in Tanzania. J Rural Commun Dev 10(2):47–62

Paul KB (2020) A second wave of forest—settlement territorialization: a case study from the Indian Sundarbans. Environ Sp Place 12(1):83–109. https://doi.org/10.5749/envispacplac.12.1.0083

Paul BK, Chatterjee S (2019) Climate change-induced environmental hazards and Aila Relief Measures undertaken to Sundarbans in Bangladesh and India. In: Sen H (ed) The Sundarbans: a disaster-prone eco-region. Springer, Cham, pp 469–490. https://doi.org/10.1007/978-3-030-00680-8_16

Pramanik M, Szabo S, Pal I, Udmale P, Pongsiri M, Chilton S (2021) Population health risks in multi-hazard environments: action needed in the Cyclone Amphan and COVID-19–hit Sundarbans region, India. Clim Dev. https://doi.org/10.1080/17565529.2021.1889948

Rahman MA, Mustafa MG, Barman BK (2011) Impacts of aquaculture extension activities on female fish farmers in different areas of Bangladesh. Bangladesh J Zool 39(2):213–221

Sahana M, Sajjad H (2019) Vulnerability to storm surge flood using remote sensing and GIS techniques: a study on Sundarban Biosphere Reserve, India. Remote Sens Appl Soc Environ 13:106–120. https://doi.org/10.1016/j.rsase.2018.10.008

Sarkhel P (2015) Flood risk, land use and private participation in embankment maintenance in Indian Sundarbans. Ecol Econ 118:272–284. https://doi.org/10.1016/j.ecolecon.2015.08.003

Shyne Anand PS, Balasubramanian CP, Lalramchhani C, Panigrahi A, Gopal C, Ghoshal TK, Vijayan KK (2018) Comparison of mudcrab-based brackishwater polyculture systems with different finfish species combinations in Sundarban, India. Aquac Res 49(9):2965–2976. https://doi.org/10.1111/are.13755

Singh A, Bhattacharya P, Vyas P, Roy S (2010) Contribution of NTFPs in the livelihood of mangrove forest dwellers of Sundarban. J Hum Ecol 29(3):191–200

Sundaray JK, Chakrabarti PP, Mohapatra BC, Das A, Hussan A, Ghosh A, Hoque F (2019) Freshwater aquaculture in Sundarbans India. In: Sen H (ed) The Sundarbans: a disaster-prone eco-region. Springer, Cham, pp 295–319. https://doi.org/10.1007/978-3-030-00680-8_10

Suresh VR, Mandhir SK, Sajina AM, Manna RK, Mukherjee J, Banik SK et al (2021) Reproductive potential and stock status of the anadromous shad, *Tenualosa ilisha* (Hamilton, 1822): implications for managing its declining fisheries in northern Bay of Bengal bordering India. Reg Stud Mar Sci 44:101753. https://doi.org/10.1016/j.rsma.2021.101753

Chapter 5
Role of Iron Fertilization on the Changes of Chlorophyll Concentration and Fish Production in the Brackish Water Ponds of Indian Sundarbans

Sourav Das

Abstract In the present study, chlorophyll-*a* and nutrients were studied in the pond water collected from three brackish water ponds of Namkhana, Indian Sundarbans. Within these three ponds, one pond is treated with iron fertilizer and another one is with mangrove litter. The third pond acted as a control pond. The maximum value of chlorophyll concentration in the iron fertilized pond indicates the higher concentration of phytoplankton due to iron enhancement. Moreover, low nutrient concentrations accorded with the maximum chl-*a* concentration in the iron-fertilized pond. This observation indicated that the nutrients were used by phytoplankton during the higher rate of photosynthesis in the presence of higher iron value. The present study can be used to enhance the fish production of aquaculture ponds of Indian Sundarbans as well as link up with the future adaptation policies related to increasing the income of the vulnerable communities of Indian Sundarban.

Keywords Iron fertilization · Higher chlorophyll concentration · Increase fish production · Aquaculture pond · Indian Sundarban

5.1 Introduction

Phytoplankton structure the basic food chain of aquatic ecosystems (all types). The information on phytoplankton is important to understand the significant features of the surface aquatic systems and the consequence of the physico-chemical parameters on the phytoplankton assemblage. Besides many other aquatic physicochemical parameters, phytoplankton can serve as a proxy of the water quality of a pond (Crossetti and Bicudo 2008). These groups of organisms are extremely sensitive to environmental variables and play a crucial role in regulating the biogeochemistry of lentic ecosystems like ponds (Çelekli et al. 2014). Chlorophyll-*a* is a unique indicator of phytoplankton abundance in the surface water (Mitra et al. 2012). The other interrelated parameters, such as bio-volume, cell size and physiology of phytoplankton are also important for aquatic ecosystem studies (Chisholm 1992). Chlorophyll-a also

S. Das (✉)
School of Oceanographic Studies, Jadavpur University, Kolkata, India

S. Das et al. (eds.), *Pond Ecosystems of the Indian Sundarbans*, Water Science and Technology Library 112, https://doi.org/10.1007/978-3-030-86786-7_5

acts as an indicator of the biological productivity of any aquatic ecosystem (Huot et al. 2007).

Indian Sundarbans is positioned at the southernmost landmass of West Bengal, India. The Indian Sundarbans comprises the largest mangrove cover (~4264 km^2) in the world (Sánchez-Triana et al. 2016). The term Sundarbans came from "*Sunderbunds*" which means the forests of *Heritiera fomes*, locally called "*Sundri*" (Mitra et al. 2004). The Indian Sundarbans comprise a group of 102 islands out of which 54 are inhabited (Mukherjee and Siddique 2018). The region is a transitional zone between freshwaters coming from the Hooghly River and the saline water from the Bay of Bengal (Akhter et al. 2018). Besides the mangrove forest and an array of diverse flora and fauna, the Indian Sundarbans hosts a highly dense population of 4.4 million people (Census 2011). Though agriculture is the principal means of earning a livelihood in this region, aquaculture fish farming practice is also very common among the people of Sundarbans (Dubey et al. 2016). Ponds are integral to the daily life activities of the people of Sundarbans; however, the fishes reared in the household ponds are mostly meant for self-consumption. Lately, this region has witnessed a surge in the commercialization of fish farming through aquaculture ponds (DasGupta et al. 2019; Thakur et al. 2021). The fish farmers of this region, even in the present date, mostly rely on their traditional fishing practice, and seldom take into account the modern scientific innovations to enhance the yield and productivity of fishes (Chand et al. 2012; Dubey et al. 2016). In addition, the ongoing climate change has significantly disrupted the already poor infrastructure of aquaculture ponds in this region (Dubey et al. 2017). Under such conditions, several initiatives like the integrated multi-trophic aquaculture concept have been put under lenses and are in the experimental phase with the main intent to this sector sustainable and at the same increase the profit margin of the fish farmers (Biswas et al. 2019).

Phytoplankton abundance or chlorophyll-a concentration in aquaculture pond has a major impact on fish production (Sara 2007a). Thus, chlorophyll concentration would be an essential tool to acquire a perfect knowledge for the use of pond habitations (Sara 2007b). The aquaculture is secondarily affected by the changes in water physicochemical environments that can control the phytoplankton abundance or chlorophyll concentration (Jewel et al. 2003) in aquaculture pond. These environments are significantly changed due to the variations of the nutrient also.

In this chapter, we have estimated Chlorophyll-*a* pigment and nutrients of three brackish water ponds to study the growth of phytoplankton under different treatments in the Namkhana Block of Indian Sundarbans. One pond (aquaculture) was kept as control (only riverine water stored in it). The other two ponds were treated with iron salt (e.g. $FeSO_4$) and mangrove litter, respectively.

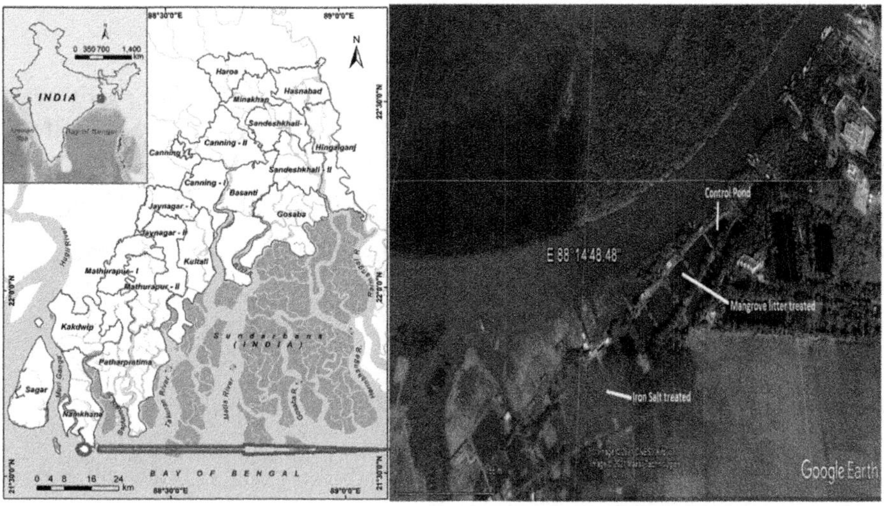

Fig. 5.1 Study area map showing the location of treated ponds in Indian Sundarbans

5.2 Materials and Methods

5.2.1 Study Area

A major part of the Sundarbans delta (mangrove-dominated estuarine delta) lies in Bangladesh and rest of the mangrove forest is within the India (Banerjee et al. 2012). There are several brackish water ponds fed by Indian Sundarbans estuaries (Mitra et al. 2004). We have selected three such brackish water ponds in the Namkhana region (Fig. 5.1). These ponds are treated differently to examine the impact of iron salt adding on the chlorophyll concentration of the pond water during the pre-monsoon season of 2016.

5.2.2 Salinity

The in-situ salinity was measured using a Multikit (WTW Multi 340i Set; Merck, Germany) close-fitting with the probe WTW Tetracon 325 and it was additionally measured argentometrically by modified Mohr–Knudsen titration method to authenticate the probe data (Grasshoff et al. 1999).

5.2.3 Dissolved Iron

10 L plastic bottles were used to collect the pond water from the three ponds. Water samples were filtered through Nucleopore filter papers (pore size 0.4 μm) within 2 h of sample collection. The aliquot was acidified with HNO_3 (sub-boiling temperature) to a pH ~2. The acidified aliquot was stored in polyethylene bottles. According to Danielsson et al. (1978), Fe (dissolved in condition) was separated and extracted from brackish pond water. Extract was analyzed by Atomic Absorption Spectrophotometer (Perkin Elmer: Model 3030).

5.2.4 Nutrient Analyses

To analyze the nutrients of pond water, the pond water samples were filtered (by 0.45 μm cellulose acetate membranes), then 0.3 ml 35 g/l $HgCl_2$ solutions were added in each 100 ml sample (Kattner 1999). Dissolved inorganic nitrogen (DIN containing nitrate + nitrite ions), dissolved inorganic phosphate (DIP containing phosphate ions), and dissolved inorganic silicate (DSi containing silicate ions) were measured by spectrophotometric (Shimadzu UV–Visible 1600 double-beam spectrophotometer) methods (DIN-cadmium reduction method, DIP and DSi-molybdate colorimetric method) defined by Grasshoff et al. (1999). The detection limits (±relative error of estimation) of nitrate, nitrite, phosphate and silicate ions were 0.25 (±2.8%), 0.08 (±4.2%), 0.05 (±3.4%) and 0.09 (±6.0%) μM, respectively. Distilled water (double) was used during the analysis.

5.2.5 Result and Discussion

The mean salinity of the Control brackish water pond, Mangrove litter treated pond and $FeSO_4$ treated pond was 4.1, 4.4, and 4.3 respectively. The Chl-*a* concentrations were maximum in the iron salt-treated pond (90.2 mg/m^3), followed by the mangrove litter treated pond (42.3 mg/m^3) and control pond (30.5 mg/m^3) (Fig. 5.2a). The nitrate concentration trend was: control pond (20.0 μM) > mangrove litter treated pond (12.5 μM) > iron salt-treated pond (6.2 μM) (Fig. 5.2b). Silicate and Phosphate also showed similar trends with the maximum values in the control ponds (82.5 μM and 10.2 μM respectively), followed by mangrove litter pond (5.3 μM and 44.2 μM respectively) and iron salt-treated pond (2.7 μM and 23.7 μM respectively) (Fig. 5.2c, d).

According to Chisholm and Morel (1991), there is a very common paradox in the surface water of different oceanic regions, i.e. High Nitrate Low Chlorophyll (HNLC). Several factors control the utilization of nitrate, the amount of CO_2 consumed by phytoplankton, and the quantity of C exported from the ocean surface

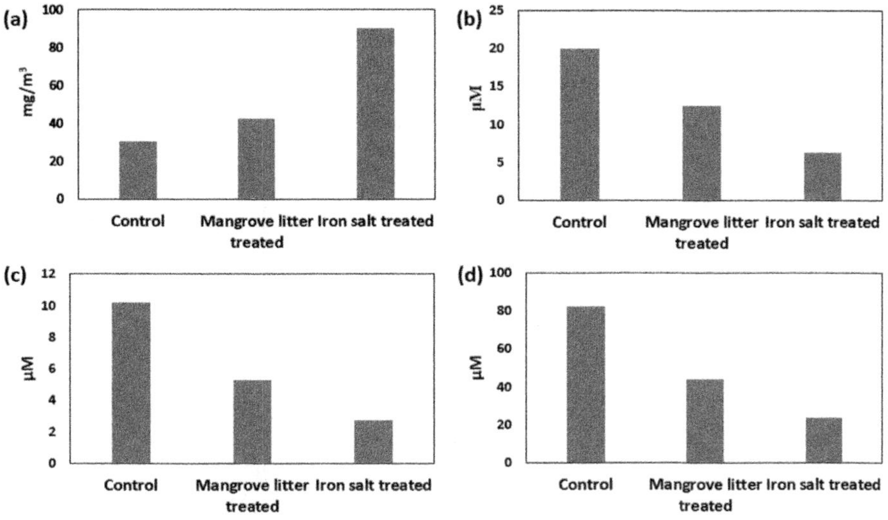

Fig. 5.2 Pond water quality **a** Chl-a, **b** nitrate, **c** phosphate and **d** silicate during different treatment

(Siegenthaler 1986). Price et al. (1991) described that grazing pressure applied on phytoplankton by quickly replicating microzooplankton and the deficiency of micronutrients (i.e. iron) may play the role together in the HNLC water. There are several studies on the role of micronutrients on the phytoplankton community, biovolume, etc. (Landry et al. 1997; Price et al. 1994; Coale 1991; Martin et al. 2013; De Baar et al. 1990). In the present study, we have noticed a considerable increase of chlorophyll concentration (i.e. higher growth of phytoplankton) in iron sulfate treated pond along with decreasing concentration of nitrate, phosphate, and silicates (Fig. 5.2). On the contrary, the control pond did not show any such type of peak of chlorophyll in this period.

All the hydrological variables (i.e. Chlorophyll-*a*, nitrate, phosphate, and silicate) exhibited a unique response to iron fertilization after 10 days of experimental period. Compared to the control pond (only brackish water), the litter (mangrove) treated brackish water pond and iron salt-treated pond indicated a 41.25% and 196.23% increase in Chl-*a*, respectively. The nitrate concentration decreased by 37.5% in litter-treated (mangrove) pond and 69.0% in the iron salt-treated pond compared to the control pond. A similar tendency was also detected for the other two nutrients. The phosphate decreased by 49.01% in the mangrove litter treated pond and by 73.52% in the iron sulfate treated pond. The silicate reduced by 46.42% in the litter (mangrove) treated brackish water pond and by 71.27% in the iron salt-treated brackish water pond (Fig. 5.2). Zaman et al. (2014) also found similar results while working in the Indian Sundarban. Kumar et al. (1995) also found a similar trend while working in the other part of the world.

From the literature review, it is established that an increase in the abundance of phytoplankton enhances the rate of CO_2 uptake from the ambient air. Many pieces

Table 5.1 Total fish production of the studied ponds during the year of 2014 (late monsoon), 2016 (late monsoon) and 2018 (late monsoon)

	Total fish production		
	Year-2014 (late monsoon) (kg)	Year-2016 (late monsoon) (kg)	Year-2018 (late monsoon) (kg)
Control pond	200	215	205
Mangrove litter treated pond	150	185	140
Iron salt treated pond	205	390	230

of evidence strongly indicate that phytoplankton productivity enhances under the enriched-iron scenario, which in turn, leads to enhanced carbon uptake; however, most of these studies were conducted in lotic ecosystems and open oceanic waters (Gervais et al. 2002; Aumont and Bopp 2006; Pollard et al. 2009; Martin et al. 2013). Iron plays a crucial role in governing the nutrient availability to the pre-existing phytoplankton community (Boyd 1997). The addition and removal of essential nutrients like nitrate, nitrite, and phosphate depend upon the availability of alternative electron acceptors like ferric ions (Boyd 1995). Thus, iron fertilization might be useful to alleviate the localized global warming issues in the Indian Sundarbans part. Apart from this, we have collected the fish catch (secondary) data from the fish farmer of these three ponds (Table 5.1). Secondary data showed that the fish production of the control pond was almost similar in the year 2014, 2016, and 2018 during the late monsoon season. The fish production increased 23% during the year of 2016 (late monsoon season) than in 2014 in the Mangrove Litter Treated Pond. Most interestingly, the fish production increased 90.24% during the year 2016 in the iron salt-treated pond, when the Chl-a also increased by 196.23% than in 2014 (late monsoon). It is worth mentioning that we have added the iron salt during the pre-monsoon season of 2016. The above-mentioned observation indicated that the iron fertilization not only increased the phytoplankton abundance but also increased the fish production of the brackish water pond of Indian Sundarban. In that way, controlled use of iron fertilizer may increase the income of fish farmers in the Indian Sundarban.

References

Akhter M, Reza MdS, Jamil NAHM, Uddin Nazim Md (2018) Assessment water quality and seasonal variations based on aquatic biodiversity of Sundarbans Mangrove Forest, Bangladesh. IOSR J Biotechnol Biochem 4(1):06–15

Aumont O, Bopp L (2006) Globalizing results from ocean in situ iron fertilization studies. Glob Biogeochem Cycles 20(2)

Banerjee K, Senthilkumar B, Purvaja R, Ramesh R (2012) Sedimentation and trace metal distribution in selected locations of Sundarbans mangroves and Hooghly estuary, Northeast coast of India. Environ Geochem Health 34:27–42
Biswas G, Kumar P, Kailasam M, Ghoshal TK, Bera A, Vijayan KK (2019) Application of integrated multi trophic aquaculture (IMTA) concept in Brackishwater ecosystem: the first exploratory trial in the Sundarban, India. J Coast Res 86(SI):49–55
Boyd CE (1995) Bottom soils, sediment, and pond aquaculture. Chapman and Hall, New York
Boyd CE (1997) Practical aspects of chemistry in pond aquaculture. Prog Fish-Cult 59(2):85–93
Çelekli A, Öztürk B, Kapı M (2014) Relationship between phytoplankton composition and environmental variables in an artificial pond. Algal Res 5:37–41
Census (2011) Census report for the state of West Bengal. Government of India
Chand BK, Trivedi RK, Biswas A, Dubey SK, Beg MM (2012) Study on impact of saline water inundation on freshwater aquaculture in Sundarban using risk analysis tools. Explor Anim Med Res 2:170–178
Chisholm SW (1992) Phytoplankton size. In: Falkowski PG, Woodhead AD (eds) Primary productivity and biogeochemical cycles in the sea. Plenum Press, New York, pp 213–237
Chisholm SW, Morel FMM (1991) What controls phytoplankton production in nutrient-rich areas of the open sea? Limnol Oceanogr 38:1507–1964
Coale KH (1991) Effects of iron, manganese, copper and zinc enrichments on productivity and biomass in the subarctic Pacific. Limnol Oceanogr 36:1865–1878
Crossetti LO, Bicudo CEDM (2008) Phytoplankton as a monitoring tool in a tropical urban shallow reservoir (Garças Pond): the assemblage index application. Hydrobiologia 610(1):161–173
Danielsson LG, Magnusson B, Westerlund S (1978) An improved metal extraction procedure for the determination of trace metals in seawater by atomic absorption spectrometry with electrothermal atomization. Anal Chem Acta 98:45–57
DasGupta R, Hashimoto S, Okuro T, Basu M (2019) Scenario-based land change modelling in the Indian Sundarban delta: an exploratory analysis of plausible alternative regional futures. Sustain Sci 14(1):221–240
De Baar HJW, Buma AGJ, Nolting RF, Cadée GC, Jacques G, Tréguer P (1990) On iron limitation of the Southern Ocean: experimental observations in the Weddell and Scotias seas. Mar Ecol Progr Ser 65:105–122
Dubey SK, Chand BK, Trivedi RK, Mandal B, Rout SK (2016) Evaluation on the prevailing aquaculture practices in the Indian Sundarban delta: an insight analysis. J Food Agric Environ 14(2):133–141
Dubey SK, Trivedi RK, Chand BK, Mandal B, Rout SK (2017) Farmers' perceptions of climate change, impacts on freshwater aquaculture and adaptation strategies in climatic change hotspots: a case of the Indian Sundarban delta. Environ Dev 21:38–51
Gervais F, Riebesell U, Gorbunov MY (2002) Changes in primary productivity and chlorophyll-a in response to iron fertilization in the Southern Polar Frontal Zone. Limnol Oceanogr 47(5):1324–1335
Grasshoff K, Kremling K, Manfred E (1999) Methods of seawater analysis. Wiley-VCH, New York, p 600
Huot Y, Babin M, Bruyant F, Grob C, Twardowski MS, Claustre H (2007) Does chlorophyll-a provide the best index of phytoplankton biomass for primary productivity studies? Biogeosci Discuss 4(2):707–745
Jewel MAS, Affan MA, Khan S (2003) Fish mortality due to cyanobacterial blooms in an aquaculture pond in Bangladesh. Pak J Biol Sci 6:1046–1050
Kattner G (1999) Storage of dissolved inorganic nutrients in seawater: poisoning with mercuric chloride. Mar Chem 67:61–66
Kumar N, Anderson RF, Mortlock RA, Froelich PN, Kubik P, Dittrich HB, Suter M (1995) Increased biological productivity and export production in the glacial Southern Ocean. Nature 378:675–680

Landry MR, Barber RT, Bidigare RR, Chai F, Coale KH, Dam HG, Lewis MR, Lindley ST, McCarthy JJ, Roman MR, Stoecker DK, Verity PG, White JR (1997) Iron and grazing constraints on primary production in the central equatorial Pacific: an EqPac synthesis. Limnol Oceanogr 42(3):405–418

Martin P, Vander Loeff MR, Cassar N, Vandromme P, d'Ovidio F, Stemmann L, Naqvi SWA (2013) Iron fertilization enhanced net community production but not downward particle flux during the Southern Ocean iron fertilization experiment LOHAFEX. Global Biogeochem Cycles 27(3):871–881

Mitra A, Banerjee K, Bhattacharyya DP (2004) The other face of Mangroves. Department of Environment, Govt. of West Bengal, India

Mitra A, Zaman S, Kanti Ray S, Sinha S, Banerjee K (2012) Interrelationship between phytoplankton cell volume and aquatic salinity in Indian Sundarbans. Natl Acad Sci Lett. https://doi.org/10.1007/s40009-012-0083-1

Mukherjee N, Siddique G (2018) Climate change and vulnerability assessment in Mousuni Island: South 24 Parganas District. Spat Inf Res. https://doi.org/10.1007/s41324-018-0168-0

Pollard RT, Salter I, Sanders RJ, Lucas MI, Moore CM, Mills RA, Zubkov MV (2009) Southern Ocean deep-water carbon export enhanced by natural iron fertilization. Nature 457(7229):577–580

Price NM, Andersen LF, Morel FM (1991) Iron and nitrogen nutrition of equatorial Pacific plankton. Deep-Sea Res 38:1361–1378

Price NM, Ahner BA, Morel FMM (1994) The equatorial Pacific Ocean: grazer-controlled phytoplankton in an iron-limited ecosystem. Limnol Oceanogr 69:520–534

Sánchez-Triana E, Ortolano L, Paul T (2016) Managing water-related risks in 481 the West Bengal Sundarbans: policy alternatives and institutions. Int J Water Resour Dev. https://doi.org/10.1080/07900627.2016.1202099

Sara G (2007a) Ecological effects of aquaculture in living and non-living suspended fractions of the water column: a meta-analysis. Water Res 41:3187–3200

Sara G (2007b) Aquaculture effects on some physical and chemical properties of the water column: a meta-analysis. Chem Ecol 23:251–262

Siegenthaler U (1986) Carbon dioxide its natural cycle and anthropogenic perturbations. In: Buat-Menard P (ed) The role of an sea exchange in geochenucal C\clim>. Reidel, pp 209–247

Thakur S, Maity D, Mondal I, Basumatary G, Ghosh PB, Das P, De TK (2021) Assessment of changes in land use, land cover, and land surface temperature in the mangrove forest of Sundarbans, northeast coast of India. Environ Dev Sustain 23(2):1917–1943

Zaman S, Bhattacharyya SB, Pramanick P, Raha AK, Chakraborty S, Mitra A (2014) Rising water salinity: a threat to mangroves of Indian Sundarbans. In: Water insecurity: a social dilemma. Emerald Group Publishing Limited

Chapter 6
CO_2 Exchange Dynamics in the Household and Abandoned Ponds of the Indian Sundarbans from the Perspective of Climate Change

Abhra Chanda and Sourav Das

Abstract Lentic ecosystems though encompass a much smaller area compared to the lotic water bodies of the world, are found to emit substantial quantities of greenhouse gases like CO_2, towards the atmosphere. The Sundarbans Biosphere Reserve (SBR) of India, besides being the abode of the world's largest mangrove forest, shelters almost 4.4 million people with a substantially high population density. The CO_2 dynamics from several compartments of this biosphere reserve is studied in the recent past; however, the ponds are yet to receive any attention as such. The present chapter reports the variability of the partial pressure of CO_2 in water [pCO_2(water)] and the air–water CO_2 flux from four different types of ponds situated within the SBR. One of these selected ponds is abandoned, and not used for any human purpose and another pond is well-maintained and not at all used for any human purpose. The rest of the two ponds are typical homestead ponds with varying degrees of anthropogenic disturbances. The results indicated that all four ponds acted as a source of CO_2 towards the atmosphere; however, the rate of emission varied across the ponds. The most well-maintained least anthropogenically disturbed pond emitted CO_2 at the lowest rate, whereas the dilapidated and abandoned pond, which exist in a hypereutrophic state emitted the most. The other two ponds showed an intermediate range of fluxes. Biological processes played a dominant role over physical processes in governing the CO_2 fluxes. Water temperature showed a strong, positive, and statistically significant relationship with pCO_2(water). Thus, given the ongoing climate-change-induced rise in temperature can effectively enhance the emission rate of these ponds.

Keywords Pond · pCO_2(water) · Air–Water CO_2 flux · Water temperature · Primary productivity · Greenhouse gas emission · Lentic system · Indian Sundarban

---

A. Chanda (✉) · S. Das
School of Oceanographic Studies, Jadavpur University, 188 Raja S. C. Mullick Road, Kolkata 700 032, India

S. Das et al. (eds.), *Pond Ecosystems of the Indian Sundarbans*, Water Science and Technology Library 112, https://doi.org/10.1007/978-3-030-86786-7_6

6.1 Introduction

Aquatic ecosystems are perhaps one of the largest repositories of the world, which store various elements and compounds that are in dynamic equilibrium with the lithosphere and the atmosphere. The aquatic ecosystems continually exchange several gases with the ambient atmosphere. Some of these gases are extremely crucial to maintain life activities within the aquatic systems like oxygen and carbon dioxide gases. However, many of these gases when found in the atmosphere at higher concentrations leads to climatic changes that are not desirable. There are several gaseous compounds in the atmosphere, which impart a greenhouse effect on the planet Earth. In other words, the prevalence of these gases in the atmosphere in higher concentrations implies that the lesser would be the energy emission from the earth back to space, which mostly takes place in the form of longer-wavelength infrared radiation (Elrod 1999). The gas molecules, which can impart a greenhouse effect to the earth and enhance the radiative forcing within the Earth's atmosphere, are collectively referred to as 'greenhouse gases'. The different greenhouse gases present in the atmosphere have the varying capability to absorb the outgoing longer wavelength radiation from the Earth and this capability is characterized by quantifying their global warming potential (GWP) (Lashof and Ahuja 1990). These greenhouse gases also happen to vary in the atmosphere in terms of their concentration. Many of these greenhouse gases occur naturally in the atmosphere and participates in several biogeochemical processes of nature. Water vapor, carbon dioxide (CO_2), methane (CH_4), and nitrous oxide (N_2O) are some such greenhouse gases that can naturally occur in the atmosphere. Whereas, the emission of some of the greenhouse gases such as perfluorocarbons, hydro-chlorofluorocarbons, hydrofluorocarbons, and sulfur hexafluoride to the atmosphere is exclusively associated with certain anthropogenic activities like refrigeration, air-conditioning, metal manufacturing plant, etc. (Khalil 1999). Though the fluorinated greenhouse gases have much higher GWP (Purohit and Höglund Isaksson 2017), it is the conventional greenhouse gases like CO_2, CH_4, and N_2O, that remain to be the point of major concern, from the perspective of the global climate (Köhler et al. 2017).

Ever since the scientific community realized the potential of greenhouse gases in altering the global climate, research with a thrust and emphasis on understanding the role of several natural ecosystems has increased manifold. After having carried out extensive research, the global scientific community can confidently and unanimously state that the terrestrial biosphere and the open oceans act as a sink for carbon (mainly by absorbing CO_2 through autotrophic activities) with considerable spatiotemporal variability (DeVries et al. 2019; Pugh et al. 2019). Though several pieces of research indicated that the inland aquatic bodies like rivers, lakes, ponds, reservoirs, dams, play a very crucial role in regulating the global carbon cycle, the knowledge, and understanding about the behavior of the different types of inland aquatic bodies concerning being a source or a sink of greenhouse gases is still poorly constrained (Harmon 2020). According to the estimates of Drake et al. (2018) terrestrial compartments export ~5.1 Pg C to the inland aquatic bodies, out of which only

33% reaches the global oceanic waters through the rivers and estuaries. However, during the transit of carbon from the terrestrial compartments to the oceans, around 25 to 44% of the carbon is respired back to the atmosphere (Harmon 2020). Researchers usually derive such estimates from the meta-analyses of the existing data. However, among the different types of inland aquatic bodies, lotic aquatic systems have received comparatively more attention (Aufdenkampe et al. 2011; Benstead and Leigh 2012; Crosswell et al. 2017) than the lentic aquatic bodies. Again, among the various lentic water bodies, large lakes and reservoirs have received a substantially higher share of attention compared to the small ponds (Bastviken et al. 2004, 2008; Åberg et al. 2010; DelSontro et al. 2018).

The ponds owing to their small size are often neglected while drawing estimates of hydrological carbon budgets but their contribution in acting as a link between the terrestrial environment and water cycle has been recognized long ago (Torgersen and Branco 2008). As far as the processing of terrestrial carbon in the pond ecosystems is concerned, several schools of thought exist at present. Overall, the small ponds usually act as a consistent source of CH_4 round the year, and both source and sink for CO_2 and N_2O; depending upon several, other biogeochemical factors that regulate these gas exchanges (Fig. 6.1) (Macrae et al. 2004; Ferrón et al. 2012). Though the air–water greenhouse gas exchange from the ponds mostly recorded a source character, Smith et al. (2002) along with many others argued that pond ecosystems are most likely a net sink of carbon due to having a very high rate of sedimentation (Chen et al. 2018). Downing et al. (2006) estimated that small ponds having a surface area less than 0.1 km^2 occupy an area of about 6,90,000 km^2 throughout the world with an additional 12,000 km^2 area covered by man-made impoundments. These figures indicate that there is no reason to neglect such a crucial aquatic ecosystem while estimating the carbon budgets and these aquatic bodies can play a significant role in combating or promoting climate change.

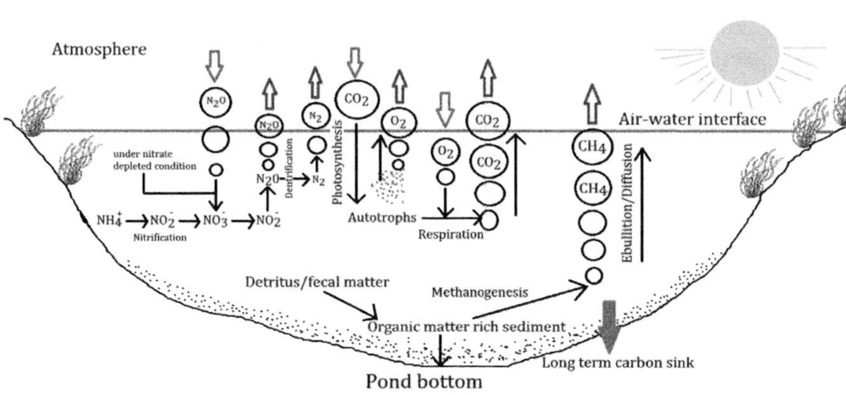

Fig. 6.1 A schematic diagram showing the pathways that lead to exchanges of several important gases between the pond system and the immediate atmosphere (modified after Glaz et al. 2016)

6.2 Pond Ecosystem Biogeochemistry

Compared to medium-sized and large lakes and ponds, the small ponds are present in much more abundance than the former classes throughout the world. These small ponds usually have a shallow water column and exhibit a wide range of biogeochemical variability over space and time (Martinsen et al. 2019). Owing to their small size and shallow depths, the terrestrial setting surrounding these ponds exert a substantial influence on these water bodies (Christensen et al. 2013), which in turn results in some unique biogeochemical conditions in such ponds. Water temperature happens to be one of the key physicochemical parameters of any aquatic system that regulates several biogeochemical processes and the biodiversity of the water column (Yvon-Durocher et al. 2012). All the atmospheric parameters that regulate the ambient temperature near the earth's surface play a crucial role in regulating the water temperature of the ponds near the air–water interface. This includes the ambient temperature itself, the degree of insolation, wind speed near the air–water interface of the pond, relative humidity, and the cloud cover (Woolway et al. 2015, 2016). The shallow ponds due to land effect typically experience rapid heating and subsequent cooling during a diurnal cycle and thus exhibits a wide range of diel variability in water temperature, which in turn regulates the stratification and mixing in the water column (Boehrer and Schultze 2008; Woolway et al. 2016).

Several researchers used to believe that shallow lakes and ponds, under their small water volume, remain homogeneously mixed, however, recent estimates indicate that such perception is wrong (Andersen et al. 2017). Contrary to such common perception, shallow ponds often undergo clear stratification during the daytime, especially if these ponds have submerged aquatic vegetation, and during the nighttime, the mixing of the water column takes place. This not only leads to high diel variability of the water temperature as mentioned above, but also in pH levels, dissolved oxygen (DO), and CO_2 concentrations (Andersen et al. 2016). Usually, the shallow ponds experience vertical stratification during the daytime, influenced by the heating of the air–water interface depending on the attenuation of light and morphometry of the lake (Fee et al. 1996). During the nighttime, the cooling of the water mass coupled with wind shear usually destabilizes the stratified water column and consequences to vertical mixing (Spigel et al. 1986). When the water column remains stratified, the ponds can be conceptually demarcated into three prominent zones from the air–water interface to the pond bottom: (i) the turbulent and disturbed epilimnion (upper layer), (ii) the metalimnion where the density sharply changes with depth (middle layer), and lastly (iii) the cold hypolimnion with almost no turbulence (bottom layer) (Kalff 2002) (Fig. 6.2). However, the degree of vertical stratification during a diel cycle might vary across the ponds of a certain region depending on multiple micrometeorological factors like shades offered by the tree canopy surrounding the ponds, hindrances to wind flow, presence of submerged aquatic vegetation like macrophytes, and so forth (Branco and Torgersen 2009; Markfort et al. 2010).

The shallow tropical ponds usually experience high turbidity, especially where land runoffs are very prominent (Sarnelle et al. 1998). Such high turbidity enables the

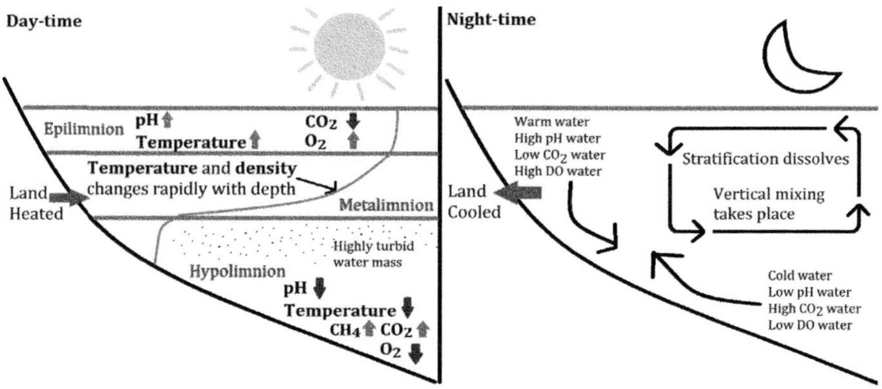

Fig. 6.2 A schematic diagram showing the vertical stratification and mixing of shallow pond waters during the day and night time. The upward and downward blue arrows indicate an increase and decrease respectively. The black arrows indicate the movement of water masses. The red arrows indicate heat exchange between land and pond water

water column to capture and sustain heat (Condie and Webster 2002) and at the same time promotes net heterotrophy as it obstructs the penetration of the photosynthetically active radiation to the deeper layers. Most of the freshwater ponds in tropical regions have zero salinity with few ponds having measurable electrical conductivity, which is situated, close to marine regions. Salt-water intrusion from shallow aquifers often enhances the salinity of some natural ponds in the deltaic and coastal regions and sometimes cyclonic events associated with storm surges and coastal flooding lead to salinization of freshwater ponds (Chand et al. 2012). In some cases like that of brackish aquaculture ponds, saltwater is deliberately introduced to some of the manmade ponds to facilitate the aquaculture of selected species, mainly shrimps (Biswas et al. 2019). Besides this, the local people to meet the water demand for several household activities use many of the small ponds, which are situated in less developed regions like that of Sundarbans. This usage of pond water for activities like bathing, washing utensils, and clothes coupled with runoffs from adjacent agricultural fields treated with fertilizers, in turn, significantly alters the biogeochemistry of such ponds (Mukhopadhyay et al. 2004).

6.3 Nutrient Dynamics in the Pond Ecosystem

There is a group of elements and compounds, which plays a crucial role in the primary autotrophic process, despite being present in minute quantities. We collectively refer to these substances as essential micronutrients. These micronutrients remain in the soil substratum and the terrestrial plants can absorb these through root absorption. Similarly, micronutrients are essential for the primary autotrophs of an aquatic ecosystem as well. Nitrogen, phosphorus, and silica are some of the most vital micronutrients that play a central role in regulating the functioning and the species composition of this base-level floristic community. Not only the absolute quantity of these dissolved nutrients (mostly in the inorganic form) but also their proportion to each other, can significantly govern the phytoplankton dynamics in any aquatic body (Elser et al. 2005; Leflaive et al. 2008). The entire microbial diversity constituting both the microscopic and macroscopic species are, in turn, dependent on the concentrations of these nutrients, especially nitrogen and phosphorus (Torsvik et al. 2002; Groszkopf and Soyer 2016). Pond ecosystems are also no exception, in this regard.

Like many other aquatic ecosystems, ponds also host a suite of microbial species that are dependent on these nutrients to carry out all the essential activities required for thriving. The different kinds of taxonomic groups or individual species have varying capabilities to utilize the respective components of these nutrients (Nelson and Carlson 2011; Corman et al. 2016). Such differences arise out of varying metabolic activities and ecological strategies of these microbes to survive in a particular type of environment (Carbonero et al. 2014). Some microbes can survive in a given state of nutrients, whereas others cannot. Changes in nutrient levels and ratios due to natural or anthropogenic reasons can often lead to only minute changes. However, pieces of evidence show that it can also bring about a paradigm shift in the species composition, coupled with major changes in species richness and evenness (Hewson et al. 2003; Van Horn et al. 2011; Soininen and Meier 2014). The capability of nutrient alteration to modify the biodiversity of a particular system, in turn, plays a major influence on the productivity of the same system. Particularly, the ponds that are small and shallow are more susceptible to such changes, as evaporation takes place at a rapid rate in such ponds and with the fast change in the volume of water, the nutrient concentration also changes (Lee et al. 2017).

Essential nutrients like nitrogen and phosphorus end up accumulating in the ponds through several pathways. Land runoff plays one of the most pivotal roles in regulating the concentration of these substances in the ponds. The ponds in the urbanized sector such as retention ponds and storm-water ponds have received substantial attention in this regard (Yang and Lusk 2018; Taguchi et al. 2020); however, the understanding of rural ponds like that of Sundarbans are still limited. However, in the Indian setup, where many of the ponds are ill-managed and indiscriminately used for various domestic purposes, aquatic pollution takes place due to activities like washing utensils and clothes, bathing, and waste disposal (Kant et al. 2019). These activities, in turn, lead to enhanced nutrient input in these waters and deteriorates the water quality. Excessive presence of nutrients leads to harmful impacts like eutrophication, which in turn, consequences to undesirable manifestations like bad odor, harmful algal blooms, and fish mortality (Howarth and Paerl 2008; Conley et al. 2009). The occurrence of such situations demands intervention in the form of pond management, however, the absence of such initiative compels the local people to abandon such ponds, and in this way, we lose the vital ecosystem services of such aquatic ecosystems.

Once introduced to the ponds, nitrogen, and phosphorus goes through several biogeochemical reactions by which these elements change their chemical speciation and find their way out of the system. Nitrogen gas is abundant in the atmosphere; however, most of the autotrophs and other life forms cannot make use of nitrogen in the gaseous form. There are certain microbes, which specialize in fixing atmospheric nitrogen to the soluble and bioavailable form of dissolved inorganic nitrogen like nitrate, nitrite, and ammonia. This process, known as nitrogen fixation, is one of the most important natural processes, which regulates the nitrogen content of any lentic aquatic body, like that of ponds (Howarth et al. 1988; Scott et al. 2008; Newell et al. 2016). Similarly, a microbial-mediated process that transforms nitrogen-bearing inorganic nutrient compounds to molecular nitrogen, commonly referred to as the denitrification process effectively removes nitrogen from the system (Seitzinger et al. 2006; Groffman et al. 2009; Collins et al. 2010; Bettez and Groffman 2012).

Like nitrogen, phosphorus also exists in different chemical forms. All these varying chemical forms have different mobilizing patterns. Some are easily accessible for aquatic plants, whereas some forms are biochemically inert. The ironbound and labile organic phosphorus is one of the most abundant forms that the aquatic biotas can readily utilize (Hansen et al. 2003; Søndergaard et al. 2003; James 2011). The physicochemical properties of the pond sediments, like the aluminum, sulfur, calcium, and iron content and dissolution potential play a crucial role in regulating the proportion of the different forms of phosphorus (Hupfer and Lewandowski 2008). Compared to nitrogen, the concentration of phosphorus prevails at much less magnitudes in natural ponds; however, a slight alteration in phosphorus concentration can lead to eutrophication (Frost et al. 2019). Like denitrification, phosphorus removal from any closed aquatic body also has to rely on biotic assimilation of phosphorus carried out by a suite of microorganisms, mainly algae (Sañudo-Wilhelmy et al. 2004; Barat et al. 2011; Zhimiao et al. 2016). A delicate balance maintains the equilibrium of these nutrients' concentration in the pond ecosystem, disturbing which can lead to far-reaching impacts, including the destruction of this ecosystem and eventual conversion to a terrestrial landform.

6.4 Eutrophication and Hypoxia

Eutrophication is the excessive growth of a suite of algal organisms and aquatic plants that take place under favorable conditions accompanied by an abundance of limiting factors essential for photosynthesis, like sunlight, CO_2, and primary nutrients (Schindler 2006; Chislock et al. 2013). Lentic systems accumulate sediments in the bottom and gradually become shallow. The lakes and ponds, in this way, undergo eutrophication over the natural course of the cycle, which takes more than a hundred years (Carpenter 1981). However, various anthropogenic activities have enhanced the rate of eutrophication. Human beings are responsible for enhancing the rate of nitrogen and phosphorus input to these shallow lentic bodies, which directly reflects an anomalous growth of macroalgae and other unwanted aquatic vegetation (Carpenter et al. 1998). We refer to this phenomenon as cultural eutrophication. During the late twentieth century, many scholars throughout the world associated cultural eutrophication with a range of industrial, domestic, and fishery-based activities (Schindler 1974). The eutrophication leads to the formation of an algal mat in the pond-air interface, which restricts the exchange of essential gases like oxygen with the ambient atmosphere. The lowering of dissolved oxygen levels in the aquatic bodies beyond a certain threshold (mostly <2 mg l^{-1}) leads to hypoxia. Several aquatic organisms find it challenging to survive under such low oxygen levels, and eventually dies. Thus, eutrophication can lead to fish mortality and enhances heterotrophy. Microbial degradation of dead remains of various flora and fauna supersedes the autotrophic CO_2 fixation. The degradation of overall water quality and foul odor are some of the common consequences of eutrophication, which leaves the aquatic bodies in an unusable state. The ill effects of this phenomenon incur substantial financial loss, as we cease to reap the benefits of several ecosystem services (like drinking water supply and recreational use) from these shallow aquatic bodies (Dodds et al. 2009).

Characterizing the trophic state of an inland lentic body is essential to understand the structure and function of such ecosystems and to project future inclination under incessantly altering environmental and climatic conditions (Zhang et al. 2018). We can quantify the trophic state of a pond based on the concentrations of several parameters like total nitrogen, total phosphorus, chlorophyll-*a*, and the degree of transparency. The combined concentrations of these indicators enable us to determine whether a pond is oligotrophic, mesotrophic, eutrophic, or hypertrophic. Minimal concentrations of nutrients accompanied by modest levels of chlorophyll-a denote the oligotrophic state, whereas the excessive presence of total N and P with very high chlorophyll-a levels indicates a hypereutrophic state (Carlson 1977; Kratzer and Brezonik 1981; Wetzel 2001). Besides the nutrient levels, this trophic state depends substantially on the degree of photosynthesis. Photosynthesis and aerobic respiration leading to the production of oxygen and carbon dioxide, respectively, are the principal metabolic pathways through which the production as well as the destruction of organic matter takes place in nature (Cole et al. 2000), and maintains the gross metabolic balance in any ecosystem (Howarth et al. 1996). Removal of organic matter from any aquatic system exclusively depends on the microbial community strength and supply of enough oxygen to decompose the organic matter (Sobek et al. 2009). However, inland aquatic bodies like lakes and ponds often receive a substantially high load of organics (Tranvik et al. 2009) and lead to an enhanced rate of respiration in degrading this organic matter (Vaquer-Sunyer and Duarte 2008). Such respiratory processes consequences of overutilization of oxygen leaving the adjacent water column deprived of the necessary oxygen levels required for many of the life forms to thrive and promoting anaerobic conditions (Conrad et al. 2011). Hypoxia can cause naturally over time with prolonged aging of any pond or lake; however, anthropogenic inputs to such water bodies accelerate the rate and intensifies the degree of hypoxic conditions (Marotta et al. 2012). In the tropical climate, warmer temperatures favor the deterioration of hypoxic conditions to total anoxia, which in turn leads to deleterious consequences to the biological life (Marotta et al. 2012).

6.5 Pond Ecosystem Productivity

The primary productivity of any ecosystem is integral to the well-being and overall health of that system. The lentic ecosystems like lakes and ponds are no exception. To restore and conserve the ponds of any region, one must properly evaluate the intrinsic algal dynamics and the nature of primary production (Mayer 2020). A suite of biotic and abiotic factors govern the ecological food chain of the ponds, and hence the net ecosystem productivity (Kitchell and Carpenter 1993). The algal species assemblage and count often follows a top-down control, driven by biotic factors like zooplankton grazing and ingestion by fish. Similarly, abiotic factors like light availability and nutrient concentrations regulate the community structure through bottom-up control (Menezes et al. 2010). Besides light penetrability, the residence time of water, the depth of the ponds, the water level, the mixing rate, and the flushing frequency also play a critical role in regulating productivity. The tropical shallow ponds are particularly susceptible to all these parameters. The ponds of the tropics are usually small and experience warmer temperatures and a high rate of evaporation, which in turn fluctuates the water level of the ponds throughout the year. The ratio of surface area to volume is substantially high with ample scope of interaction between pond bottom and the above-lying water mass. All these characteristics play a significant role in governing the primary productivity of these ponds (Zohary et al. 2010; Jeppesen et al. 2015).

The pond substratum and the nature and quality of sediments can indirectly regulate the rate of primary production and shape the algal community structure that thrives therein. The vertical movement of the loose sediments controls the degree of transparency of the water column as these particulate matters impart turbidity. The higher the concentration of these turbid materials the lower would be the light penetration, and hence, it can potentially compromise the rate of photosynthesis. The bottom-churning induced sediment resuspension takes place in particular in the shallow ponds, as wind-driven shear can disturb the pond bottom (Talling 2001). Rapid changes in water level in the seasonal time scale can exacerbate the situation, as, under the circumstances, sediments tend to leave the bottom and remains suspended in the water column (Rodrigues et al. 2016). However, there could be some positive feedbacks of sediment suspension towards enhancing primary productivity, as they often release dissolved nutrient matter to the water, thereby promoting photosynthesis (Dantas et al. 2019).

Among the biotic factors, zooplankton and fish dynamics cast a significant impact on the phytoplankton community structure and the associated gross primary production. Zooplankton grazing can directly control the phytoplankton counts. However, the grazing rate depends on the size and ingestion capability of the zooplankton (Iglesias et al. 2011). Fish population, especially in the tropics, can both promote as well as negate the primary productivity by controlling the phytoplankton community composition (Attayde et al. 2010). The consumption of phytoplankton by the fishes reduces the productivity potential of the ponds. However, by preferential consumption of zooplankton, fishes occasionally relive the phytoplankton from grazing pressure (Torres et al. 2016). Besides, some of the bottom feeders disturb the pond bottom in search of food and release nutrients, which in turn helps in enhancing productivity (Starling et al. 2002). Thus, we can infer that several biotic options can alter and control the primary productivity of the ponds.

6.6 Ponds and Greenhouse Gases

Downing (2009) strongly emphasized that inland aquatic bodies like lakes and ponds, though neglected due to their small size and discrete distribution throughout the land surface, play a crucial role in regulating the greenhouse gases in the atmosphere. Downing et al. (2006) established that these shallow aquatic bodies encompass substantially large areas compared to previous estimates. Contrary to the pre-existing belief, many scholars proved that the smaller ponds and lakes emitted higher magnitudes of greenhouse gases towards the atmosphere (Kortelainen et al. 2006; Juutinen et al. 2009). The production of greenhouse gases, its removal, transformation, and emission towards the atmosphere depends on several factors like temperature, size and depth, hydrology, maintenance, and usage types, and many others (van Hulzen et al. 1999; Rantakari and Kortelainen 2005; Battin et al. 2009; Duc et al. 2010; Kosten et al. 2010). Atmospheric variables like ambient temperature, relative humidity, solar insolation, wind speed change from minutes to years to decades, and play a crucial role in regulating the greenhouse gas fluxes. However, it is difficult to ascertain the role of climate change in changing the flux patterns from shallow ponds, due to the absence of long-term data and repeated measurements in the same spots (Natchimuthu et al. 2014).

Besides the phytoplankton, all the other life forms breathe out CO_2 in the aquatic column of the ponds. If the rate of CO_2 production by the autotrophs supersedes the rate of CO_2 respiration by the rest of the biotic community in a diurnal cycle, the system altogether becomes net autotrophic, and usually acts as a sink of CO_2. However, in most cases, the reverse scenario prevails, which makes the system net heterotrophic and hence, a source of CO_2. As ponds often receive substantial quantities of organic load and suffer from hypoxia due to human-induced eutrophication, the ponds experience anaerobic conditions, especially near the bottom. This type of condition activates the methanogens, and instead of CO_2, leads to the production of a more potent greenhouse gas, CH_4. The input of nitrogenous materials leads to the generation of N_2O under hypoxic conditions during denitrification (Codispoti 2010). The global warming potential of N_2O is much higher than CO_2 as well as CH_4, and research shows that small water bodies can emit substantial quantities of N_2O (Gorsky et al. 2019). The enhanced use of nitrogenous fertilizers and domestic waste through land runoff ends in the lentic ecosystems, whereby, these substances undergo denitrification and emit N_2O towards the atmosphere (Blaszczak et al. 2018).

In this chapter, we have dealt with only CO_2 emissions from the homestead ponds of the Sundarbans region.

6.7 Household Ponds of Sundarbans and Their Characteristic Features

Not only in the Sundarbans but also in many rural parts of this country, ponds are an integral part of many domestic activities. In the Sundarbans, a small pond within the household of a kaccha or even pukka houses is a very common sight. Mandal et al. (2015) reported that in the rural sectors of Sundarban, the majority of the households possess a pond in its adjacency, having an average surface of 400 to 500 m^2. The landowners own most of these ponds legally as these ponds are essentially a part of the entire land area purchased by an individual or passed on through a family heirloom. Many households with a larger area under their share have more than one pond of varying size. These ponds vary in size and shape. However, household ponds larger than 1000 m^2 are scarce. Most of these ponds exist within a distance of 10–15 m from the dwelling. It is a common practice in this part of the world to dig ponds and utilize the unearthed soils to increase the basement of the dwelling or even construct the house itself (the kaccha houses). Besides these small ponds, which are in the true sense personal property of the landowner, there are some large ponds as well. These large ponds are less in number; however, exist in one or two numbers across a village. This type of ponds are public properties and local villagers can access these ponds to meet several daily life activities.

These ponds are mostly perennial. During the non-monsoon dry months, the water level in many of these ponds goes down leaving a limited amount of water. The rain during the monsoon season recharges these ponds up to the brim. The depth of these ponds varies from a meter to three meters. Most of these are manually dug ponds, and these have a typical shape of steadily inclining edges with the highest depth around the center. Mandal et al. (2015) observed that the average age of the ponds in the Sundarbans is around 60 years. Some of the ponds are as old as hundred years, whereas, very few ponds came into operation in the last decade. The older the ponds the higher is the rate of sedimentation, which affects the primary productivity of these ponds (Boyd 2012). These ponds are in a way, lifelines for the rural population in the Sundarbans. They rely on these ponds for several services. Fish rearing is a very common practice in these ponds. Local people grow a wide variety of freshwater fishes in these homestead ponds, which mainly includes *Labeo rohita, Labeo catla, Catla catla, Cirrhinus cirrhosis, Puntius sophore, Hypophthalmichthys molitrix, Mystus tengara, Lates calcarifer, Oreochromis niloticus, Amblypharyngodon mola, Probarbus jullieni, Clarias batrachus,* and many more (Mandal et al. 2015). The local population does not rear fish for commercial purposes. They harvest these fishes regularly from the ponds for their daily meals. They can rear fish for almost eight to nine months, excluding the dry months when the water column becomes extremely shallow, and hence, unfavorable for the fishes to survive. They also do not practice any scientific management for the well-being of these ponds, mostly because they cannot afford to do so. Hence, most of these ponds usually do not undergo conventional pond management practices like periodic dredging, lime treatment, and de-silting. The fish rearing practiced in these ponds also does not follow any stringent aquaculture protocols. The traditional practice just includes the introduction of seedlings at different times of the year. However, the local people do not care to take into account the conventional aquaculture parameters like stocking density, and fish composition. They also do not provide specific feeds for the fishes. The fishes mostly grow on their own depending on the natural foods that end up in these ponds, mostly out of primary autotrophic processes. Mandal et al. (2015) observed that the local people could enhance the aquaculture potential of these ponds; however, the financial constraints bar them from doing so.

Besides being a source of fish, the pond water finds its utility in growing fruits and vegetables in the homestead gardens for household consumption. Mandal et al. (2015) mentioned in their study that local households of Sundarbans deploy about 120–240 m^2 of their land plot to grow a wide range of vegetables throughout the year. The adjacent ponds serve as the sole source of water to sustain this miniature form of agriculture. Besides growing vegetables and rearing fishes, the local people also use this water for bathing and washing purposes. Many people prefer to take a dip and swim across the ponds as a daily routine. Washing utensils along the edge of the ponds is also a very common sight in this region.

6.8 Materials and Methods

Though we understand the importance of these small lentic bodies in regulating the greenhouse gas concentration, there are almost no studies that focused on the air–water exchange of the conventional greenhouse gases in these ponds of Sundarbans. The ongoing climate change is significantly altering the ambient temperature and the rainfall pattern. Sundarbans experience the wrath of climate change mainly in the form of tropical cyclones, which have exhibited an alarming increase both in terms of frequency as well intensity (Dubey et al. 2017). Storm surges and relative sea-level rise have posed a severe threat to the functioning of several aquaculture ponds situated near the coastline due to a change in the overall salinity regime of the aquatic column. All these factors can potentially alter the CO_2 dynamics of these small ecosystems as well. To characterize the air–water CO_2 exchange across the pond interface, we have sampled in few of the representative ponds within the Indian Sundarbans Biosphere Reserve. In the present date, there are several established protocols for measuring the air–water CO_2 fluxes. Eddy covariance technique has become extremely popular among meteorologists, and it has replaced all the measurement protocols because of its robustness and long-term flux monitoring capability. However, this method requires a sophisticated set of instruments and proper stationing, which was beyond the ambit of the author's reach. We implemented the bulk density method for spot measurements and estimated the partial pressure of CO_2 in water [pCO_2(water)]. The concentration gradient across the air–water interface enabled us to compute the CO_2 fluxes. We have also measured a suite of other biogeochemical parameters and observed the relationship of those with the CO_2 fluxes. We have detailed the methodology adopted for the present study in the following sub-sections.

6.8.1 Location of Ponds Selected for This Study

We carried out the sampling at four selected ponds distantly located throughout the South 24 Paraganas district of the Sundarbans Biosphere Reserve. Two of these ponds are in the Namkhana Block, one in the Patharpratima Block, and the other in the Basanti Block. Figure 6.3 shows the location of the respective ponds in the study area. We tried to encompass the different varieties of ponds based on their state of health, water usage, size, and shape. Some of these are typical household ponds used for multifarious domestic activities. Some ponds receive comparatively less domestic attention, as the local people are not exclusively dependent on the services provided by these ponds. Besides, there are some water bodies, which are usually larger than the other two types of ponds mentioned in the preceding sentence, and these undergo proper maintenance, mainly for their aesthetic value and beautification purposes. Lastly, there exists a class of ponds, which are most neglected, not used for any domestic activity, and suffer from intense eutrophication. Somehow, these ponds can manage a water column for a substantially long time, as these ponds get replenishment of water from monsoonal rain.

The pond in the Basanti Block exists near a popular tourist destination named Jharkhali (hereafter referred to as P1). This pond roughly covers an area of 250 m^2. This pond is not a typical household pond as it lies within the premises of a government office. However, stationed officials make use of the pond water. Occasionally very few people bathe and wash utensils in this pond. The pond has fishes but the people surrounding this pond are not very dependent on it, and they harvest fishes rarely. Overall, this pond has limited use for domestic purposes.

The second pond selected for this study is in Ramganga (hereafter referred to as P2), which is one of the busiest places in the Patharpratima Block, as many people commute through this critical junction to go towards Kolkata and the suburban localities. The pond merely covers an area of about 400 m^2. Situated in a typical locality, this pond is almost of no use to the local people. A thin algal mat prevailed over the water surface almost for the entire duration of an annual cycle. In the monsoon season, heavy rain sometimes broke off the intact mat cover. The visual scenario of the very pond can tell us that it is abandoned and suffering from eutrophication. We selected this pond as one of the representative classes, as these types of ponds are prevalent all around the Sundarban.

The third pond sampled in this study represents the most typical type of household ponds. This pond is in the Namkhana Block (hereafter referred to as P3), situated in a typical semi-rural locality. We have noticed a substantial number of people bathing

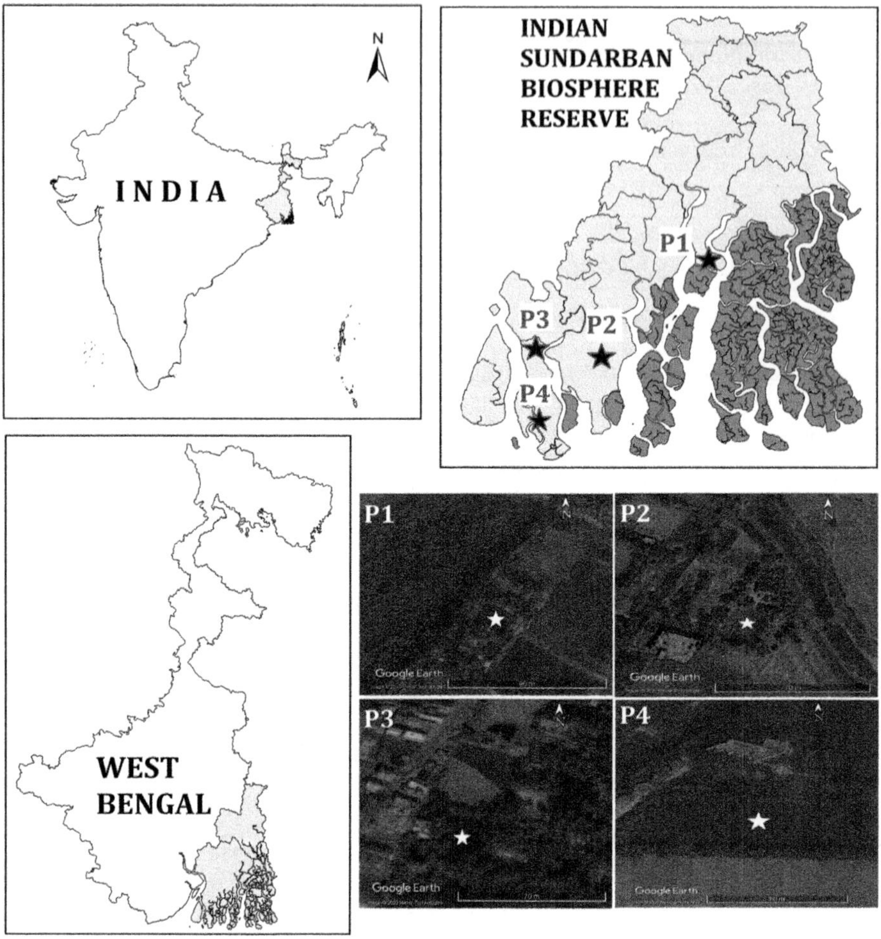

Fig. 6.3 The study area map showing the locations of the four ponds (selected for the present study) situated in the Indian Sundarbans Biosphere Reserve

in this pond all-round the year. Women from nearby houses occasionally use the pond water to wash the utensils or even clothes, sitting near the periphery of the pond. This pond occupies an area of about 380 m^2.

Lastly, we selected a pond, situated in the same Namkhana Block, but it is located within the premises of a Guest House (hereafter referred to as P4) that mainly cater to tourists from Kolkata who ventures into the nearby Bakkhali and Henry Island sea beaches. Out of the four ponds we sampled, this is the largest having an area of about 2700 m^2. The pond is full of fishes and occasionally the caretakers harvest fishes out of the pond but they essentially do not rely on these fishes for consumption. There is a strict prohibition for bathing and washing clothes or utensils in this pond. The guesthouse authority maintains this pond mainly for beautification purposes. All the four ponds had one thing in common. There are no concrete embankments in any of these ponds. Almost all the ponds have some naturally growing bushes and shrubs in their periphery.

6.8.2 Sampling Strategy

We sampled the above-mentioned four ponds during the years 2016 to 2018. The broad demarcation of seasons in this part of the world includes the pre-monsoon season (February to May), the monsoon season (June to September), and the post-monsoon season (October to January). We sampled in each of the three seasons during the three years (2016–2018). However, we carried out the sampling in different months of a particular season in the three respective years to characterize the short-term temporal variability in the physicochemical parameters as well as the CO_2 fluxes. In this way, we sampled in all the twelve months but of different calendar years. We presented the seasonal mean data taking into account the measurements carried out in all the months of that particular season. We assumed that there was no significant inter-annual variability in the measured parameters.

We sampled once every month during the daytime (between 10:00 a.m. and 12:00 p.m.) and the nighttime (between 06:00 p.m. and 08:00 p.m.) from all four ponds. We collected both the day and night samples to analyze the difference in the CO_2 flux dynamics in the presence and absence of photosynthesis. In the year 2016, we sampled in March, May, September, and November. Similarly, in the year 2017, we sampled in January, April, August, and October. Finally, in the year 2018, we sampled in February, June, July, and December. We collected three replicate samples from each of the ponds with the help of a pre-cleaned polypropylene glass bottle from the surface water. We collected the water samples from at least a meter distance from the edge of the ponds.

6.8.3 Analytical Protocol

We measured most of the parameters in situ by deploying standard probes. However, for those parameters, which require laboratory analysis, we collected the samples in specific containers, preserved the samples with the addition of necessary preservatives for each parameter, and transferred the bottles in freezing conditions to the laboratory. We took adequate care to calibrate and standardize the sensors and probes used in this study. We ensured that all the reagents used for the different laboratory analyses were of analytic grade.

6.8.3.1 Spot Measurements of Atmospheric and Biogeochemical Parameters

Most of these ponds sampled in the present study were essential of freshwater type. Thus, the total dissolved solid content was much low to have any reportable salinity. Thus, we measured the in-situ electrical conductivity of the surface water of these ponds with the help of a Multikit (Multi 340i set, WTW, Germany) fitted with a standard probe (Tetracon 325, WTW, Germany). We also monitored the in-situ water temperature with the same instrument. The sensors had a measurement resolution of 1 μS cm^{-1} and 0.1 °C, for electrical conductivity and water temperature, respectively. We estimated the analytical precision of the sensors by measuring a few of the same samples repeatedly for ten times. The precision of measurements for electrical conductivity was ± 2 μS cm^{-1} and for water temperature, it was 0.1 °C. We measured the dissolved oxygen (DO) concentration by deploying a standard probe (FiveGo portable F4 Dissolved Oxygen meter, Mettler Toledo, Switzerland) having a resolution of 0.01 mg l^{-1}. We measured the DO once during each sampling endeavor following the modified Winkler's titrimetric method. We checked the offset and calibrated the sensor's DO measurement readings with the help of the results obtained from Winkler's method. We also regularly calibrated the DO sensor with the help of boiled water (having zero DO) and fully oxygenated water (having a hundred percent oxygen saturation). The DO sensor exhibited an accuracy of ± 1 percentage. We also measured the saturation state of DO using the empirical formula of Weiss (1970).

We measured the in-situ pH with the help of a micro-pH meter equipped with standard glass electrodes (Orion PerpHecT ROSS Combination pH Microelectrode) fitted to a data logger [Thermo Scientific, U.S.A.]. The analytical resolution of the pH meter was 0.001. We calibrated the glass electrodes for pH measurements (once before the sampling endeavor begun for each month and once after the completion) using the NBS scale technical buffers (Merck, Germany) at a controlled temperature of 25 °C. We deployed a nephelometer (Eutech TN-100, Singapore) to measure the turbidity of the water samples. With the help of a standard underwater light sensor (LI-192SA, Li-Cor, USA; analytical resolution 0.1 μ mol m^{-2} s^{-1}) fitted to a data logger (Li-250A, Li-Cor, USA), we measured the underwater photosynthetically active radiation (UWPAR). We carried a portable weather station (WS-2350, La Crosse Technology, Wisconsin, USA) to the sampling spots to measure the ambient temperature, atmospheric pressure, and wind velocity. We implemented the standard light–dark bottle method by incubating the samples for 12 h from dawn to dusk and measured the gross primary productivity (GPP) and community respiration (CR) of the surface water by measuring the DO concentrations with standard probes. We deployed three replicate sets of bottles for GPP and CR measurements.

6.8.3.2 Estimation of pCO_2(Water) and Other Laboratory Measurements

We filtered 100 ml water samples through GF/C filter papers and preserved the samples for total alkalinity (TAlk) analysis (Frankignoulle et al. 1996) by poisoning each sample with 20 μl saturated $HgCl_2$ solution (7.2 g $HgCl_2$ in 100 ml distilled water) (Kattner 1999). We used an automated titrator (905 Titrando, Metrohm, Switzerland) to measure the TAlk following a closed chamber titration. We estimated pCO_2(water), dissolved inorganic carbon (DIC), and hydroxyl ion (OH^-) concentration using the water temperature, atmospheric pressure, TAlk, and pH data with the help of the software CO_2SYS.EXE (Lewis and Wallace 1998). Oceanographers and marine scientists mostly use this software to estimate pCO_2 in seawater. However, the software has a provision to estimate pCO_2(water) in absolutely fresh water. Recent studies like Chanda et al. (2020) have successfully used this software to estimate the pCO_2(water) in the urban tidal river of Hooghly, which flows by the twin cities of Kolkata and Howrah, situated in the north of the Sundarbans Biosphere Reserve. The Hooghly river water in this part had negligible salinity. Following Chanda et al. (2020) we used the dissociation constants, K_1 and K_2, of Millero (1979) for zero salinity water on the NBS scale. We used the corrections of Khoo et al. (1977) for sulfate concentrations. We used a nondispersive infrared (NDIR) sensor (Li-840A; Li-COR, USA) to measure the CO_2 concentration in the ambient air. We calibrated the instrument before each sampling campaign with certified reference standard gases of known concentrations of CO_2 (0, 300, and 600 CO_2 concentration) in N_2 as base gas (Chemtron Science Laboratories, India). We converted the mole fraction of CO_2 in ambient air to the partial pressure of air [pCO_2(air)] by using the ambient temperature and atmospheric pressure, and the virial equation of state (Weiss 1974).

We collected 1 l of a water sample from the pond surface, stored the same in an amber-colored container, preserved under ice-cold condition, and sent the samples back to the laboratory within 24 h for chlorophyll-*a* (chl-*a*) analysis. We kept the samples in the dark in a freezer until further analysis. We implemented standard spectrophotometric protocols (Parsons et al. 1992) to measure chl-*a*. The precision of the chl-*a* measurement was ± 0.02 mg m^{-3}. We used standard lyophilized chlorophyll-*a* (Sigma-Aldrich, Merck, Germany) for the preparation of standard stock solutions and calibration of the spectrophotometer.

6.8.3.3 Computation of Air–Water CO_2 Fluxes

For the sake of the CO_2 flux computation across the air–water interface, we converted the pCO_2(water) and the pCO_2(air) to concentrations of carbon dioxide in water (CO_2wc) and air (CO_2ac), respectively, following the equations given below (Weiss 1974; Anderson 2002).

$$CO_2wc = K_H \times pCO_2(water)$$

$$CO_2ac = K_H \times pCO_2(air)$$

where K_H stands for the gas partition constant of CO_2 in freshwater at the sampling temperature. We computed KH (in mole l^{-1} atm^{-1}) following the equation below.

$$\ln KH = -58.0931 + 90.5069 \times (100/T_K) + 22.294 \times \ln (T_K/100)$$

where T_K denotes the water temperature in the unit of Kelvin (Weiss 1974).

We computed the air–water CO_2 exchange rate (flux) according to the formula furnished in MacIntyre et al. (1995) as shown below.

$$\text{Air-water } CO_2\text{flux } [FCO_2] = k_x \times (CO_2wc - CO_2ac)$$

where k_x stands for the gas transfer coefficient (cm h^{-1}) that we estimated according to the formula given by Wanninkhof (1992).

$$k_x = k_{600} \times (600/S_c)^x$$

where S_c denotes the Schmidt number (for CO_2). The S_c depends on the water temperature (T, in the unit of Kelvin) according to the formula given below.

We calculated k600 using the wind speed at a height of 10 m above the pond interface (U10), according to Cole and Caraco (1998), where, the magnitude 'x' is equal to 0.5 and 0.66 for wind speeds >3 m s^{-1} and $\leq$3 m s^{-1}, respectively. We measured the wind speed with the help of a weather station at 1 m height from the pond interface and converted the speed magnitudes for 10 m height, using the empirical wind profile equations given by Kondo (2000).

6.8.4 Statistical Analyses

We used the Statistical Product and Service Solutions software (SPSS version 16.0, Inc., USA) and Microsoft Excel for Windows 2010 to carry out all the statistical analyses and prepare the graphical illustrations for this chapter. We examined whether the arithmetic means of pCO_2(water), CO_2 fluxes, and other biogeochemical parameters were significantly different or not between all the four ponds and across all the seasons by carrying out a one-way analysis of variance test (ANOVA; F-test). We conducted a posthoc Tukey's honest significant difference (HSD) test to specify the difference in mean of the biogeochemical parameters among the selected ponds if any. To analyze the inter-relationship between the pCO_2(water) and the other regulating factors, we calculated the Pearson correlation coefficient. We considered the outcomes of these tests statistically significant at $p \leq 0.05$.

6.9 Results and Discussion

6.9.1 Variability of Physicochemical Parameters

The surface water temperature in the ponds varied between 20.5 and 35.4 °C (Table 6.1). The annual mean temperature was close to 30 °C in all four ponds. However, P2 which is an abandoned pond recorded slightly higher temperature than the other

Table 6.1 The annual mean ± standard deviation along with the range of biogeochemical parameters observed in the four ponds

Parameters	P1	P2	P3	P4
Water temperature (°C)	29.6 ± 5.0 (21.0 – 34.6)	30.7 ± 4.4 (23.4 – 35.4)	29.6 ± 4.4 (22.9 – 34.2)	29.3 ± 5.1 (20.5 – 34.2)
Electrical conductivity (µS cm^{-1})	1100 ± 142 (886 – 1388)	1235 ± 254 (856 – 1563)	1129 ± 134 (974 – 1378)	1088 ± 117 (918 – 1313)
Dissolved oxygen (mg l^{-1})	5.4 ± 0.5 (4.5 – 6.4)	2.8 ± 0.4 (2.2 – 3.6)	5.9 ± 0.4 (5.1 – 6.8)	6.1 ± 0.5 (5.3 – 7.2)
Oxygen saturation (mg l^{-1})	7.0 ± 0.6 (6.4 – 8.1)	6.8 ± 0.5 (6.3 – 7.7)	7.0 ± 0.5 (6.4 – 7.8)	7.0 ± 0.6 (6.4 – 8.2)
Turbidity (NTU)	18.6 ± 3.1 (13.6 – 24.6)	63.7 ± 5.1 (55.6 – 72.3)	21.4 ± 3.1 (16.8 – 27.5)	15.5 ± 3.1 (11.5 – 21.5)
pH	7.874 ± 0.066 (7.744 – 7.985)	7.525 ± 0.072 (7.399 – 7.659)	8.106 ± 0.064 (7.998 – 8.212)	8.368 ± 0.109 (8.151 – 8.497)
Total alkalinity (µmol kg^{-1})	1297 ± 241 (894 – 1731)	1295 ± 123 (1098 – 1527)	2461 ± 166 (2219 – 2754)	1416 ± 241 (1013 – 1850)
Dissolved inorganic carbon (µmol kg^{-1})	1328 ± 247 (914 – 1773)	1375 ± 128 (1171 – 1613)	2485 ± 169 (2235 – 2779)	1409 ± 238 (1006 – 1838)
Hydroxyl ion (µmol kg^{-1})	1.1 ± 0.3 (0.6 – 1.6)	0.6 ± 0.2 (0.3 – 0.9)	1.9 ± 0.4 (1.1 – 2.6)	3.3 ± 0.9 (2.0 – 4.8)
pCO_2(water) (µatm)	1247 ± 358 (637 – 1811)	2798 ± 472 (1853 – 3567)	1378 ± 317 (911 – 1910)	433 ± 140 (202 – 731)
CO_2 flux (µmol m^{-2} h^{-1})	1666 ± 771 (431 – 2817)	4707 ± 1099 (2944 – 6812)	1915 ± 729 (929 – 3007)	69 ± 281 (-371 – 723)
UWPAR (µmol m^{-2} s^{-1})	11.6 ± 1.4 (9.9 – 14.6)	5.0 ± 1.4 (3.2 – 7.5)	6.3 ± 0.9 (5.1 – 8.1)	20.3 ± 2.5 (17.3 – 25.7)
Chl-a (mg m^{-3})	17.7 ± 3.4 (12.0 – 23.0)	81.3 ± 13.2 (61.0 – 102.0)	16.4 ± 2.7 (12.0 – 21.0)	13.3 ± 2.4 (10.0 – 19.0)
GPP (gO_2 m^{-2} d^{-1})	5.1 ± 0.8 (3.9 – 6.9)	10.7 ± 1.7 (7.8 – 13.8)	4.7 ± 0.7 (3.5 – 6.1)	11.6 ± 1.9 (8.9 – 15.8)
CR (gO_2 m^{-2} d^{-1})	11.3 ± 2.1 (8.7 – 15.2)	28.8 ± 5.8 (20.2 – 39.1)	12.1 ± 2.4 (9.1 – 16.5)	11.9 ± 3.6 (6.1 – 15.9)

three ponds consistently throughout the year though the difference was not statistically significant ($p > 0.05$). Due to a comparatively lesser volume of water in P2, the influence of land could have been slightly higher in this pond, which in turn, led to this marginal difference. Seasonally, monsoon recorded the highest temperature followed by pre-monsoon and post-monsoon in all four ponds. The surface water temperature is mainly governed by the ambient temperature. In this region, the temperature remains consistently high throughout the monsoon. Though the summer months (April and May) fall under the pre-monsoon season, the other two months of this season (February and March) experiences much lower temperatures. All four ponds were essentially freshwater types. The electrical conductivity ranged between 856 and 1563 $\mu S\ cm^{-1}$. P2, the abandoned pond showed significantly higher electrical conductivity than the three other ponds ($p > 0.05$). One of the main reasons could be the lesser volume of water coupled with the dilapidated condition of the pond, which enabled the dissociation of various litter in the pond water that enhanced the total dissolved solids. P2 not being used for any purpose was left neglected and such ponds often receive a higher degree of anthropogenic litters than the ones which are consistently in use for some purpose. Dissolved oxygen exhibited significant differences among the four ponds ($p < 0.05$). Again, the most notable difference was observed between P2 and all the other ponds. P2 almost recorded half the annual mean dissolved oxygen observed in the other three ponds. We observed an algal mat over P2 almost every month, except few monsoon months when heavy torrential downpour cleared off some portion of the algal cover. Thus, it was expected that net heterotrophy is quite persistent in this pond compared to the others. This aspect could have led to the lowered dissolved oxygen in P2. It is also worth mentioning that P1 and P4 showed a significant difference, with P4 showing the highest annual mean dissolved oxygen among the four ponds. Similar to dissolved oxygen, turbidity also showed a significant difference among the ponds, and in this case, also P2 exhibited significantly high turbidity compared to the other three ponds. The shallow depth in P2 might have enabled bottom churning and the public lettering might have added more suspended materials, which led to the enhanced turbidity. The presence of a thick algal cover might have also contributed to the enhanced turbidity. Besides P2, the other ponds also showed a subtle but significant difference in turbidity among them. P4, the one that is maintained properly, showed the least annual mean turbidity. The variation of the physicochemical parameters is illustrated in Fig. 6.4.

6.9.2 Variability of Primary Productivity Parameters

The UWPAR gives us an idea about how conducive the aquatic environment is, in terms of carrying out primary autotrophic processes. It depends on the solar insolation as well as the light penetrating ability of the water column. In this study, a significant difference in UWPAR was observed among the ponds ($p < 0.05$). P2 recorded the lowest annual mean UWPAR, whereas P4 recorded the highest annual mean UWPAR. The ponds with minimal anthropogenic disturbance (P1 and P4)

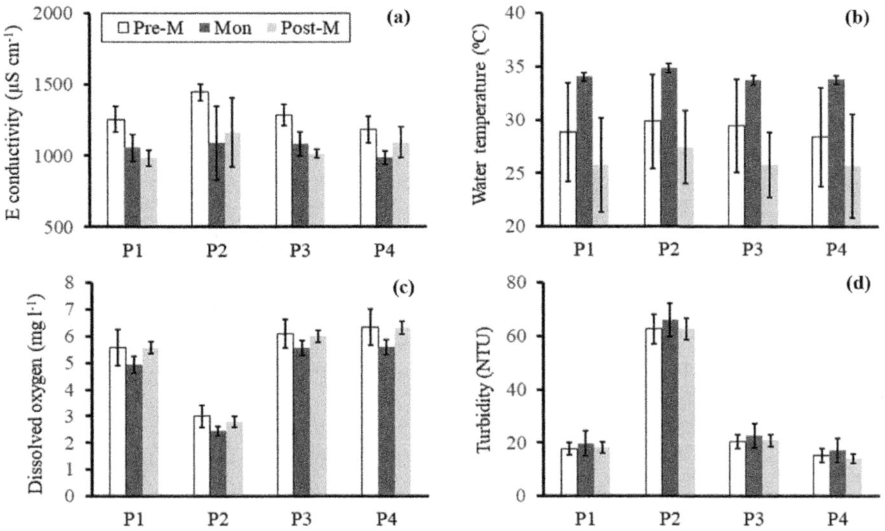

Fig. 6.4 The seasonal variability of the physicochemical parameters in the four selected ponds of the present study

exhibited comparatively higher UWPAR, which showed that proper maintenance can enhance the UWPAR in the water column and can increase the autotrophic potential of these ponds. The high turbidity in P2 could be one of the major reasons behind the low UWPAR observed in this pond. Chl-a concentration gives us an idea about the standing stock of phytoplankton biomass in an aquatic system. The results obtained from this study showed that P1, P3, and P4 did not exhibit any significant difference in chl-a concentration amongst each other. However, P2 showed significantly ($p < 0.01$) higher chl-a concentration all through the sampling period. The annual mean chl-a concentration indicated that P2 exhibited 4–5 times the concentration observed in the other three ponds. This shows that the neglected pond, P2 is prone to eutrophication and most likely exists in a hypereutrophic state. It is interesting to note that despite having the highest mean concentration of chl-a, P2 did not record the highest rate of GPP. P4 exhibited the highest GPP, followed by P2, P1, and P3. The GPP of P1 and P3 had no statistically significant difference. Similarly, P2 and P4 did not show any statistically significant difference. However, it is worth mentioning that the GPP in P1 and P3 was nearly half of what was observed in the case of P2 and P4. The reason behind high GPP in P2 could be the higher concentration of chl-a, whereas in the case of P4 despite having a modest chl-a concentration, favorable conditions of photosynthesis could have led to the higher GPP, compared to P1 and P3. CR magnitudes indicated that all four ponds acted as net heterotrophic systems. However, in P4 there was hardly any difference between the GPP and CR, which indicated that this pond was neither net autotrophic, not net heterotrophic, in the strict sense. CR followed the trend like that of chl-a. P2 exhibited significantly high CR compared to the other three ponds and P1, P3, and P4 did not exhibit any significant

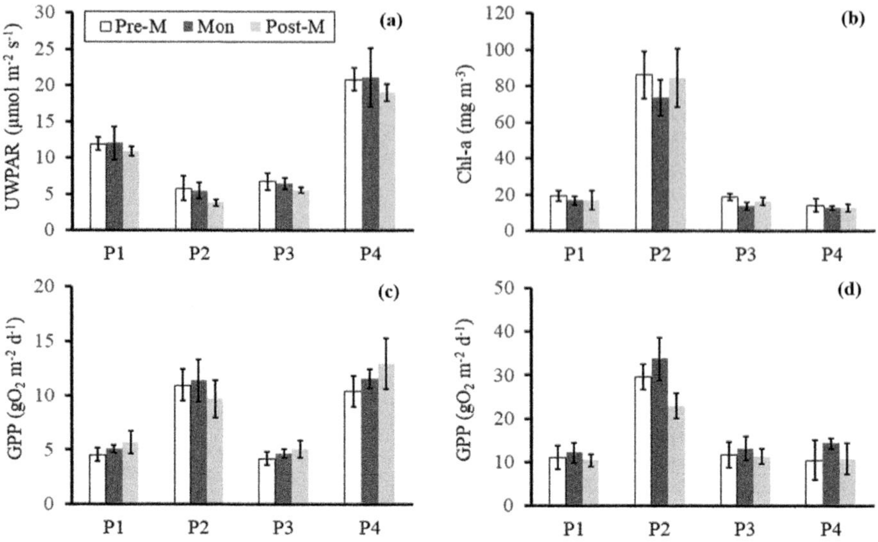

Fig. 6.5 The seasonal variation in productivity parameters in the four selected ponds of the present study

difference in CR amongst each other. This showed that excessively high chl-*a* led to hyper-eutrophic conditions, which in turn, promoted community respiration. The NPP followed the order P4 (-0.2 ± 3.0 gO_2 m^{-2} d^{-1}) > P1 (-6.2 ± 2.0 gO_2 m^{-2} d^{-1}) > P3 (-7.5 ± 2.1 gO_2 m^{-2} d^{-1}) > P2 (-18.2 ± 5.5 gO_2 m^{-2} d^{-1}). The variation of productivity parameters in the four ponds is illustrated in Fig. 6.5.

6.9.3 Variability in Carbonate Chemistry Parameters

All four ponds were alkaline throughout the year. However, P2 exhibited the lowest mean pH compared to the other three ponds. At the same time, all four ponds showed significantly different pH amongst each other. P4 recorded the highest pH (8.368 ± 0.109) among the four ponds. Overall, pH remained the lowest in the monsoon season compared to the other two seasons, except in P2. In P2, pH increased during the monsoon. Monsoonal rain usually brings with low pH water, as CO_2 remains dissolved in rainwater. However, in P2, the already low pH water was probably diluted by the rain which could have increased the pH. Total alkalinity in P1, P2, and P4 did not show any significant difference. However, the annual mean TAlk in P3 was almost double the other three ponds. This could be due to both geogenic or anthropogenic reasons (like liming); however, it could not be delineated from the present study. The TAlk concentration of the respective ponds remained more or less constant throughout the year and did not any significant seasonal variability. Estimated DIC also mirrored the variability of TAlk in all the ponds. Hydroxyl ion

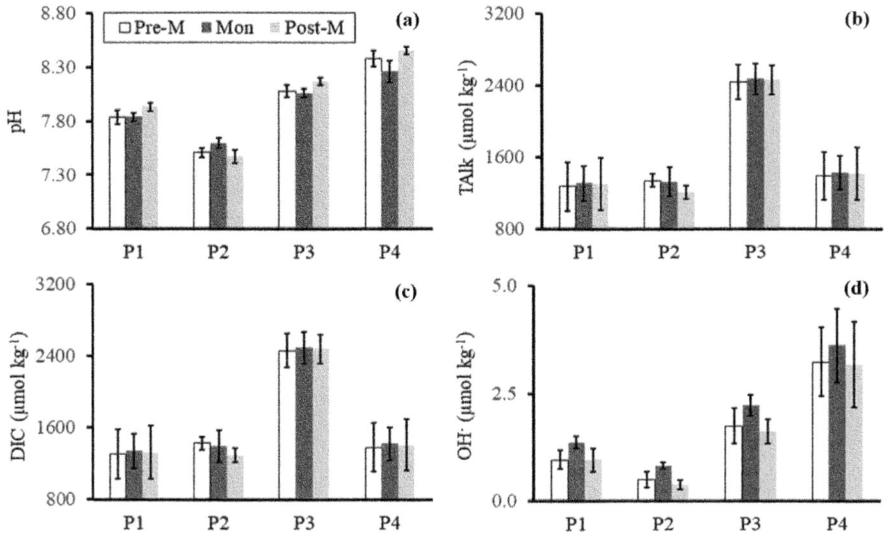

Fig. 6.6 The seasonal variation in carbonate chemistry parameters in the four selected ponds of the present study

concentration followed the same trend as that of pH. The variability of the carbonate chemistry parameters is illustrated in Fig. 6.6.

6.9.4 Variability of pCO_2(Water) and Air–Water CO_2 Flux

The present study indicated that P1, P2, and P3 were always supersaturated with CO_2 in terms of the atmospheric CO_2 concentration, as the minimum pCO_2(water) observed in these ponds was higher than the atmospheric CO_2 concentration. The atmospheric CO_2 concentration during the sampling period did not vary significantly among the ponds. Altogether, the atmospheric CO_2 concentration ranged between 402 and 413 μatm. Thus, it is clear that pCO_2(water) was mainly responsible for governing the ΔpCO_2, and hence the air–water CO_2 flux. All four ponds exhibited significant differences in pCO_2(water) amongst each other. However, P2 showed the highest mean pCO_2(water) among all four ponds (2798 ± 472 μatm). This indicated that the least favorable photosynthetic conditions coupled with a hyper-eutrophic state enhanced the pCO_2(water) to a great extent in P2. P1 and P3 also showed significantly high pCO_2(water). The only exception in this regard was P4. The pCO_2(water) in P4 was undersaturated in some months during the late half of the post-monsoon season and the early half of the pre-monsoon season. Apart from these months, the other months had pCO_2(water) higher than the atmospheric CO_2 concentration. Thus, this pond, P4, exhibited a dual character in terms of source and sink of CO_2. The air–water CO_2 flux mirrored the variability of pCO_2(water), as

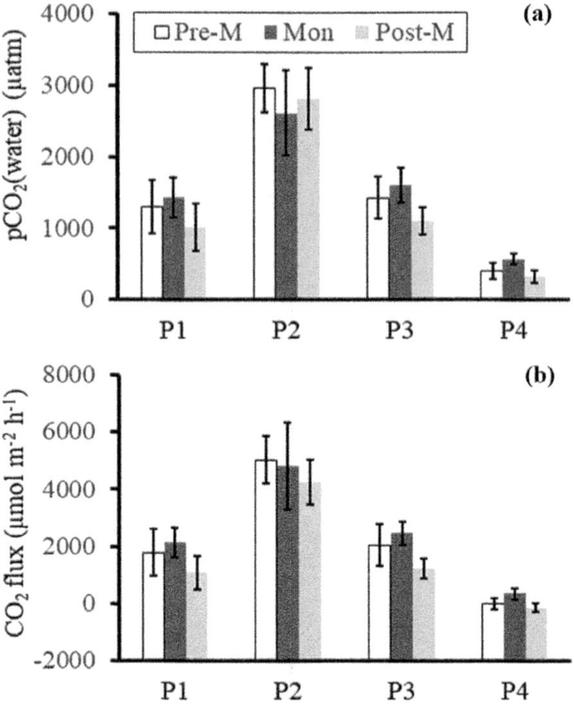

Fig. 6.7 The seasonal variability in pCO_2(water) and air–water CO_2 flux in the four selected ponds of the present study

ΔpCO_2 was mainly governed by this parameter, and pCO_2(air) exhibited marginal variation. The annual mean air–water CO_2 flux varied in the order: P2 (4707 ± 1099 μmol m^{-2} h^{-1}) > P3 (1915 ± 729 μmol m^{-2} h^{-1}) > P1 (1666 ± 771 μmol m^{-2} h^{-1}) > P4 (69 ± 281 μmol m^{-2} h^{-1}) (Fig. 6.7). Thus it is evident that the abandoned pond, which is in the most dilapidated state emits CO_2 at the highest rate and the well-maintained pond with the least anthropogenic disturbance emits CO_2 at the lowest rate. Sometimes this well-maintained pond acted as a CO_2 sink also.

6.9.5 Relationship Between pCO_2(Water) and Other Biogeochemical Variables

Water temperature, electrical conductivity, dissolved oxygen, oxygen saturation, turbidity, UWPAR, chl-a, GPP, CR, and NPP were correlated with pCO_2(water) to identify the potential regulators of the air–water CO_2 fluxes (Table 6.2). Three out of four ponds (except P2) showed a statistically significant positive relationship between water temperature and pCO_2(water). This shows that temperature plays a

Table 6.2 The relationship between pCO_2(water) and other biogeochemical parameters depicted through the Pearson correlation coefficient

Parameters	P1	P2	P3	P4
Water temperature	**r = 0.72 p < 0.01**	r = 0.05 p > 0.05	**r = 0.84 p < 0.01**	**r = 0.81 p < 0.01**
Electrical conductivity	r = 0.15 p > 0.05	r = 0.46 p > 0.05	r = 0.27 p > 0.05	r = −0.39 p > 0.05
Dissolved oxygen	r = −0.42 p > 0.05	r = 0.21 p > 0.05	r = −0.46 p > 0.05	**r = −0.81 p < 0.01**
Oxygen saturation	**r = −0.72 p < 0.01**	r = −0.06 p > 0.05	**r = −0.83 p < 0.01**	**r = −0.83 p < 0.01**
Turbidity	r = 0.21 p > 0.05	r = −0.42 p > 0.05	r = 0.30 p > 0.05	r = 0.03 p > 0.05
UWPAR	r = 0.02 p > 0.05	r = 0.40 p > 0.05	r = 0.46 p > 0.05	**r = 0.57 p < 0.05**
Chl-a	r = 0.37 p > 0.05	r = 0.41 p > 0.05	r = −0.31 p > 0.05	r = −0.04 p > 0.05
GPP	r = 0.37 p > 0.05	r = 0.30 p > 0.05	r = 0.28 p > 0.05	r = 0.18 p > 0.05
CR	**r = 0.79 p < 0.01**	r = −0.12 p > 0.05	**r = 0.80 p < 0.01**	**r = 0.87 p < 0.01**
NPP	**r = −0.68 p < 0.05**	r = 0.22 p > 0.05	**r = −0.83 p < 0.05**	**r = −0.93 p < 0.01**

crucial role in aiding and enhancing the net heterotrophy, which led to this positive relationship. Given the ongoing climate change and the increase in ambient temperature, these ponds are expected to emit more CO_2 in the near future.

Electrical conductivity, which did not vary to a great extent amongst the ponds (as all the ponds are of freshwater type) did not show any significant relationship in any of the ponds. Usually, high pCO_2(water) is accompanied by low dissolved oxygen and vice-versa, when both of these parameters are governed exclusively by biological mechanisms. However, such a negative relationship between the two parameters was only observed in the case of P4, which is a well-maintained pond with the least anthropogenic disturbance. Oxygen saturation, however, showed a negative relationship with pCO_2(water) in P1, P3, and P4. This indicated that biological mechanisms played a crucial role in the O_2-CO_2 equilibrium. Though chl-a and GPP did not show any significant relationship with pCO_2(water), CR and NPP showed a significant relationship in P1, P3, and P4. This showed that the pCO_2(water) dynamics in these three ponds were mainly regulated by the biological net heterotrophy. This is why temperature and pCO_2(water) also could have shown a positive relationship because increased temperature facilitates the degradation of organic matter, and enhances net heterotrophy. It is also worth mentioning and quite intriguing that P2 did not show any significant relationship with any of the parameters. This led us to conclude that under hyper-eutrophic and abandoned state, the dynamics of pCO_2(water) depends on a combination of factors like net heterotrophy, photosynthesis inhibition in the water column, a higher rate of degradation, and so forth, and it is often not observed through a simple linear correlation between the respective parameters.

6.10 Conclusion

Overall, the ponds of Indian Sundarbans are found to be net heterotrophic. The ponds where daily household activities are carried out or are anthropogenically disturbed, emitted more CO_2 than the ones that experience lesser anthropogenic intervention. Well-maintained ponds without any human intervention like bathing, washing utensils, and clothes, can even act as sinks for CO_2 during some time of the year when conducive conditions for net autotrophy prevails. The results indicate that biological mechanisms mainly govern the CO_2 exchange and physical forcing plays a lesser role. Water temperature, dissolved oxygen saturation, community respiration, and net primary productivity showed a significant relationship with pCO_2(water), and hence, air–water CO_2 flux. These observations indicate that biological organisms responsible for photosynthesis mainly regulates the pCO_2(water). Thus, maintaining conditions favorable for photosynthesis in these ponds can provide negative feedback to climate change and vice-versa. This study also inferred that abandoned hypereutrophic ponds are the worst as these ponds emit significantly higher CO_2 than the ordinary homestead and multipurpose ponds, where the regular human intervention takes place.

References

Åberg J, Jansson M, Jonsson A (2010) Importance of water temperature and thermal stratification dynamics for temporal variation of surface water CO_2 in a boreal lake. J Geophys Res 115:G02024. https://doi.org/10.1029/2009JG001085

Andersen MR, Kragh T, Sand-Jensen K (2017) Extreme diel dissolved oxygen and carbon cycles in shallow vegetated lakes. Proc R Soc B Biol Sci 284(1862):20171427. https://doi.org/10.1098/rspb.2017.1427

Andersen MR, Sand-Jensen K, Iestyn Woolway R, Jones ID (2016) Profound daily vertical stratification and mixing in a small, shallow, wind-exposed lake with submerged macrophytes. Aquat Sci 79:395–406. https://doi.org/10.1007/s00027-016-0505-0

Anderson CB (2002) Understanding carbonate equilibria by measuring alkalinity in experimental and natural systems. J Geosci Educ 50:389–403. https://doi.org/10.5408/1089-9995-50.4.389

Attayde JL, van Nes EH, Araujo AI, Corso G, Scheffer M (2010) Omnivory by planktivores stabilizes plankton dynamics, but may either promote or reduce algal biomass. Ecosystems 13(3):410–420. https://doi.org/10.1007/s10021-010-9327-4

Aufdenkampe AK, Mayorga E, Raymond PA, Melack J, Doney SC, Alin SR, Aalto RE, Yoo K (2011) Riverine coupling of biogeochemical cycles between land, oceans, and atmosphere. Front Ecol Environ 9:53–60. https://doi.org/10.1890/100014

Barat R, Montoya T, Seco A, Ferrer J (2011) Modelling biological and chemically induced precipitation of calcium phosphate in enhanced biological phosphorus removal systems. Water Res 45(12):3744–3752. https://doi.org/10.1016/j.watres.2011.04.028

Bastviken D, Cole J, Pace M, Tranvik L (2004) Methane emissions from lakes: dependence of lake characteristics, two regional assessments, and a global estimate. Glob Biogeochem Cycl 18(4). https://doi.org/10.1029/2004GB002238

Bastviken D, Cole JJ, Pace ML, VandeBogert MC (2008) Fates of methane from different lake habitats: connecting whole-lake budgets and CH_4 emissions. J Geophys Res Biogeosci 113:G02024. https://doi.org/10.1029/2007JG000608

Battin TJ, Luyssaert S, Kaplan LA, Aufdenkampe AK, Richter A, Tranvik LJ (2009) The boundless carbon cycle. Nat Geosci 2(9):598–600. https://doi.org/10.1038/ngeo618

Benstead JP, Leigh DS (2012) An expanded role for river networks. Nat Geosci 5:678–679. https://doi.org/10.1038/ngeo1593

Bettez ND, Groffman PM (2012) Denitrification potential in stormwater control structures and natural riparian zones in an urban landscape. Environ Sci Technol 46(20):10909–10917. https://doi.org/10.1021/es301409z

Biswas G, Kumar P, Kailasam M, Ghoshal TK, Bera A, Vijayan KK (2019) Application of Integrated Multi Trophic Aquaculture (IMTA) concept in brackishwater ecosystem: the first exploratory trial in the Sundarban, India. J Coast Res 86(SI):49–55. https://doi.org/10.2112/SI86-007.1

Blaszczak JR, Steele MK, Badgley BD et al (2018) Sediment chemistry of urban stormwater ponds and controls on denitrification. Ecosphere 9(6):e02318. https://doi.org/10.1002/ecs2.2318

Boehrer B, Schultze M (2008) Stratification of lakes. Rev Geophys 46:RG2005. https://doi.org/10.1029/2006RG000210

Boyd CE (ed) (2012) Bottom soils, sediment, and pond aquaculture. Springer Science & Business Media, UK

Branco BF, Torgersen T (2009) Predicting the onset of thermal stratification in shallow inland waterbodies. Aquat Sci 71:65–79. https://doi.org/10.1007/s00027-009-8063-3

Carbonero F, Oakley BB, Purdy KJ (2014) Metabolic flexibility as a major predictor of spatial distribution in microbial communities. PLoS ONE 9:e85105. https://doi.org/10.1371/journal.pone.0085105

Carlson RE (1977) A trophic state index for lakes. Limnol Oceanogr 22(2):361–369. https://doi.org/10.4319/lo.1977.22.2.0361

Carpenter SR (1981) Submersed vegetation: an internal factor in lake ecosystem succession. Am Nat 118(3):372–383

Carpenter SR, Caraco NF, Correll DL, Howarth RW, Sharpley AN, Smith VH (1998) Nonpoint pollution of surface waters with phosphorus and nitrogen. Ecol Appl 8(3):559–568. https://doi.org/10.1890/1051-0761(1998)008[0559:NPOSWW]2.0.CO;2

Chand BK, Trivedi RK, Biswas A, Dubey SK, Beg MM (2012) Study on impact of saline water inundation on freshwater aquaculture in Sundarban using risk analysis tools. Explor Anim Med Res 2:170–178

Chanda A, Das S, Bhattacharyya S et al (2020) CO_2 effluxes from an urban tidal river flowing through two of the most populated and polluted cities of India. Environ Sci Pollut Res 27:30093–30107. https://doi.org/10.1007/s11356-020-09254-6

Chen Y, Dong S, Bai Y, Xu S, Yang X, Pan Z (2018) Carbon budgets from mariculture ponds without a food supply. Aquac Environ Interact 10:465–472. https://doi.org/10.3354/aei00279

Chislock MF, Doster E, Zitomer RA, Wilson AE (2013) Eutrophication: causes, consequences, and controls in aquatic ecosystems. Nat Educ Knowl 4(4):10. https://www.nature.com/scitable/knowledge/library/eutrophication-causes-consequences-and-controls-in-aquatic-102364466/

Christensen J, Sand-Jensen K, Staehr PA (2013) Fluctuating water levels control water chemistry and metabolism of a charophyte-dominated pond. Freshw Biol 58:1353–1365. https://doi.org/10.1111/fwb.12132

Codispoti LA (2010) Interesting times for marine N_2O. Science 327(5971):1339–1340. https://doi.org/10.1126/science.1184945

Cole JJ, Caraco NF (1998) Atmospheric exchange of carbon dioxide in a low-wind oligotrophic lake measured by the addition of SF_6. Limnol Oceanogr 43(4):647–656. https://doi.org/10.4319/lo.1998.43.4.0647

Cole JJ, Pace ML, Carpenter SR, Kitchell JF (2000) Persistence of net heterotrophy in lakes during nutrient addition and food web manipulations. Limnol Oceanogr 45:1718–1730. https://doi.org/10.4319/lo.2000.45.8.1718

Collins KA, Lawrence TJ, Stander EK et al (2010) Opportunities and challenges for managing nitrogen in urban stormwater: a review and synthesis. Ecol Eng 36(11):1507–1519. https://doi.org/10.1016/j.ecoleng.2010.03.015

Condie SA, Webster IT (2002) Stratification and circulation in a shallow turbid waterbody. Environ Fluid Mech 2:177–196. https://doi.org/10.1023/A:1019898931829

Conley DJ, Paerl HW, Howarth RW et al (2009) Controlling eutrophication: nitrogen and phosphorus. Science 323:1014–1015. https://doi.org/10.1126/science.1167755

Conrad R, Noll M, Claus P, Klose M, Bastos WR, Enrich-Prast A (2011) Stable carbon isotope discrimination and microbiology of methane formation in tropical anoxic lake sediments. Biogeosciences 8:795–814. https://doi.org/10.5194/bg-8-795-2011

Corman JR, Poret-Peterson AT, Uchitel A, Elser JJ (2016) Interaction between lithification and resource availability in the microbialites of Rio Mesquites, Cuatro Ciénegas, México. Geobiology 17:176–189. https://doi.org/10.1111/gbi.12168

Crosswell JR, Anderson IC, Stanhope JW, Van Dam B, Brush MJ, Ensign S, Piehler MF, McKee B, Bost M, Paerl HW (2017) Carbon budget of a shallow, lagoonal estuary: transformations and source-sink dynamics along the river-estuary ocean continuum. Limnol Oceanogr 62:S29–S45. https://doi.org/10.1002/lno.10631

Dantas DDF, Rubim PL, de Oliveira FA, da Costa MRA, Moura CGB, Teixeira LH, Attayde JL (2019) Effects of benthivorous and planktivorous fish on phosphorus cycling, phytoplankton biomass, and water transparency of a tropical shallow lake. Hydrobiologia 829:31–41. https://doi.org/10.1007/s10750-018-3613-0

DelSontro T, Beaulieu JJ, Downing JA (2018) Greenhouse gas emissions from lakes and impoundments: upscaling in the face of global change. Limnol Oceanogr Lett 3:64–75. https://doi.org/10.1002/lol2.10073

DeVries T, Le Quéré C, Andrews O et al (2019) Decadal trends in the ocean carbon sink. Proc Natl Acad Sci 116(24):11646–11651. https://doi.org/10.1073/pnas.1900371116

Dodds WK, Bouska WW, Eitzmann JL et al (2009) Eutrophication of US freshwaters: analysis of potential economic damages. Environ Sci Technol 43(1):12–19. https://doi.org/10.1021/es801217q

Downing JA (2009) Global limnology: up-scaling aquatic services and processes to planet Earth. Verh Internat Verein Limnol 30(8):1149–1166. https://doi.org/10.1080/03680770.2009.11923903

Downing JA, Prairie YT, Cole JJ et al (2006) The global abundance and size distribution of lakes, ponds, and impoundments. Limnol Oceanogr 51:2388–2397. https://doi.org/10.4319/lo.2006.51.5.2388

Drake TW, Raymond PA, Spencer RGM (2018) Terrestrial carbon inputs to inland waters: a current synthesis of estimates and uncertainty. Limnol Oceanogr Lett 3:132–142. https://doi.org/10.1002/lol2.10055

Dubey SK, Trivedi RK, Chand BK, Mandal B, Rout SK (2017) Farmers' perceptions of climate change, impacts on freshwater aquaculture and adaptation strategies in climatic change hotspots: a case of the Indian Sundarban delta. Environ Dev 21:38–51. https://doi.org/10.1016/j.envdev.2016.12.002

Duc N, Crill P, Bastviken D (2010) Implications of temperature and sediment characteristics on methane formation and oxidation in lake sediments. Biogeochemistry 100(1):185–196. https://doi.org/10.1007/s10533-010-9415-8

Elrod MJ (1999) Greenhouse warming potentials from the infrared spectroscopy of atmospheric gases. J Chem Educ 76(12):1702–1705. https://doi.org/10.1021/ed076p1702

Elser JJ, Schampel JH, Kyle M et al (2005) Response of grazing snails to phosphorus enrichment of modern stromatolitic microbial communities. Freshw Biol 50:1826–1835. https://doi.org/10.1111/j.1365-2427.2005.01453.x
Fee E, Hecky R, Kasian S, Cruikshank D (1996) Effects of lake size, water clarity, and climatic variability on mixing depths in Canadian Shield lakes. Limnol Oceanogr 41:912–920. https://doi.org/10.4319/lo.1996.41.5.0912
Ferrón S, Ho DT, Johnson ZI, Huntley ME (2012) Air-water fluxes of N_2O and CH_4 during microalgae (*Staurosira* sp.) cultivation in an open raceway pond. Environ Sci Technol 46(19):10842–10848. https://doi.org/10.1021/es302396j
Frankignoulle M, Bourge I, Wollast R (1996) Atmospheric CO_2 fluxes in a highly polluted estuary (the Scheldt). Limnol Oceanogr 41(2):365–369. https://doi.org/10.4319/lo.1996.41.2.0365
Frost PC, Prater C, Scott AB, Song K, Xenopoulos MA (2019) Mobility and bioavailability of sediment phosphorus in urban stormwater ponds. Water Resour Res 55:3680–3688. https://doi.org/10.1029/2018WR023419
Glaz P, Bartosiewicz M, Laurion I, Reichwaldt ES, Maranger R, Ghadouani A (2016) Greenhouse gas emissions from waste stabilization ponds in Western Australia and Quebec (Canada). Water Res 101:64–74. https://doi.org/10.1016/j.watres.2016.05.060
Gorsky AL, Racanelli GA, Belvin AC, Chambers RM (2019) Greenhouse gas flux from stormwater ponds in southeastern Virginia (USA). Anthropocene 28:100218. https://doi.org/10.1016/j.ancene.2019.100218
Groffman PM, Butterbach-Bahl K, Fulweiler RW et al (2009) Challenges to incorporating spatially and temporally explicit phenomena (hotspots and hot moments) in denitrification models. Biogeochemistry 93(1–2):49–77. https://doi.org/10.1007/s10533-008-9277-5
Groszkopf T, Soyer OS (2016) Microbial diversity arising from thermodynamic constraints. ISME J 10:2725–2733. https://doi.org/10.1038/ismej.2016.49
Hansen J, Reitzel K, Jensen HS, Andersen FØ (2003) Effects of aluminum, iron, oxygen, and nitrate additions on phosphorus release from the sediment of a Danish softwater lake. Hydrobiologia 492:139–149. https://doi.org/10.1023/A:1024826131327
Harmon TC (2020) Carbon gas flux to and from inland waters: support for a global observation network. Limnology 21:429–442. https://doi.org/10.1007/s10201-020-00623-1
Hewson I, Vargo GA, Fuhrman JA (2003) Bacterial diversity in shallow oligotrophic marine benthos and overlying waters: effects of virus infection, containment, and nutrient enrichment. Microb Ecol 46:322–336. https://doi.org/10.1007/s00248-002-1067-3
Howarth R, Paerl HW (2008) Coastal marine eutrophication: control of both nitrogen and phosphorus is necessary. Proc Natl Acad Sci USA 105:E103. https://doi.org/10.1073/pnas.0807266106
Howarth RW, Marino R, Lane J, Cole JJ (1988) Nitrogen fixation in freshwater, estuarine, and marine ecosystems. 1. Rates and importance. Limnol Oceanogr 33(4 part 2):669–687
Howarth RW, Schneider R, Swaney D (1996) Metabolism and organic carbon fluxes in the tidal freshwater Hudson River. Estuaries 19:848–865. https://doi.org/10.2307/1352302
Hupfer M, Lewandowski J (2008) Oxygen controls the phosphorus release from lake sediments—a long-lasting paradigm in limnology. Int Rev Hydrobiol 93:415–432. https://doi.org/10.1002/iroh.200711054
Iglesias C, Mazzeo N, Meerhoff M et al (2011) High predation is of key importance for dominance of small-bodied zooplankton in warm shallow lakes: evidence from lakes, fish exclosures, and surface sediments. Hydrobiologia 667:133–147. https://doi.org/10.1007/s10750-011-0645-0
James WF (2011) Variations in the aluminum: Phosphorus binding ratio and alum dosage considerations for Half Moon Lake, Wisconsin. Lake Reser Manage 27:128–137. https://doi.org/10.1080/07438141.2011.572232
Jeppesen E, Brucet S, Naselli-Flores L et al (2015) Ecological impacts of global warming and water abstraction on lakes and reservoirs due to changes in water level and related changes in salinity. Hydrobiologia 750:201–227. https://doi.org/10.1007/s10750-014-2169-x

Juutinen S, Rantakari M, Kortelainen P et al (2009) Methane dynamics in different boreal lake types. Biogeosciences 6(2):209–223. https://doi.org/10.5194/bg-6-209-2009

Kalff J (2002) Limnology: inland water ecosystems. Prentice-Hall, New Jersey

Kant S, Mandal NK, Dubey MK (2019) Impact of anthropogenic activities on physicochemical properties of water and sediment soil of a perennial pond of Godda District (Santal Pargana), Jharkhand. Indian J Environ Sci 23(2):69–74

Kattner G (1999) Storage of dissolved inorganic nutrients in seawater: poisoning with mercuric chloride. Mar Chem 67(1–2):61–66. https://doi.org/10.1016/S0304-4203(99)00049-3

Khalil MAK (1999) Non-CO_2 greenhouse gases in the atmosphere. Ann Rev Energy Environ 24(1):645–661

Khoo KH, Ramette RW, Culberson CH, Bates RG (1977) Determination of hydrogen ion concentrations in seawater from 5 to 40 °C: standard potentials at salinities from 20 to 45‰. Anal Chem 49(1):29–34

Kitchell JF, Carpenter SR (1993) Cascading trophic interactions. In: Carpenter SR, Kitchell JF (eds) The trophic cascade in Lakes. Cambridge University Press, New York, pp 1–14

Köhler P, Nehrbass-Ahles C, Schmitt J, Stocker TF, Fischer H (2017) A 156 kyr smoothed history of the atmospheric greenhouse gases CO_2, CH_4, and N_2O and their radiative forcing. Earth Syst Sci Data 9(1):363–387. https://doi.org/10.5194/essd-9-363-2017

Kondo J (2000) Atmospheric science near the ground surface. University of Tokyo, Japan

Kortelainen P, Rantakari M, Huttunen JT, Mattsson T, Alm J, Juutinen S, Larmola T, Silvola J, Martikainen PJ (2006) Sediment respiration and lake trophic state are important predictors of large CO_2 evasion from small boreal lakes. Glob Change Biol 12(8):1554–1567. https://doi.org/10.1111/j.1365-2486.2006.01167.x

Kosten S, Roland F, Da MM, Van Nes EH, Mazzeo N, Sternberg LdSL, Scheffer M, Cole JJ (2010) Climate-dependent CO_2 emissions from lakes. Glob Biogeochem Cycle 24(2):GB2007. https://doi.org/10.1029/2009GB003618

Kratzer CR, Brezonik PL (1981) A Carlson-type trophic state index for nitrogen in Florida lakes. J Am Water Resour Assoc 17(4):713–715. https://doi.org/10.1111/j.1752-1688.1981.tb01282.x

Lashof DA, Ahuja DR (1990) Relative contributions of greenhouse gas emissions to global warming. Nature 344(6266):529–531. https://doi.org/10.1038/344529a0

Lee ZMP, Poret-Peterson AT, Siefert JL et al (2017) Nutrient stoichiometry shapes microbial community structure in an evaporitic shallow pond. Front Microbiol 8:949. https://doi.org/10.3389/fmicb.2017.00949

Leflaive J, Danger M, Lacroix G, Lyautey E, Oumarou C, Ten-Hage L (2008) Nutrient effects on the genetic and functional diversity of aquatic bacterial communities. FEMS Microbiol Ecol 66:379–390. https://doi.org/10.1111/j.1574-6941.2008.00593.x

Lewis E, Wallace DWR (1998) Program developed for CO_2 system calculations. ORNL/CDIAC–105. Carbon Dioxide Information Analysis Center, Oak Ridge National Laboratory, U.S. Department of Energy, Oak Ridge, Tennessee

MacIntyre S, Wanninkhof R, Chanton JP (1995) Trace gas exchange across the air-water interface in freshwater and coastal marine environments. In: Matson PA, Harriss RC (eds) Biogenic trace gases: measuring emissions from soil and water. Blackwell Science, Oxford, pp 52–97

Macrae ML, Bello RL, Molot LA (2004) Long-term carbon storage and hydrological control of CO_2 exchange in tundra ponds in the Hudson Bay Lowland. Hydrol Processes 18(11):2051–2069. https://doi.org/10.1002/hyp.1461

Mandal S, Burman D, Sarangi SK, Bandyopadhyay BK, Maji B (2015) Homestead production systems in Sundarbans region of West Bengal, India–Current status and opportunities. CGIAR-CPWF

Markfort CD, Perez AL, Thill JW, Jaster DA, Porte-Agel F, Stefan HG (2010) Wind sheltering of a lake by a tree canopy or bluff topography. Water Resour Res. https://doi.org/10.1029/2009wr007759

Marotta H, Fontes MLS, Petrucio MM (2012) Natural events of anoxia and low respiration index in oligotrophic lakes of the Atlantic Tropical Forest. Biogeosciences 9:2879–2887. https://doi.org/10.5194/bg-9-2879-2012

Martinsen KT, Andersen MR, Sand-Jensen K (2019) Water temperature dynamics and the prevalence of daytime stratification in small temperate shallow lakes. Hydrobiologia 826(1):247–262. https://doi.org/10.1007/s10750-018-3737-2

Mayer T (2020) Interactions of fish, algae, and abiotic factors in a shallow, tropical pond. Hydrobiologia 847(20):4145–4160. https://doi.org/10.1007/s10750-020-04375-y

Menezes RF, Attayde JL, Vasconcelos FR (2010) Effects of omnivorous filter-feeding fish and nutrient enrichment on the plankton community and water transparency of a tropical reservoir. Freshw Biol 55:767–779. https://doi.org/10.1111/j.1365-2427.2009.02319.x

Millero FJ (1979) Thermodynamics of the carbon dioxide system in the oceans. Geochim Cosmochim Acta 59(4):661–677. https://doi.org/10.1016/0016-7037(94)00354-O

Mukhopadhyay G, Mondal D, Biswas P, Dewanji A (2004) Water quality monitoring of tropical ponds: location and depth effect in two case studies. Acta Hydrochim Hydrobiol 32(2):138–148. https://doi.org/10.1002/aheh.200300524

Natchimuthu S, Selvam BP, Bastviken D (2014) Influence of weather variables on methane and carbon dioxide flux from a shallow pond. Biogeochemistry 119(1–3):403–413. https://doi.org/10.1007/s10533-014-9976-z

Nelson CE, Carlson CA (2011) Differential response of high-elevation planktonic bacterial community structure and metabolism to experimental nutrient enrichment. PLoS ONE 6:e18320. https://doi.org/10.1371/journal.pone.0018320

Newell SE, McCarthy MJ, Gardner WS, Fulweiler RW (2016) Sediment nitrogen fixation: a call for re-evaluating coastal N budgets. Estuar Coast 39(6):1626–1638. https://doi.org/10.1007/s12237-016-0116-y

Parsons TR, Maita Y, Lalli CM (1992) A manual of chemical and biological methods for sea water analysis. Pergamon, New York, pp 101–103

Pugh TA, Lindeskog M, Smith B, Poulter B, Arneth A, Haverd V, Calle L (2019) Role of forest regrowth in global carbon sink dynamics. Proc Natl Acad Sci 116(10):4382–4387. https://doi.org/10.1073/pnas.1810512116

Purohit P, Höglund Isaksson L (2017) Global emissions of fluorinated greenhouse gases 2005–2050 with abatement potentials and costs. Atmos Chemi Phys 17:2795–2816. https://doi.org/10.5194/acp-17-2795-2017

Rantakari M, Kortelainen P (2005) Interannual variation and climatic regulation of the CO_2 emission from large boreal lakes. Glob Change Biol 11(8):1368–1380. https://doi.org/10.1111/j.1365-2486.2005.00982.x

Rodrigues M, da Costa A, Attayde JL, Becker V (2016) Effects of water level reduction on the dynamics of phytoplankton functional groups in tropical semi-arid shallow lakes. Hydrobiologia 778:75–89. https://doi.org/10.1007/s10750-015-2593-6

Sañudo-Wilhelmy SA, Tovar-Sanchez A, Fu FX, Capone DG, Carpenter EJ, Hutchins DA (2004) The impact of surface-adsorbed phosphorus on phytoplankton Redfield stoichiometry. Nature 432(7019):897–901. https://doi.org/10.1038/nature03125

Sarnelle O, Cooper SD, Wiseman S, Mavuti KM (1998) The relationship between nutrients and trophic-level biomass in turbid tropical ponds. Freshw Biol 40(1):65–75. https://doi.org/10.1046/j.1365-2427.1998.00332.x

Schindler DW (1974) Eutrophication and recovery in experimental lakes:implications for lake management. Science 184(4139):897–899. https://doi.org/10.1126/science.184.4139.897

Schindler DW (2006) Recent advances in the understanding and management of eutrophication. Limnol Oceanogr 51:356–363. https://doi.org/10.4319/lo.2006.51.1_part_2.0356

Scott JT, McCarthy MJ, Gardner WS, Doyle RD (2008) Denitrification, dissimilatory nitrate reduction to ammonium, and nitrogen fixation along a nitrate concentration gradient in a created freshwater wetland. Biogeochemistry 87(1):99–111. https://doi.org/10.1007/s10533-007-9171-6

Seitzinger S, Harrison JA, Böhlke JK et al (2006) Denitrification across landscapes and waterscapes: a synthesis. Ecol Appl 16(6):2064–2090. https://doi.org/10.1890/1051-0761(2006)016[2064:DALAWA]2.0.CO;2

Smith SV, Renwick WH, Bartley JD, Buddemeier RW (2002) Distribution and significance of small, artificial water bodies across the United States landscape. Sci Total Environ 299:21–36. https://doi.org/10.1016/S0048-9697(02)00222-X

Sobek S, Durisch-Kaiser E, Zurbrugg R, Wongfun N, Wessels M, Pasche N, Wehrli B (2009) Organic carbon burial efficiency in lake sediments controlled by oxygen exposure time and sediment source. Limnol Oceanogr 54:2243–2254. https://doi.org/10.4319/lo.2009.54.6.2243

Soininen J, Meier S (2014) Phytoplankton richness is related to nutrient availability, not to pool size, in a subarctic rock pool system. Hydrobiologia 740:137–145. https://doi.org/10.1007/s10750-014-1949-7

Søndergaard M, Peder Jensen J, Jeppesen E (2003) Role of sediment and internal loading of phosphorus in shallow lakes. Hydrobiologia 506:135–145. https://doi.org/10.1023/B:HYDR.0000008611.12704.dd

Spigel RH, Imberger J, Rayner KN (1986) Modeling the diurnal mixed layer. Limnol Oceanogr 31(3):533–556. https://doi.org/10.4319/lo.1986.31.3.0533

Starling F, Lazzaro X, Cavalcanti C, Moreira R (2002) Contribution of omnivorous tilapia to eutrophication of a shallow tropical reservoir: evidence from a fish kill. Freshw Biol 47:2443–2452. https://doi.org/10.1046/j.1365-2427.2002.01013.x

Taguchi VJ, Olsen TA, Natarajan P, Janke BD, Gulliver JS, Finlay JC, Stefan HG (2020) Internal loading in stormwater ponds as a phosphorus source to downstream waters. Limnol Oceanogr Lett 5(4):322–330. https://doi.org/10.1002/lol2.10155

Talling JF (2001) Environmental controls on the functioning of shallow tropical lakes. Hydrobiologia 458:1–8. https://doi.org/10.1023/A:1013121522321

Torgersen T, Branco B (2008) Carbon and oxygen fluxes from a small pond to the atmosphere: temporal variability and the CO_2/O_2 imbalance. Water Res Res 44(2):W02417. https://doi.org/10.1029/2006WR005634

Torres SG, Silva LHS, Rangel LM, Attayde JL, Huszar VLM (2016) Cyanobacteria are controlled by omnivorous filter-feeding fish (Nile tilapia) in a tropical eutrophic reservoir. Hydrobiologia 765:115–129. https://doi.org/10.1007/s10750-015-2406-y

Torsvik V, Øvreås L, Thingstad TF (2002) Prokaryotic diversity–magnitude, dynamics, and controlling factors. Science 296:1064–1066. https://doi.org/10.1126/science.1071698

Tranvik LJ, Downing JA, Cotner JB et al (2009) Lakes and reservoirs as regulators of carbon cycling and climate. Limnol Oceanogr 54:2298–2314. https://doi.org/10.4319/lo.2009.54.6_part_2.2298

Van Horn D, Sinsabaugh R, Takacs-Vesbach C, Mitchell K, Dahm C (2011) Response of heterotrophic stream biofilm communities to a gradient of resources. Aquat Microb Ecol 64:149–161. https://doi.org/10.3354/ame01515

van Hulzen JB, Segers R, van Bodegom PM, Leffelaar PA (1999) Temperature effects on soil methane production: an explanation for observed variability. Soil Biol Biochem 31(14):1919–1929. https://doi.org/10.1016/S0038-0717(99)00109-1

Vaquer-Sunyer R, Duarte CM (2008) Thresholds of hypoxia for marine biodiversity. Proc Nat Acad Sci 105:15452–15457. https://doi.org/10.1073/pnas.0803833105

Wanninkhof R (1992) Relationship between wind speed and gas exchange over the ocean. J Geophys Res 97(CS5):7373–7382. https://doi.org/10.1029/92JC00188

Weiss RF (1970) The solubility of nitrogen, oxygen, and argon in water and seawater. Deep-Sea Res Oceanogr Abstr 17(4):721–735. https://doi.org/10.1016/0011-7471(70)90037-9

Weiss RF (1974) Carbon dioxide in water and seawater: the solubility of non-ideal gas. Mar Chem 2(3):203–215. https://doi.org/10.1016/0304-4203(74)90015-2

Wetzel RG (2001) Limnology: lake and river ecosystems. Gulf Professional Publishing

Woolway RI, Jones ID, Hamilton DP et al (2015) Automated calculation of surface energy fluxes with high frequency lake buoy data. Environ Model Softw 70:191–198. https://doi.org/10.1016/j.envsoft.2015.04.013

Woolway RI, Jones ID, Maberly SC et al (2016) Diel surface temperature range scales with lake size. PLoS ONE 11:e0152466. https://doi.org/10.1371/journal.pone.0152466

Yang YY, Lusk MG (2018) Nutrients in urban stormwater runoff: current state of the science and potential mitigation options. Curr Pollut Rep 4(2):112–127. https://doi.org/10.1007/s40726-018-0087-7

Yvon-Durocher G, Caffrey JM, Cescatti A et al (2012) Reconciling the temperature dependence of respiration across timescales and ecosystem types. Nature 487:472–476. https://doi.org/10.1038/nature11205

Zhang Y, Zhou Y, Shi K, Qin B, Yao X, Zhang Y (2018) Optical properties and composition changes in chromophoric dissolved organic matter along trophic gradients: implications for monitoring and assessing lake eutrophication. Water Res 131:255–263. https://doi.org/10.1016/j.watres.2017.12.051

Zhimiao Z, Xinshan S, Yanping X et al (2016) Influences of seasons, N/P ratios, and chemical compounds on phosphorus removal performance in algal pond combined with constructed wetlands. Sci Total Environ 573:906–914. https://doi.org/10.1016/j.scitotenv.2016.08.148

Zohary T, Padisak J, Naselli-Flores L (2010) Phytoplankton in the physical environment: beyond nutrients, at the end, there is some light. Hydrobiologia 639:261–269. https://doi.org/10.1007/s10750-009-0032-2

Chapter 7
CH_4 Emission from Household and Abandoned Ponds of the Indian Sundarbans: Positive Feedback to Climate Change

Abhra Chanda and Sourav Das

Abstract Small aquatic ecosystems like ponds and lakes have been found to emit a significant amount of CH_4 towards the atmosphere and their role is worth inclusion in delineating the global CH_4 budget. The Sundarbans Biosphere Reserve (SBR) of India, besides being the abode of the world's largest mangrove forest, shelters almost 4.4 million people with a substantially high population density. The CH_4 dynamics from several compartments of this biosphere reserve is studied in the recent past; however, the ponds are yet to receive any attention as such. The present chapter reports the variability of the partial pressure of CH_4 in water [pCH_4(water)] and the subsequent air–water CH_4 fluxes from four different types of ponds situated within the SBR. One of these selected ponds is abandoned, and not used for any human purpose and another pond is well-maintained and not at all used for any human purpose. The rest of the two ponds are typical homestead ponds with varying degrees of anthropogenic disturbances. The results indicated that all four ponds acted as a source of CH_4 towards the atmosphere; however, the rate of emission varied across the ponds. The most well-maintained least anthropogenically disturbed pond emitted CH_4 at the lowest rate, whereas the dilapidated and abandoned pond, emitted the most. The other two ponds showed an intermediate range of fluxes. Water temperature showed a strong, positive, and statistically significant relationship with pCH_4(water). The results indicate that the ongoing climate-change-induced rise in temperature can effectively enhance the CH_4 emission rate from these ponds. However, the dissolved oxygen levels exhibited a significant negative relationship with pCH_4(water), which indicates that if autotrophic conditions can be maintained through proper pond management practices, methanotrophy can negate the dominance of methanogens to a large extent.

Keywords Pond · pCH_4(water) · Air–water CH_4 flux · Water temperature · Net heterotrophy · Methanogenesis · Methanotrophy · Lentic system · Indian sundarbans

---

A. Chanda (✉) · S. Das
School of Oceanographic Studies, Jadavpur University, 188 Raja S. C. Mullick Road, Kolkata 700032, India

S. Das et al. (eds.), *Pond Ecosystems of the Indian Sundarbans*, Water Science and Technology Library 112, https://doi.org/10.1007/978-3-030-86786-7_7

7.1 Introduction

Methane (CH_4) is one of the most crucial greenhouse gases and its concentration has increased substantially since the industrial revolution. A substantial amount of CH_4 emission takes place from the natural wetlands of this planet. However, a suite of anthropogenic activities, like incomplete combustion of fossil fuels and ruminant CH_4 production from livestock farming has increased the atmospheric CH_4 concentration. CH_4 comes immediately next to carbon dioxide (CO_2) in terms of the significance of their greenhouse effects. The global warming potential of CH_4 is almost 28 times more than that of CO_2 (Peters et al. 2011; Azar and Johansson 2012). This implies that the emission of one molecule of CH_4 can impart a greenhouse effect equivalent to 28 molecules of CO_2 emission. This fact gives rise to concerns regarding the rising CH_4 concentration, despite having significantly less atmospheric concentration when compared to CO_2. In the pre-industrial era, atmospheric CH_4 concentration was close to 680 ppbv (parts per billion volume). To the present date, the CH_4 concentration in the lower part of the troposphere has increased to 1800 ppbv (Kirschke et al. 2013). Myhre et al. (2013) argued that this enhancement of CH_4 concentration to the present-day levels contributed to almost one-fourth of the rise in the overall greenhouse effect in Mother Earth, since the year 1750. Paleo records suggest that during the last glacial maximum (~21,000 years ago), atmospheric CH_4 concentration was 375 ppbv, which increased to 680 ppbv just before the onset of the Industrial Revolution (Hopcroft et al. 2017). Though the reason behind the increase in CH_4 concentration, which took place before the industrial revolution, remains an area of active research, the scientific community ubiquitously accepts that CH_4 concentration was principally dependent on natural processes. Similarly, characterizing the enhancement in the atmospheric CH_4 concentration in the post-industrial revolution era requires more focused endeavors; however, changes in biogenic emissions and biomass burning play a pivotal role in regulating the CH_4 levels (Schaefer et al. 2016; Rigby et al. 2017; Worden et al. 2018). Besides, there remains considerable uncertainty in the residence time of CH_4 in the atmosphere. According to some pieces of research, CH_4 remains reactive in the atmosphere for almost nine years, and it can indirectly alter the concentrations of other important greenhouse gases like nitrous oxide (Prather et al. 2012; Hopcroft et al. 2017).

The temporal dynamics of the CH_4 level in the atmosphere exhibited interesting observations in the past decades. The years 1999 to 2006 witnessed almost no enhancement in atmospheric CH_4 concentration (Kai et al. 2011; Nisbet et al. 2014). Since the year 2007, we again came to notice an alarming rate of increase in CH_4 level; however, the scientific community could not reach a consensus for the reasons behind such an increase (Allen 2016; Schwietzke et al. 2016; Turner et al. 2017). Rice paddy cultivation accounts for 25–36% of the global CH_4 emissions (Saunois et al. 2016; Zhang et al. 2020). Similarly, livestock rearing contributes almost 5% of the global CH_4 emissions (Kirschke et al. 2013; Greening et al. 2019). Besides these sectors, inland aquatic bodies also play a significant role in regulating the atmospheric CH_4 concentration. The estuarine and wetland sediments rich in organic matter leads to substantial CH_4 emissions from the water column (Stamp et al. 2013; DelSontro et al. 2015). A suite of biogenic activities in the estuaries throughout the world leads to an emission of almost 1.8–3.0 Tg CH_4 yr^{-1} (Middelburg et al. 2002). Rivers and estuaries drew the attention of environmentalists long ago; however, in the recent past, inland lentic ecosystems, like lakes and ponds, have come under the lenses owing to their potential in emitting CH_4 (Matthews et al. 2020). Due to the ongoing global-warming-induced rise in temperature, these aquatic bodies manifested higher emissions in the recent past, owing to the temperature enhancement of the water column and the elongation in the thaw period, especially for the lakes in the northern latitudes (IPCC 2014; AMAP 2017). Globally the lakes and ponds emit around 37 to 112 Tg CH_4 yr^{-1} (Saunois et al. 2016). Most of these landlocked aquatic bodies exist in the northern hemisphere (Verpoorter et al. 2014); however, the lakes and ponds of tropical belts, despite covering only 27% of the global total lake area, contributes almost 49% of the global CH_4 emissions from such water bodies (Bastviken et al. 2011; Holgerson and Raymond 2016). A disequilibrium between two primary processes, namely methanogenesis, and methanotrophy, gives rise to the net CH_4 emission from any aquatic system. Methanogens produce CH_4 as a by-product when they feed upon organic carbon substrates under anaerobic conditions (West et al. 2016), and methanotrophs utilize CH_4 and oxidize it to CO_2 in oxygen-rich environments (Martinez-Cruz et al. 2015). CH_4 generated within the aquatic systems by methanogenesis eventually effluxes towards the atmosphere following a suite of pathways, like ebullition, release of stored carbon (storage flux), diffusion, and flux mediated by emergent plants (Bastviken et al. 2004). Besides climate warming, several other phenomena like eutrophication and improper management of lakes and ponds can give rise to varying CH_4 emissions (Yvon-Durocher et al. 2014; Aben et al. 2017; Schilder et al. 2017; Davidson et al. 2018). The CH_4 emission from these lakes and ponds has also exhibited large-scale spatiotemporal variability depending on the local regulating factors. Compared to the significance of this issue, studies in this field remain constrained by a lack of data. Thus, characterizing the CH_4 emission scenario from different variety of ecosystems has become an urgent need of the hour, as this would enable us to understand how these systems are going to respond in the future to the ongoing climate change (Negandhi et al. 2016; Audet et al. 2017). Keeping in view this background, the present chapter has focussed on delineating the CH_4 flux dynamics from the household ponds of the Indian Sundarbans Biosphere Reserve.

7.2 Physicochemical Dynamics of Ponds

Unlike the lotic ecosystems like rivers, streams, estuaries, and open oceanic waters, lentic ecosystems, like ponds, lakes, and reservoirs exhibit some unique biogeochemical properties. Temperature-induced stratification, occasionally salinity-driven stratification, high turbidity, shallow depths, organic matter-rich environment, and high abundance of nutrients are some of the characteristic features of a typical tropical pond. Thermal stratification is a common phenomenon in lakes and ponds, which in turn, regulates several principal biogeochemical processes within these aquatic systems (Crawford and Collier 2007; Boehrer et al., 2009). Prolonged stratification seldom takes place in temperate lakes, as seasonal variability in the heat exchange between the lake surface and the immediate atmosphere occasionally leads to complete mixing of the water column. However, tropical lentic systems, especially the shallow ones, mostly remain completely mixed throughout the year (Katsev et al. 2010). Some of the oligohaline or mesohaline ponds exhibit stable layers of water regulated by varying salinity (Schmid et al. 2005; Boehrer and Schultze 2008). The stratification can indirectly regulate the CH_4 emission from ponds towards the atmosphere. CH_4 production usually takes place in the bottom of the ponds, where anoxic conditions prevail, and a thermally stratified water column prevents the methane so produced to reach the upper oxygenated layers. Moreover, this restriction provides the methanotrophs ample time to act upon the CH_4 and reduce it to CO_2, thus lowering the net CH_4 fluxes towards the atmosphere (Encinas Fernandez et al. 2014; McGinnis et al. 2015). However, observations contrary to this established understanding are also prevalent. Tang et al. (2016) argued that the enhanced rate of eutrophication and climate warming favors the growth of cyanobacteria, which in turn, facilitates methane production in the upper oxic layers. They further inferred that a suite of microorganisms is capable of producing CH_4 in the oxygenated upper layers of ponds by implementing specialized enzymes, which can deliberately negate the effect of oxygen during methanogenesis.

Among the various physicochemical parameters, salinity and sulfate levels can significantly regulate methanogenesis (Poffenbarger et al. 2011). Usually, the presence of sulfate enables the sulfate-reducing bacteria to outcompete the methanogens to utilize the organic substrates (DeLaune et al. 1983; Wang et al. 1996). Seawater having high salinity are usually rich in sulfate levels. Besides providing competitors in the form of electron acceptors like sulfate ions for the substrate availability (Sun et al. 2013), saline waters also impart ion stress to the methanogens, which in turn, compels these microorganisms to limit the CH_4 production (Chambers et al. 2013; Neubauer et al. 2013). Thus, several studies have a lower rate of CH_4 production from saline waters. This phenomenon is equally true for brackish water ponds, where an admixture of saline water and freshwater takes place (Yang et al. 2018). However, this relationship is not ubiquitous, as local factors often lead to contrary observations (Megonigal et al. 2004; Weston et al. 2011). Sulfate oxidation capability varies across sites, and this can enable the methanogens to act upon the organic substrates freely, despite having a saline environment (Treude et al. 2005). Moreover, the rate of CH_4 oxidation has exhibited significant spatial variability, which in turn, is capable of consuming a substantial portion of CH_4 generated through methanogenesis (Barbosa et al. 2018). Besides salinity, pH also plays an indirect role in governing the magnitudes of CH_4 emission by regulating the methanogen community composition (Bertolet et al. 2019). pH regulates the metabolic rate among the different species under the class of methanogens, and thus, varying pH levels lead to an intraspecific competition to thrive, which in turn, decides the rate of CH_4 production (Jin and Kirk 2018). Xiong et al. (2012) further asserted that the combination of pH and salinity gave rise to the colonization of different methanogenic communities in the bottom sediment profiles of lakes.

Essential micronutrients like dissolved inorganic nitrogen and phosphorus along with reactive silica are the principal limiting factors of photosynthesis in the tropical lentic ecosystems where sunlight, temperature, and all other factors that favor photosynthesis remain abundant all through the year. The presence of excessive nutrients in the water column accelerates the rate of primary productivity and eventually makes such waterbodies unsuitable for any faunal lifeforms to thrive and degrades the ecosystem altogether. We refer to this phenomenon as eutrophication. Eutrophication occurs in the lakes and ponds due to natural causes, like land runoff, aging, and shallowing of the ponds. However, a suite of anthropogenic activities introduces nutrients to these water bodies and worsens the scenario of eutrophication. We refer to manmade eutrophication as cultural eutrophication. Eutrophication hampers the biodiversity as well as the biogeochemical cycles of lentic waters (Moss et al. 2011; Schilder et al. 2017), which in turn, regulates the oxygen and organic substrate availability. These factors govern the processes of both methanogenesis and methanotrophy, and hence the air–water CH_4 fluxes from the lakes and ponds (Sepulveda-Jauregui et al. 2018). The ongoing climate warming tends to raise the temperature of the water column of such closed lentic waters, which favors deoxygenation and leads to an anaerobic environment, thereby encouraging methanogenesis (Moss et al. 2011). With an increase in anthropogenic inputs led by the growing population, most of the lakes and ponds have witnessed eutrophication in the past decades. These lentic waterbodies throughout the globe have exhibited negligible resistance to such an alarming rise in nutrient levels (Jenny et al. 2016). As per projections, the global population can rise by 37% by the year 2050 (with respect to the population in the year 2000), and by 50% by the year 2100 (Samir and Lutz 2017). This population expansion would inevitably lead to enhanced nutrient discharge through sewage and agricultural fertilizer use (Caraco and Cole 1999; Cordell et al. 2009). Moreover, climate change has significantly altered the global precipitation pattern with increased occurrences and frequencies of storms, which in turn, could significantly augment the land runoff (Jeppesen et al. 2009; Sinha et al. 2017; Pacheco et al. 2014). Thus, the expected severity of eutrophication in the days to come can significantly increase the CH_4 emission from these water bodies (Adrian et al. 2016).

7.3 Methanogenesis

Community respiration is one of the essential phenomena of any ecosystem that drives several biogeochemical processes. Respiration that takes place under aerobic conditions facilitates the oxidization of an organic substrate to carbon dioxide (CO_2) accompanied by the reduction of oxygen to water. On the contrary, respiration under anaerobic conditions leads to hydrogenotrophic methanogenesis (Ferry and Kastead

2007). In this respiratory pathway, hydrogen oxidizes to the hydrogen ion, and CO_2 reduces to CH_4. Thus, in the trade-off between aerobic and anaerobic respiration, the two greenhouse gases replace each other. However, owing to the higher greenhouse gas potential of CH_4 over CO_2, anaerobic respiration provides more positive feedback to climate change. Unlike the other common respiratory pathways, methanogenesis has a typical low energy yield with the synthesis of less than one adenosine triphosphate per molecule of CH_4 produced (Lyu et al. 2018). The distinct group of microorganisms that are capable of opting for the anaerobic pathways is methanogens. Methanogens mostly comprise archaeal prokaryotic freshwater plankton, which acts as obligate methane producers (Casamayor and Borrego 2009; Adam et al. 2017). These methanogens are essentially anaerobes and do not thrive under oxygen-rich environments, as they do not opt for other alternative electron acceptors to respire. The limnologists believed that only a group of prokaryotes could carry out methanogenesis; however, recent advances revealed that methanogenesis is one of the primordial processes on earth and the methanogens comprise a wide range of biodiversity (Lyu and Liu 2018).

According to the latest estimates, 600 Tg CH_4 release towards the atmosphere takes place each year. Methanogenesis accounts for about 70% of this total emission (Lyu et al. 2018). The anaerobic pathway of methanogenesis makes use of the oxidized form of carbon, i.e. CO_2, as an alternative electron acceptor. A competition among electron acceptors in any terrestrial and aquatic habitat is a common phenomenon. The relative proportion of these electron acceptors exhibit substantial spatial variability. In habitats deficient in other electron acceptors, like ferric ion (Fe^{3+}), nitrate ion (NO_3^-), gaseous oxygen (O_2), and sulfate ions (SO_4^{2-}); methanogenesis takes place freely. Regions characterized with high primary production, where consumption of other electron acceptors takes place at a rapid rate, leave the organic substrates for the methanogens to utilize. Thus, methanogens exist in a suite of anaerobic environments, like hydrothermal vents situated at the abyssal plain, permafrost soils of the Polar Regions, oligohaline to mesohaline lakes and ponds sediments, wetlands like peats, bogs, and swamps, and rice fields (Lyu et al. 2018). Earlier, the scientific community had a perception that methanogens essentially grow and thrive in strictly anaerobic environments. However, the latest advancements in this domain revealed that various methanogens could carry out methane production even in the presence of oxygen. These methanogens have developed physiological mechanisms to tolerate oxidative stress.

The substrate utilization pathways vary among the methanogens. The methanogen biodiversity comprises three broad groups: hydrogenotrophic, methylotrophic, and aceticlastic (Lyu et al. 2018). The hydrogenotrophic methanogens are widely abundant and this pathway substantially contributes to the total biogenic CH_4 emissions. These methanogens occur in the gastrointestinal tracts of several higher life forms. The cow is one such organism, which alone accounts for almost 72 Tg CH_4 emissions per year. This type of methanogen acts upon hydrogen and basic alcoholic groups accompanied by the reduction of CO_2 to CH_4. Methylotrophic methanogens, on the other hand, make use of a suite of organic compounds with methyl groups. The contribution of this subgroup towards the global CH_4 emission is the least among the three broad groups. Aceticlastic methanogens are perhaps the most dominant class, which thrive in a wide range of habitats across wetlands and rice fields and contribute to almost 66% of the biogenic CH_4 emissions. These methanogens typically flourish in regions where the performance of hydrogenotrophic methanogens is poor and does not generate the required amount of acetate to accomplish methane production. Methanogens have lately become an active area of research owing to their potential in driving energy through the global carbon cycle as well as their role in enhancing global warming. Several pieces of research are at present ongoing that are trying to capitalize the methanogens by putting these microbes into commercial use. Generation of CH_4 as biogas by utilizing the anaerobic digestion pathways of methanogens has successfully led to the production of energy (Mayumi et al. 2016). Artificial microbial electro-synthesis with the involvement of methanogens have also been successful in the past decades (Enzmann et al. 2018). Genetically engineered methanogens have been put to use to manufacture CH_4 from simpler organic substantiates. On the other hand, research has gained impetus on reducing the activities of methanogens in the gastrointestinal tracts of ruminants. Thus, methanogenesis as a topic will continue to remain under the lenses for a long time to come, given its importance in regulating the aquatic as well as atmospheric biogeochemistry.

7.4 Methanotrophy

The CH_4 produced through methanogenesis and anthropogenic activities interacts with several reactive ions and radicals in the atmosphere, and thereby undergoes oxidation in the presence of light (Saarnio et al. 2009). However, biogenic oxidation of CH_4 substantially reduces the CH_4 load in the atmosphere as well as in the pedosphere and sediment–water interface, and hence plays a critical role in governing the global CH_4 dynamics (Serrano-Silva et al. 2014). The group of microbial organisms that are capable of oxidizing CH_4 is methanotrophs. Based on the type of environment, where these microbes perform the oxidation, the methanotrophs can be of two broad groups: one being the aerobic methanotrophs and the other is a conglomeration of anaerobic archaea and bacteria. The latter can perform CH_4 oxidation under oxygen-deficient conditions (Ettwig et al. 2010). These microorganisms are abundant in a wide variety of soil types, like wetlands, peats, bogs, swamps, lake sediments, agricultural paddy fields, marine sediments, acidic and alkaline soils, and cold environments (Semrau et al. 2011). However, these microbes perform best under negligible salinity, moderate temperature, and neutral pH environments (Le Mer and Roger 2001). In all these types of sites, methanotrophs prefer to thrive in the critical junction between the anoxic and environments (Wendlandt et al. 2010).

According to recent estimates, globally around 80% of the CH_4 generated in the pedosphere undergoes oxidation led by methanotrophic bacteria. This bacterial community utilizes the CH_4, as a source of carbon and hence, energy (Conrad et al. 2008). Moist sediments and especially, rice fields are active sources of CH_4. However, a substantial proportion of this emitted CH_4 upon consumption by the methanotrophs leads to much lower net CH_4 emission from these agricultural fields (Conrad 2009). Depending on the CH_4 concentration that prevails in the topsoil layer, the methanotrophic mechanism follows two distinct types, which vary in the rate kinetics (Nayak et al. 2007). The scientific community refers to these two processes as low affinity and high-affinity mechanisms. The low-affinity process takes place at almost all field conditions where a substantial amount of CH_4 production occurs through methanogenesis (Reay 2003). The high-affinity pathway is preferable under a very low concentration of CH_4 in the sediment surface (Lau et al. 2007). Overall, the methanotrophic bacterial community comprises three genes, which include the subunits of soluble methane-monooxygenase (MMO) (*mmoX*), particulate MMO (*pmoA*), and methanol dehydrogenase (*mxaF*), respectively (Shukla et al. 2009). The metabolic pathway of CH_4 oxidation follows a sequence of steps and involves three intermediate reaction by-products, namely methanol (CH_3OH), formaldehyde (HCHO), and formate ion ($HCOO^-$). MMO initiates the oxidation reaction and converts CH_4 to CH_3OH. The enzymatic complex methanol dehydrogenase facilitates the oxidation of CH_3OH to HCHO. Finally, HCHO undergoes oxidation to $HCOO^-$, and then to CO_2, which marks the end of methanotrophic oxidation of CH_4 (Hanson and Hanson 1996; Chistoserdova et al. 2004).

A suite of environmental and biogeochemical factors regulate the abundance of the methanotrophs and hence, the degree of CH_4 oxidation. Though the exact mechanism remains uncertain, the enhanced concentration of atmospheric CO_2 hampers the CH_4 oxidation by methanotrophs (Dubbs and Whalen 2010). Phillips et al. (2001) argued that an increase in atmospheric CO_2 concentration triggers the competition for accessing and utilizing oxygen among the methanotrophs, which dampens the rate of methanotrophy. Ambient temperature has the potential to regulate almost any and every biochemical reaction mediated by microbes. Methanotrophy is no exception; however, several authors emphasized that the relation between temperature and methanogens is far more prominent than that observed in the case of methanotrophs (Borken et al. 2006). At temperatures below 10 °C, methane oxidation reduced substantially; however, above 10 °C the vice-versa effect was not significant (Castro et al. 1995). Sediment surface with substantially high CH_4 exhibited a significant relationship between methanotrophic oxidation rates and temperature (Kallistova et al. 2005; Jäckel et al. 2005). Horz et al. (2005) observed significant changes in microbial community composition with alterations in temperature. Soil water content also plays a critical role in regulating CH_4 oxidation (Jugnia et al. 2008). Optimum moisture condition in the upper sediment profile is beneficial towards methanotrophy. However, with an increase in soil moisture beyond a certain threshold, the CH_4 oxidation rate exhibits significant reduction, most probably due to the lower degree of solubility of both CH_4 and oxygen (Del Grosso et al. 2000; Werner et al. 2006). Usually higher moisture content in the sediment surface favors methanogenesis, but observations of net CH_4 oxidation exist at soil moisture beyond 60%. These rare observations are responsible for the activity of aerobic microsites or anaerobic CH_4 oxidation pathways (Khalil and Baggs 2005).

Besides the atmospheric parameters discussed in the preceding paragraph, sediment parameters can also play a pivotal role in governing the CH_4 oxidation dynamics. The sediment texture and the bulk density regulate the diffusion of both CH_4 and oxygen in the top layer of the pedosphere. Higher diffusivity of the sediments allows the CH_4 and oxygen to interact in the presence of the methanotrophs. Thus, sediments with higher pore size exhibit enhance methanotrophy (Castaldi et al. 2006; Templeton et al. 2006). Availability of oxygen itself is a crucial factor. In this regard also, soil porosity plays a significant role in the diffusion of oxygen in the upper permeable soil profile (Mancinelli 1995). The role of soil pH needs more elaborate study, as CH_4 oxidation by methanotrophs appears to take place in highly acidic as well as highly alkaline regions (Benstead and King 2001; Saari et al. 2004). However, in soils having neutral pH, CH_4 oxidation was highest (Mosier and Delgado 1997). Soil salinity provides negative feedback to methanotrophy, as enhanced salinity increases the osmotic pressure in the soils, which in turn, hampers the microbial activity (Price et al. 2004). Inorganic fertilizers, especially ammonia-rich ones, reduce CH_4 oxidation rates. Bykova et al. (2007) argued that ammonium ions bind with the enzymatic complex MMO and thus, inhibits the initiation of methanotrophy. However, contradicting results are also available (Jacinthe and Lal 2006). The approach of different methanotrophs towards inorganic nutrients was the main reason behind such contrasting observations (Mohanty et al. 2006). Some studies indicated that a suite of herbicides and pesticides also inhibited methanotrophy (Priemé and Ekelund 2001). Thus, on the whole, we can infer that like methanogenesis, methanotrophy also has various shades that require further study. From the perspective of the changing climate, methanotrophy plays a crucial role in delimiting net CH_4 emissions towards the atmosphere. Hence, nurturing this issue is of utmost importance to understand, and at the same time, combat global CH_4 emissions.

7.5 Methane Emission Pathways

Within an aquatic system, especially, a lentic system, CH_4 generation principally takes place at the sediment–water interface. Comparatively, a much lower proportion of CH_4 generates in the water column. As long as the CH_4 remains within the sediment interface or dissolved in the water column, it does not pose any threat to the atmosphere. However, as soon as the CH_4 leaves the air–water interface, it imparts the effect of radiative forcing in the atmosphere. However, the escape of CH_4 from the sediments via the water column follows several physical pathways. Principally CH_4 ends up in the atmosphere, namely through four pathways: diffusion, ebullition, storage flux, and flux mediated through aquatic vegetation (Sanches et al. 2019). The proportion of CH_4 fluxes among the four pathways from lentic ecosystems depend on the characteristic features of the aquatic body (Bastviken et al. 2004). However, not all these four pathways received equal attention from the scientific community, while characterizing the global estimates (Bastviken et al. 2011). Studies on diffusive CH_4 fluxes are disproportionately higher than that on ebullition fluxes. Studies on the other two pathways are scarce throughout the world.

7.5.1 Diffusive Flux

The CH_4 molecules produced by the methanogens that can bypass the methanotrophy end up in the aquatic phase of the lentic bodies. These CH_4 molecules remain dissolved in the water and enhance the partial pressure of CH_4 in water [pCH_4(water)]. When these dissolved CH_4 molecules reach the air–water interface they leave the water surface and escapes to the atmosphere depending on the partial pressure gradient. According to the gradient principle, the molecules diffuse from the site where it is in higher proportion to the site with lower strength. The pCH_4(water) in the upper surface of lakes and ponds usually remains substantially high, and much higher than the partial pressure of CH_4 in the air [pCH_4(air)]. This difference in pCH_4 allows the methane to diffuse through the air–water interface towards the atmosphere. In some rare cases, when pCH_4(water) remains lower than pCH_4(air), the aquatic body can act sink for CH_4. Several authors believe that diffusive flux comes next to the ebullition flux, in terms of the total CH_4 emission from aquatic bodies. Holgerson and Raymond (2016) inferred that the small ponds having an area of fewer than 0.001 km^2 occupies only 8.6% of the global lentic bodies but contributes to 41% of the total diffusive CH_4 fluxes. Sanches et al. (2019) inferred that diffusive CH_4 fluxes exhibited significant positive correlations with the concentration of dissolved organic carbon of several lakes throughout the world. Li et al. (2020) analyzed diffusive CH_4 flux data from 744 lakes situated in different corners of the world and observed that the higher the depth of the lakes the lesser is the diffusive CH_4 emission. They argued that the deeper lakes furnished ample scope of CH_4 oxidation compared to the shallower ones. Holgerson and Raymond (2016), in their study, estimated that the global lakes and ponds emit 16 Tg CH_4 per year through the diffusive pathway.

7.5.2 Ebullition Flux

Ebullition of CH_4 from aquatic bodies refers to the upward movement of CH_4 bubbles produced at the sediment–water junction towards the air–water interface and eventually escapes to the atmosphere. Pieces of evidence indicate that this pathway is responsible for most CH_4 emissions from global lentic bodies (Bastviken et al. 2011). There are several established protocols to measure the diffusive fluxes across the air–water interface; however, monitoring ebullition fluxes remains a challenge (Gålfalk et al. 2013; Kosten et al. 2020). Compared to the other greenhouse gases like CO_2 and N_2O that originate within the aquatic system, CH_4 is much less soluble. This low solubility enhances the likelihood of bubble formation in the case of CH_4 and this makes ebullition a dominant component among the four types of fluxes, especially for CH_4 (van Bergen et al. 2019). Quantifying ebullition flux requires long-term monitoring, since the bubbles that originate in the sediment–water interface come up to the air–water interface occasionally, which makes it difficult to characterize during short-term measurements (Wik et al. 2016). The existing record of ebullition flux measurements indicates that this type of flux usually exhibits significant spatial variability even within a small lentic system (DelSontro et al. 2016). The center of the lakes and ponds, which is often associated with a thick sediment layer leads to higher ebullition fluxes compared to the edges (van Bergen et al. 2019). The presence of fish in closed shallow lentic systems can significantly reduce CH_4 ebullition due to the disturbance posed by the fishes that prevent bubble formation (Oliveira Junior et al. 2019). This is especially true for bottom dwellers, i.e., those fishes that tend to thrive in the sediment–water interface. Bioturbation by fish also enhances CH_4 oxidation rate by enhancing the redox potential (Joyni et al. 2011), and thus, reduces chances of ebullition.

7.5.3 *Storage Flux*

The term storage flux refers to the emission of CH_4, which takes place due to the accumulation of CH_4 for a long period within the aquatic phase of the lakes and ponds followed by its sudden emission through diffusive pathways. This type of emission takes place mostly due to two phenomenon that essentially occurs in lentic ecosystems, one being the stratification and the other is ice formation (that occurs in the high latitude lakes and ponds). Strong stratification prevents the CH_4 produced in the bottom anoxic layers to travel upwards and similarly, ice formation on the air–water interface bars the accumulated CH_4 from escaping towards the atmosphere. Seasonal overturn (Michmerhuizen et al. 1996; Riera et al. 1999) and ice-melting (Phelps et al. 1998; López Bellido et al. 2009) lead to the release of the CH_4 after prolonged accumulation. This type of CH_4 flux has received much less attention compared to diffusive and ebullition fluxes. Sanches et al. (2019), in their meta-analysis on global lakes, reported that the share of storage flux and diffusive flux are almost equal. They further observed that precipitation played a crucial role in regulating storage flux. Annual mean precipitation exhibited a significant negative correlation with storage flux all around the world. Like in the case of ebullition, storage flux also occurs for a small period and that too occasionally, which makes it difficult to quantify. However, this type of flux does not take place from all the lakes and ponds of the world, as ice-formation and stratification do not occur in all these lentic bodies. Thus, the importance of ebullition flux, which happens to be a common phenomenon in almost all the lakes and ponds, far supersedes that of storage flux (Downing et al. 2006).

7.5.4 *Flux Mediated Through Aquatic Vegetation*

Among the four types of pathways through which CH_4 escapes towards the atmosphere, this type has received the least attention. One of the reasons could be because this type of flux does not originate from the aquatic column directly, in the strictest sense. Different types of vegetation often accompany several lakes and ponds, especially near the periphery. These vegetative structures play a crucial role in trapping the CH_4 that generates in the sediment–water interface and eventually, the trapped CH_4 reaches the atmosphere through this vegetation (Laanbroek 2010). Aquatic macrophytes usually adapt themselves to an oxygen-deficient environment as they mostly thrive under waterlogged conditions. This physiological adaptation enables the macrophytes to transcend O_2 through their body towards the roots to carry out some essential life activities. It is through the pores from which O_2 leaves the plant bodies, CH_4 molecules enter the vegetative structure and the plants exhale the absorbed CH_4 back to the atmosphere (Van der Nat and Middelburg 1998). However, existing research findings also indicate that the oxygen that transcends down through the plant bodies can play a crucial role in methane oxidation, thereby, reducing the net methanogenesis and CH_4 emission.

7.6 Characteristic Features of the Ponds of the Sundarbans

Ponds are a common land-use class in the entire rural sector of India. Villagers rely substantially on the household ponds to carry out multifarious daily life activities. The Sundarbans are also not an exception in this regard. Almost every household that encompasses a minimum sizable land plot invariably consists of a small pond. The size of these ponds varies in shape and size. However, both kuccha, as well as pukka houses, essentially comprise a pond in the Sundarbans, with of course few exceptions. Mandal et al. (2015) observed that in the rural sectors of Sundarbans, most of the households enjoy a pond in its immediate adjacency. The surface area of these ponds varies from 400 to 500 m^2. The typical rural villagers of this region own a piece of land and these land plots consist of a pond. Thus, most of these ponds are private property by nature. Well-off families often, occupy a larger land area that constitutes more than one ponds of varying size. However, household ponds larger than 1000 m^2 are scarce. To utilize the pond water fruitfully, these ponds situate very close to the dwelling within the land plot of each household. It is a common practice in this part of the world to dig ponds and utilize the unearthed soils to increase the basement of the dwelling or even construct the house itself (the kuccha houses). Apart from these small ponds, which are in the true sense personal property of the landowner, there are some large ponds as well. These large ponds are less in number; however, exist in one or two numbers across a village. This type of ponds are public properties and local villagers can access these ponds to meet several daily life activities.

All these ponds mostly remain full of water all through the year; however, the water level fluctuates seasonally. During the dry months (non-monsoon seasons like summer and winter), the volume of water reduces significantly in many of these ponds. The monsoon-induced rain recharges the ponds to the full and sometimes due to heavy rain in some of the years, the water from the ponds overflow. Most of these ponds are not a part of the natural landscape. The local population unearth the soils and prepare a pond of the required size. These ponds usually have a steadily inclined slope and the center of the pond has the highest depth. The depth of these ponds varies from a meter to three meters. The ponds of Sundarbans appeared increased hand in hand with human in-migration, which began in the late 1960s and 70s. Very few ponds in this area are more than 100 years old. Mandal et al. (2015) observed that the average age of the ponds in the Sundarbans is around 60 years. New ponds, which are a decade old, are scarce in number. The rate of sedimentation increases with the age of the ponds, which affects the primary productivity (Boyd 2012).

The ponds provide a multitude of services to the rural population of Sundarbans. In other words, these ponds are an asset to the local villages. Fish rearing is a very common practice in these ponds. Local people grow a wide variety of freshwater fishes in these homestead ponds, which mainly includes *Labeo rohita, Labeo catla, Catla catla, Cirrhinus cirrhosis, Puntius sophore, Hypophthalmichthys molitrix, Mystus tengara, Lates calcarifer, Oreochromis niloticus, Amblypharyngodon mola, Probarbus jullieni, Clarias batrachus,* and many more (Mandal et al. 2015). The local population does not rear fish for commercial purposes. They regularly harvest these fishes from the ponds for their daily meals. They can rear fish for almost eight to nine months, excluding the dry months when the water column becomes extremely shallow, and hence, unfavorable for the fishes to survive. They also do not practice any scientific management for the well-being of these ponds, mostly because they cannot afford to do so. Hence, most of these ponds usually do not undergo conventional pond management practices like periodic dredging, lime treatment, and de-silting. The fish rearing practiced in these ponds also does not follow any stringent aquaculture protocols. The traditional practice just includes the introduction of seedlings at different times of the year. However, the local people do not care to take into account the conventional aquaculture parameters like stocking density, and fish composition. They also do not provide specific feeds for the fishes. The fishes mostly grow on their own depending on the natural foods that end up in these ponds, mostly out of primary autotrophic processes. Mandal et al. (2015) observed that the local people could enhance the aquaculture potential of these ponds; however, the financial constraints bar them from doing so.

Besides being a source of fish, the local villagers also utilize this pond water for growing fruits and vegetables in the homestead gardens for household consumption. Mandal et al. (2015) reported in their study that local households of Sundarbans deploy about 120 to 240 m^2 of their land plot to grow a wide range of vegetables throughout the year. The adjacent ponds act as the sole source of water to sustain this miniature form of agriculture. Besides growing vegetables and rearing fishes, the local people also rely on this water for bathing and washing purposes. They also use this water for washing utensils and they usually wash the utensils on the edge of the pond. This practice directly introduces several inorganic nutrients to the pond water that often leads to cultural eutrophication.

7.7 Methodology

The role of small ponds in the global methane emission budget is unquestionable. However, the ponds of Sundarbans have not received any attention as such, in this regard. The significance of CH_4 emission from the tropical ponds of diverse regions needs elaborate study to diminish the uncertainties in flux estimates. Sundarbans, in this regard, deserve special attention due to several factors. Firstly, the population density in the Sundarbans Biosphere Reserve is very high. This fact can make us expect that the number of ponds in this region per unit area should be also substantially high. Moreover, the Sundarbans is one of the few critical sites of this country, where the wrath of ongoing climate change has taken a toll. A rise in ambient temperature, relative sea-level rise, drastic changes in precipitation pattern, increased frequency of tropical cyclones, and storm surges are some of the factors that have posed an immense threat to the entire region (Dubey et al. 2017). Storm-induced saline water intrusion has also become a recurrent problem in the Sundarbans, especially in the last two decades. All these factors can potentially alter the CH_4 dynamics of the small lentic ecosystems, which is put under the lenses in this chapter. To characterize the air–water CH_4 exchange across the pond interface, we have sampled in few of the representative ponds within the Indian Sundarbans Biosphere Reserve. We mentioned about four types of CH_4 fluxes in the background of this chapter; however, we could measure only the diffusive CH_4 flux from these ponds. Measuring the rest of the three types of fluxes was beyond the ambit and capability of the authors. In the present date, there are several established protocols for measuring the diffusive air–water CH_4 fluxes. Eddy covariance technique has become extremely popular among meteorologists, and it has replaced all the measurement protocols because of its robustness and long-term flux monitoring capability. However, this method requires a sophisticated set of instruments and proper stationing, which was not possible for us to deploy. We implemented the bulk density method for spot measurements and estimated the partial pressure of CH_4 in water [pCH_4(water)] as well as in ambient air just above the pond surface [pCH_4(air)]. The concentration gradient across the air–water interface enabled us to compute the CH_4 fluxes. We have also measured an assembly of other biogeochemical parameters and analyzed their inter-relationship with the CH_4 fluxes. We mentioned the nitty–gritty details of the measurement protocol in the following sub-sections.

7.7.1 Sampling Locations

We carried out the air–water CH_4 flux measurements in the same ponds, where we monitored the air–water CO_2 fluxes (mentioned in Chap. 6). We selected four ponds situated in three administrative blocks of the South 24 Paraganas district of the Sundarbans Biosphere Reserve. Two of these ponds are in the Namkhana Block, one in the Patharpratima Block, and the other in the Basanti Block. Figure 7.1 shows the

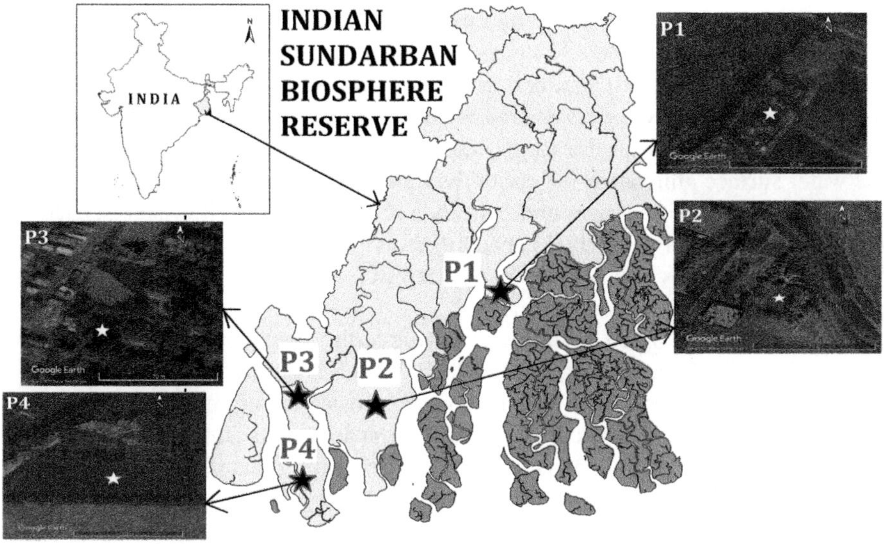

Fig. 7.1 The study area map showing the locations of the four ponds (selected for the present study) situated in the Indian Sundarbans Biosphere Reserve

location of the respective ponds in the study area and their present state (courtesy: Google Earth, Google Inc.). The main intention behind sampling site selection was to cover the common types of ponds, in terms of their state of health, water usage, size, and shape. The selected ponds include a typical household pond used for multifarious domestic activities. One of these ponds receives comparatively less domestic attention, as the local people are not exclusively dependent on the services provided by these ponds. In Sundarbans, we often come across some waterbodies, which are usually larger than the typical household ponds and these undergo proper maintenance, mainly for their aesthetic value and beautification purposes. Besides, we also often encounter some ponds, which remain in a dilapidated and neglected state. This type of pond is not at all suitable for any kind of domestic activity and suffers from intense eutrophication. Though the local people are least bothered about this type of ponds and simply abandon these water bodies, these ponds can manage a water column for a substantially long time, as like all other ponds, these get replenishment of water from monsoonal rain.

We selected a pond in the Basanti Block near a popular tourist destination, named Jharkhali (hereafter referred to as P1). It encompasses an area of around 250 m^2. This waterbody is not one of the typical household ponds as it lies within the premises of a government office (Jhar Bungalow). However, stationed officials seldom use the pond water for miscellaneous purposes. Occasionally very few people bathe and wash utensils in this pond. The pond has fishes but the people surrounding this pond are not very dependent on it. Rarely do they harvest fishes. Overall, compared to typical domestic ponds, this pond has very limited use.

The second pond selected for this study is in Ramganga (hereafter referred to as P2), which is one of the busiest places in the Patharpratima Block, as many people commute through this critical junction to go towards Kolkata and the suburban localities. This pond merely covers an area of about 400 m^2. Situated in a typical locality, this pond is almost of no use to the local people. A thin algal mat prevailed over the water surface almost for the entire duration of an annual cycle. In the monsoon season, heavy rain sometimes broke off the intact mat cover. The visual scenario of the very pond can tell us that it is abandoned and suffering from eutrophication. We selected this pond as one of the representative classes, as these types of ponds are prevalent all around the Sundarbans.

The third pond sampled in this study represents the most typical type of household ponds. This pond is in the Namkhana Block (hereafter referred to as P3), situated in a typical semi-rural locality. We have noticed a substantial number of people bathing in this pond all-round the year. Women from nearby houses occasionally use the pond water to wash the utensils or even clothes, sitting near the periphery of the pond. This pond occupies an area of about 380 m^2.

Lastly, we selected a pond, situated in the same Namkhana Block, but it is located within the premises of a Guest House (hereafter referred to as P4) that mainly cater to tourists from Kolkata who venture into the nearby Bakkhali and Henry Island sea beaches. Out of the four ponds we sampled, this is the largest having an area of about 2700 m^2. The pond is full of fishes and occasionally the caretakers harvest fishes out of the pond but they essentially do not rely on these fishes for consumption. There is a strict prohibition for bathing and washing clothes or utensils in this pond. The guesthouse authority maintains this pond mainly for beautification purposes. All the four ponds had one thing in common. There are no concrete embankments in any of these ponds. Almost all the ponds have some naturally growing bushes and shrubs in their periphery.

7.7.2 Sampling Plan and Strategy

The seasonality in eastern India has three broad demarcations, which include the pre-monsoon season (February to May), the monsoon season (June to September), and the post-monsoon season (October to January). We sampled in each of the three seasons over the years 2016, 2017, and 2018. However, we could not carry out sampling in all the months of a calendar year. Instead, we sampled in different months of a particular season in the three respective years to characterize the short-term temporal variability in the physicochemical parameters as well as the CO_2 fluxes. In this way, we sampled in all the twelve months, but of different calendar years. We presented the seasonal mean data taking into account the measurements carried out in all the months of that particular season. We assumed that there was no significant inter-annual variability in the measured parameters.

We sampled in all four ponds once a month during the daytime (between 10:00 a.m. and 12:00 p.m.) and the nighttime (between 06:00 p.m. and 08:00 p.m.). We collected both the day and night samples to analyze the difference in the CO_2 flux dynamics in the presence and absence of photosynthesis. In the year 2016, we sampled in March, May, September, and November. Similarly, in the year 2017, we sampled in January, April, August, and October. Finally, in the year 2018, we sampled in February, June, July, and December. We collected three replicate samples from each of the ponds with the help of a pre-cleaned polypropylene glass bottle from the surface water. We collected the water samples from at least a meter distance from the edge of the ponds.

7.7.3 *Physicochemical Measurements*

We measured most of the parameters in situ by deploying standard probes. However, for those parameters, which require laboratory analysis, we collected the samples in specific containers, preserved the samples with the addition of necessary preservatives for each parameter, and transferred the bottles in freezing conditions to the laboratory. We took adequate care to calibrate and standardize the sensors and probes used in this study. We ensured that all the reagents used for the different laboratory analyses were of analytic grade.

7.7.3.1 In-Situ Measurements

Wind speed (precision: 0.1 m s^{-1}) and air temperature (precision: 0.1 °C) were measured by a weather station (WS–2350, La Crosse Technology, USA). Electrical conductivity (EC) (precision: 1 μS cm^{-1}) and water temperature (precision: 0.1 °C) were measured by a digital salinity meter (Thermo Scientific, Eutech, Germany). pH (precision: 0.001) was measured using Orion PerpHecT ROSS Combination pH Micro Electrode coupled with a micro–pH data logger (Thermo Scientific, USA). The glass electrodes were calibrated on each day of sampling with NBS scale standard buffers at constant temperature (25 °C). Dissolved oxygen (DO) (precision: 0.01 mg l^{-1}) was measured with a FiveGo portable F4 Dissolved Oxygen meter (Mettler Toledo, Switzerland). Winkler's titrimetric method was used once in each day of sampling to calibrate the DO sensor's readings. The turbidity (precision: 1 NTU) of the water samples was measured by a nephelometer (Eutech TN–100, Singapore). Underwater photosynthetically active radiation (UWPAR) was measured using a standard sensor (LI–192SA, LiCor, USA, precision: 0.1 μmol m^{-2} s^{-1}) and a data logger (Li–250A, LiCor, USA). Chlorophyll–*a* (Chl–*a*; precision: 0.01 mg m^{-3}) was measured according to standard spectrophotometric procedures. The gross primary productivity (GPP) and community respiration (CR) in both the ponds were measured using three replicates of a light bottle and dark bottle method. 24 h incubation was performed on each day of sampling and the changes in DO concentration were monitored. BOD was also measured by taking three replicate samples were also collected from each pond during each sampling day to measure $^{27\,°C}BOD_{3days}$. Chl-*a*, GPP, CR, and BOD were measured according to standard protocols detailed in APHA (2005).

7.7.3.2 PCH$_4$(Water) and pCH$_4$(Air) Measurement Protocol

Surface water samples were collected in 40 ml glass vials each fitted with a rubber septum and leaving no headspace. The water samples were poisoned by adding 20 μl of 8% $HgCl_2$ solution to stop all the biological activities. Using a pre-cleaned syringe, 20 ml water was removed from the glass vial and the remaining 20 ml water sample was purged with nitrogen gas (99.99% pure) in the laboratory. The glass vials were kept undisturbed for 2 h to equilibrate. 5 ml of the headspace gas was collected by a Hamilton syringe and CH_4 concentration was analyzed using a gas chromatograph (GC) (Systronics GC–8205) having a mean relative uncertainty of ±2.9%. Pure nitrogen gas was used as the carrier gas and the retention time for CH_4 was 0.623 min. The moisture was removed by raising the injector temperature to ~105 °C. Certified standard methane gas (0 and 20 ppm) was used for calibration of the GC. Ambient air samples above the water surface were collected in glass bulbs fitted to a battery-operated air pump. The stopcock of the glass bulbs was sealed by parafilm to prevent leakage. Air samples were also brought to the laboratory within 24 h of sampling and the CH_4 concentration was analyzed in GC by using the above-mentioned injection method.

7.7.3.3 Estimation of Air–Water CH$_4$ Exchange

pCH$_4$(water) and pCH$_4$(air) were converted to concentration of methane in water (CH$_4$wc) and air (CH$_4$ac) according to the following equations (Morel 1983; Lide 2007):

$$\mathrm{CH_4wc} = K_\mathrm{H} \times \mathrm{pCH_4(water)}$$

$$\mathrm{CH_4ac} = K_\mathrm{H} \times \mathrm{pCH_4(air)}$$

$$\ln K_\mathrm{H} = -115.6477 + 155.5756/(\mathrm{T_K}/100) + 65.2553 \times \ln(\mathrm{T_K}/100) - 6.1698 \times (\mathrm{T_K}/100)$$

where K_H denotes the gas partition constant of CH_4 in water at sampling temperature, expressed in mole l^{-1} atm^{-1}, and T_K denotes the temperature in Kelvin. The air–water CH_4 flux was calculated according to equation (MacIntyre et al. 1995):

$$\mathrm{CH_4Flux}\left(\mathrm{mg\ m^{-2}h^{-1}}\right) = \mathrm{k_x(CH_4wc{-}CH_4ac)}$$

where, k_x stands for the mass transfer coefficient (cm h^{-1}) and it is computed according to equation (Wanninkhof 1992):

$$k_\mathrm{x} = k_{600} \times (\mathrm{S_c}/600)^{-\mathrm{x}}$$

where S_c is the Schmidt number for CH_4 and it is dependent on water temperature according to the equation:

$$S_c = 1897.8 - 114.28 \times T + 3.290 \times T^2 - 0.039061 \times T^3$$

k_{600} is computed from the wind speed (U_{10}) and 'x' is equal to 0.66 for wind speed ≤ 3 m s^{-1} and is equal to 0.5 for wind speed > 3 m s^{-1} (Cole and Caraco 1998).

$$k_{600} = 2.07 + \left(0.215 \times U_{10}^{1.7}\right)$$

7.7.4 Statistical Analyses

We used the Statistical Product and Service Solutions software (SPSS version 16.0, Inc., USA) and Microsoft Excel for Windows 2010 to carry out all the statistical analyses and prepare the graphical illustrations for this chapter. We examined whether the arithmetic means of pCH_4(water), CH_4 fluxes, and other biogeochemical parameters were significantly different or not between all the four ponds and across all the seasons by carrying out a one-way analysis of variance test (ANOVA; F-test). We conducted a posthoc Tukey's honest significant difference (HSD) test to specify the difference in mean of the biogeochemical parameters among the selected ponds if any. To analyze the inter-relationship between the pCH_4(water) and the other regulating factors, we calculated the Pearson correlation coefficient. We also performed a multivariate analysis namely Principal component analysis (PCA) to examine the effect of the plausible regulatory parameters on the variation of pCH_4(water), and hence the air–water CH_4 fluxes. We considered the outcomes of these tests statistically significant at $p \leq 0.05$.

7.8 Results and Discussion

7.8.1 The Dynamics of Physicochemical Parameters

The mean surface water temperature was highest during the monsoon season followed by the pre-monsoon and the post-monsoon season (Table 7.1). There was no significant difference in the mean temperature amongst the four ponds throughout the year ($p > 0.05$). However, pond P2 recorded a slightly higher temperature compared to the other three ponds that could be due to its abandoned state and comparatively lower volume of water (Fig. 7.2). The land effect could have played a role in enhancing the temperature of this water body, however, a statistically significant

Table 7.1 The mean ± standard deviation of the physicochemical parameters observed in the four ponds in all three seasons

	P1	P2	P3	P4
EC (μS cm^{-1})				
Pre-M	1259 ± 90	1449 ± 59	1287 ± 74.4	1185 ± 95.3
Mon	1056 ± 95	1092 ± 260	1085 ± 82.9	986 ± 44.6
Post-M	986 ± 55	1164 ± 244	1014 ± 29.1	1093 ± 105.4
pH				
Pre-M	7.843 ± 0.065	7.508 ± 0.046	8.083 ± 0.054	8.382 ± 0.075
Mon	7.841 ± 0.037	7.595 ± 0.046	8.063 ± 0.038	8.264 ± 0.101
Post-M	7.938 ± 0.040	7.472 ± 0.061	8.173 ± 0.036	8.458 ± 0.032
Water temperature (°C)				
Pre-M	28.9 ± 4.6	29.8 ± 4.4	29.4 ± 4.4	28.4 ± 4.7
Mon	34.0 ± 0.4	34.9 ± 0.4	33.7 ± 0.4	33.8 ± 0.4
Post-M	25.8 ± 4.4	27.5 ± 3.4	25.8 ± 3.0	25.7 ± 4.8
DO (mg l^{-1})				
Pre-M	5.6 ± 0.7	3.0 ± 0.4	6.1 ± 0.5	6.3 ± 0.7
Mon	4.9 ± 0.3	2.5 ± 0.2	5.6 ± 0.3	5.6 ± 0.3
Post-M	5.6 ± 0.2	2.8 ± 0.2	6.0 ± 0.2	6.3 ± 0.2
Turbidity (NTU)				
Pre-M	17.8 ± 2.4	62.5 ± 5.6	20.5 ± 2.6	15.2 ± 2.5
Mon	19.8 ± 4.6	65.9 ± 6.2	22.7 ± 4.4	17.2 ± 4.4
Post-M	18.2 ± 2.2	62.8 ± 3.9	21.0 ± 2.3	14.0 ± 1.8
BOD (mg l^{-1})				

(continued)

Table 7.1 (continued)

	P1	P2	P3	P4
Pre-M	4.58 ± 0.43	11.48 ± 0.46	5.13 ± 0.59	2.25 ± 0.13
Mon	5.33 ± 0.10	12.35 ± 0.39	5.70 ± 0.39	2.38 ± 0.17
Post-M	4.65 ± 0.39	11.33 ± 1.05	4.90 ± 0.22	2.03 ± 0.26

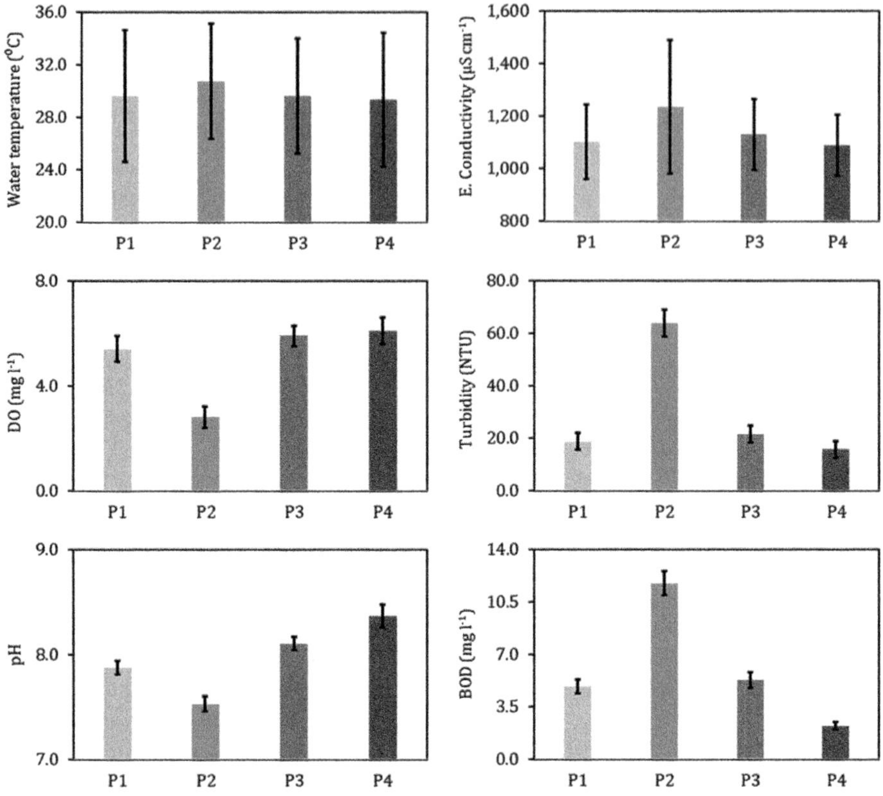

Fig. 7.2 The spatial variability of the annual mean physicochemical parameters (water temperature, electrical conductivity, dissolved oxygen, turbidity, pH, and BOD) in the four selected ponds of the present study

difference with the other three ponds was not observed. The annual mean temperature was around 30 °C in all the ponds. The surface water temperature mainly depends on the ambient air temperature. This part of the world experiences a typical hot and humid climate almost throughout the year, except for the winter months of December and January. The summer months of April and May come under the pre-monsoon season, however, the monsoon season (June to September) experiences a consistently high temperature. This is why the surface temperature in the monsoon months remained highest followed by the pre-monsoon months (which included the summer months) and the lowest was observed during the post-monsoon months (which included the winter months). All four ponds selected in this study belong to the freshwater category. The electrical conductivity varied from 856 to 1563 $\mu S\ cm^{-1}$ in these ponds. The lowest mean electrical conductivity was observed in the monsoon and the post-monsoon seasons due to the effect of the monsoon-induced rainfall. On the contrary, the electrical conductivity was marginally higher in the pre-monsoon season. This could be due to the effect of hot summer that led to the evaporation of water and enhanced the electrical conductivity. P2 was the only pond that exhibited significantly high electrical conductivity throughout the year ($p < 0.05$). P2, not being used for a specific purpose acted as a dump yard where indiscriminate anthropogenic littering took place. The lower volume of water coupled with high temperature might have facilitated the degradation of various substances which in turn enhanced the total dissolved solids, and hence the electrical conductivity of P2. On the whole, the dissolved oxygen concentration was found the highest in the post-monsoon season, followed by the pre-monsoon season, and the lowest was observed in the monsoon season. The post-monsoon season is the most favorable from the perspective of photosynthetic activities as the sky remains clear and the temperature remains optimum. During the monsoon, the dark rain clouds lead to a drop in photosynthetically active radiation, which in turn affects the autotrophic potential of these lentic ecosystems and thus leads to reduced dissolved oxygen levels. The pre-monsoon months receive ample sunlight, however, the high temperature during the summer months often enables the degassing of oxygen from the water surface. The dissolved oxygen levels showed statistically significant differences among the ponds. However, the most notable difference was observed in the case of P2. The dissolved oxygen levels in P2 were substantially lower compared to the other three ponds ($p < 0.05$). In some of the monsoon months, the oxygen levels reached hypoxic conditions rendering the water body unfit for any kind of biological activities. It is worth mentioning that P2 maintained an algal mat cover throughout the year, which is a sign of a hypereutrophic state. This algal mat inevitably led to consistent heterotrophy which led to lower oxygen levels. The turbidity was manifold high in this pond P2, compared to the other three ponds ($p < 0.05$), which in turn, further justifies the net heterotrophy of this pond. The high turbidity could have been due to the algal mat as well, however, it barred the penetration of sunlight to its shallow depth. The algal covering on the other hand prevented the free gaseous exchange with the atmosphere. Thus, these factors imparted a synergistic effect to reduce the oxygen levels in pond P2. On the contrary, pond P4, which happens to the most well-maintained ponds and where almost all sorts of anthropogenic activities are prohibited exhibited the least turbidity and the highest dissolved oxygen levels ($p < 0.05$).

The pH magnitudes indicated that all four ponds were alkaline throughout the year. P2 exhibited the lowest mean pH and P4 the highest mean all through the year. The difference in mean pH amongst the four ponds was statistically significant. Seasonally, the post-monsoon months exhibited the highest pH followed by the pre-monsoon months, and the least was observed during the monsoon. P2 was the only exception in this regard, which exhibited an increase in pH during the monsoon season. Monsoonal rain usually brings with low pH water, as CO_2 remains dissolved in rainwater. However, in P2, the already low pH water was probably diluted by the rain which could have increased the pH. Like turbidity, the BOD values were also significantly different among the four ponds, and the highest mean BOD was observed in P2. P4, the most well-maintained pond recorded the lowest BOD. BOD usually acts as a proxy of organic matter that can be biologically oxidized. The dilapidated state of P2 and the public littering ensured the abundance of such organic matter in substantial quantity, which is evident from the high BOD values. Higher

BOD values also indicate the chances of a higher degree of heterotrophy. In P1 and P3, which varying degrees of anthropogenic disturbances, the BOD values varied in between that observed in P2 and P4 ($p < 0.05$).

7.8.2 *The Dynamics of Primary Productivity Parameters*

The UWPAR magnitudes indicate the penetrability of photosynthetically active radiation below the water surface that can be utilized by the autotrophs to carry out primary autotrophic processes. It is dependent on the intensity of the solar insolation and at the same time on the transparency of the water column. The UWPAR exhibited a statistically significant difference amongst the four selected ponds ($p < 0.05$) (Table 7.2). The highest annual mean UWPAR was observed in P4, whereas the lowest was observed in P2. The ponds with a comparatively lesser degree of anthropogenic disturbance (P1 and P4) showed comparatively higher UWPAR, which indicates that appropriate management practices can enhance the UWPAR in the water column that in turn can enhance the photosynthetic potential of these ponds.

The significantly low UWAPR observed in P2 could be attributed to the excessively high turbidity compared to the other ponds. Chl-a concentration acts as a proxy of the standing stock of phytoplankton biomass in any aquatic system. There was no statistically significant difference in the mean chl-a concentrations of P1, P3,

Table 7.2 The mean ± standard deviation of the primary productivity parameters observed in the four ponds in all three seasons

	P1	P2	P3	P4
UWPAR (μmol m^{-2} s^{-1})				
Pre-M	11.9 ± 0.9	5.8 ± 1.7	6.7 ± 1.1	20.9 ± 1.6
Mon	12.0 ± 2.3	5.5 ± 1.1	6.5 ± 0.8	21.1 ± 4.0
Post-M	10.8 ± 0.7	3.8 ± 0.5	5.6 ± 0.4	19.0 ± 1.2
Chl-a (mg m^{-3})				
Pre-M	19.3 ± 3.0	86.0 ± 12.7	18.8 ± 1.7	14.3 ± 3.6
Mon	16.8 ± 2.2	73.5 ± 9.9	14.0 ± 2.2	12.8 ± 1.0
Post-M	17.0 ± 5.0	84.5 ± 16.0	16.5 ± 2.1	12.8 ± 2.2
GPP (gO_2 m^{-2} d^{-1})				
Pre-M	4.6 ± 0.6	11.0 ± 1.5	4.2 ± 0.6	10.4 ± 1.4
Mon	5.1 ± 0.4	11.4 ± 1.9	4.7 ± 0.4	11.6 ± 0.9
Post-M	5.7 ± 1.0	9.7 ± 1.7	5.1 ± 0.8	13.0 ± 2.4
CR (gO_2 m^{-2} d^{-1})				
Pre-M	11.2 ± 2.7	29.8 ± 2.9	11.8 ± 2.9	10.5 ± 4.6
Mon	12.2 ± 2.2	33.8 ± 4.9	13.2 ± 2.8	14.4 ± 1.2
Post-M	10.5 ± 1.3	23.0 ± 2.9	11.4 ± 1.7	10.8 ± 3.6

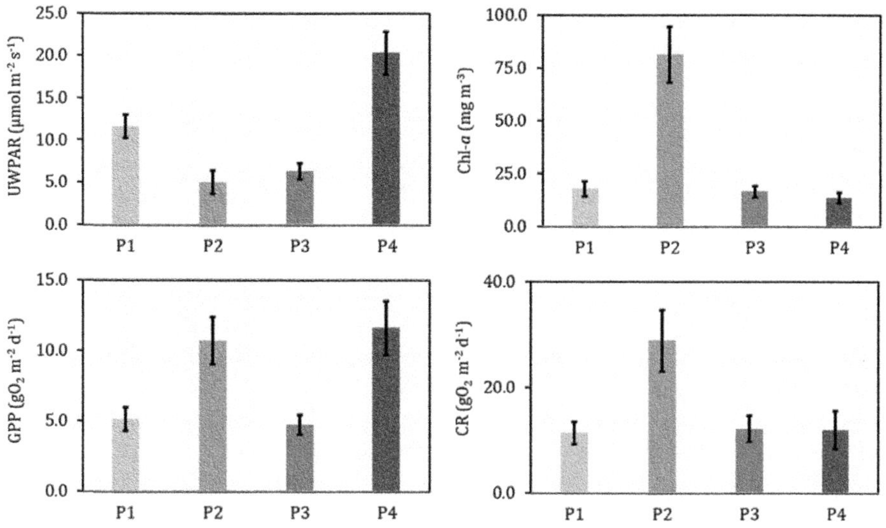

Fig. 7.3 The spatial variability of the annual mean primary productivity parameters (UWPAR, Chl-a, GPP, and CR) in the four selected ponds of the present study

and P4. However, P2 showed significantly ($p < 0.01$) higher chl-a concentration all through the sampling period (Fig. 7.3). The annual mean chl-a concentration in P2 was 5 times higher than the concentration observed in the other three ponds. This shows that the neglected pond, P2 is prone to eutrophication and most likely exists in a hypereutrophic state. Though P2 had the highest chl-a concentration, the highest GPP was observed in P4 followed by P2, P1, and P3. The GPP of P1 and P3 had no statistically significant difference. Similarly, P2 and P4 did not show any statistically significant difference. However, it is worth mentioning that the GPP in P1 and P3 was nearly half of what was observed in the case of P2 and P4. The reason behind high GPP in P2 could be the higher concentration of chl-a, whereas in the case of P4 despite having a modest chl-a concentration, favorable conditions of photosynthesis could have led to the higher GPP, compared to P1 and P3. It was evident from the mean CR magnitudes that all four ponds acted as net heterotrophic systems. In the case of P4, the mean GPP and CR were almost the same, which indicates that this pond was neither net autotrophic, not net heterotrophic. CR followed the trend like that of chl-a. P2 exhibited significantly high CR compared to the other three ponds and P1, P3, and P4 did not exhibit any significant difference in CR amongst each other. This observation indicated that excessively high chl-*a* led to hyper-eutrophic conditions, which in turn, promoted a higher rate of community respiration. The net primary productivity (NPP) was obtained from the difference in GPP and CR. The NPP followed the order P4 (-0.2 ± 3.0 gO_2 m^{-2} d^{-1}) > P1 (-6.2 ± 2.0 gO_2 m^{-2} d^{-1}) > P3 (-7.5 ± 2.1 gO_2 m^{-2} d^{-1}) > P2 (-18.2 ± 5.5 gO_2 m^{-2} d^{-1}).

7.8.3 The Dynamics of pCH_4(Water) and Air–Water CH_4 Flux

The present study indicated that all four ponds remained supersaturated with CH_4 in comparison to atmospheric CH_4 concentration and hence, acted as a source of CH_4 towards the atmosphere. The atmospheric CH_4 concentration varied over a small range of 1.859 to 1.954 μatm. On the contrary, pCH_4(water) varied between 2.030 and 15.230 μatm. This observation indicates that the gradient of partial pressure of CH_4 between air and water (ΔpCH_4) is principally regulated by pCH_4(water) and hence, the air–water CH_4 fluxes. However, the mean pCH_4(water) differed significantly among the four ponds. P2, the abandoned pond, exhibited the highest mean pCH_4(water) among all four ponds (11.670 ± 2.349 μatm). This observation indicated that the hyper-eutrophic state, shallow depth, and the dilapidated condition facilitated the enhancement in pCH_4(water). P3 and P1 exhibited a mean pCH_4(water) of 5.898 ± 0.711 and 3.816 ± 0.613 μatm, respectively (Table 7.3). Between P3 and P1, P3 is comparatively more anthropogenically disturbed and it exhibited a higher rate of efflux than P1. Thus, it can be inferred that the more the degree of anthropogenic intervention the more would be the rate of CH_4 production, and hence emission from these ponds. This notion could be further justified from the observations made in P4 which exhibited the lowest mean pCH_4(water) of 2.880 ± 0.535 μatm (Fig. 7.4). P4 is the most well-maintained pond among the selected four ponds with the lowest degree of anthropogenic intervention. The air–water CH_4 flux followed the same trend as that of pCH_4(water), as ΔpCH_4 was mainly governed by pCH_4(water), and pCH_4(air) exhibited marginal variation. The annual mean air–water CH_4 flux varied in the order: P2 (38.3 ± 18.7 mg m^{-2} h^{-1}) > P3 (14.3 ± 6.1 mg m^{-2} h^{-1}) > P1 (7.2 ± 4.0 mg m^{-2} h^{-1}) > P4 (3.8 ± 2.7 mg m^{-2} h^{-1}). Thus it is evident that the abandoned pond, which is in the most dilapidated state emits CH_4 at the highest rate and the well-maintained pond with the least anthropogenic disturbance exhibited the lowest rate of CH_4 emission. However, none of the ponds acted as a sink for atmospheric CH_4 on any of the occasions.

Table 7.3 The mean ± standard deviation of pCH_4(water) and air–water CH_4 fluxes observed in the four ponds in all three seasons

	P1	P2	P3	P4
pCH_4(water) (μatm)				
Pre-M	3.51 ± 0.63	11.56 ± 2.15	5.54 ± 0.66	2.55 ± 0.50
Mon	4.43 ± 0.24	13.97 ± 1.15	6.69 ± 0.25	3.44 ± 0.25
Post-M	3.51 ± 0.38	9.48 ± 0.80	5.47 ± 0.30	2.65 ± 0.30
Air–water CH_4 flux (mg m^{-2} s^{-1})				
Pre-M	5.96 ± 3.84	35.70 ± 16.43	13.01 ± 5.54	2.54 ± 2.39
Mon	11.44 ± 1.57	57.87 ± 8.70	20.97 ± 2.12	6.86 ± 1.24
Post-M	4.33 ± 1.91	21.27 ± 5.70	8.97 ± 1.83	2.11 ± 1.28

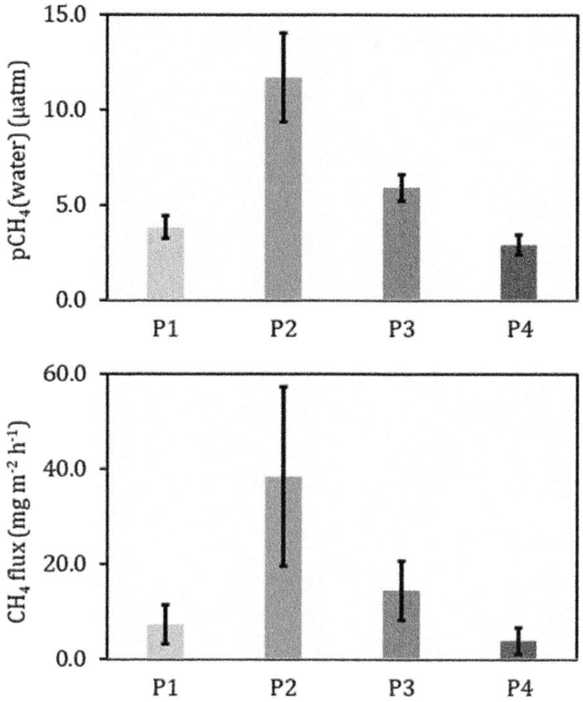

Fig. 7.4 The spatial variability of the annual mean pCH_4(water) and air–water CH_4 flux in the four selected ponds of the present study

7.8.4 Relationship Between pCH_4(Water) and Other Biogeochemical Variables

Water temperature, electrical conductivity, dissolved oxygen, oxygen saturation, turbidity, UWPAR, chl-a, GPP, CR, NPP, and BOD were correlated with pCH_4(water) to identify the potential regulators of the air–water CH_4 fluxes (Table 7.4). None of the ponds exhibited any statistically significant relationship between pCH_4(water) and electrical conductivity indicating that the minute variation in the conductivity of the water column due to the runoff and monsoon-induced rain did not affect regulating the dissolved CH_4 concentration. However, all four ponds exhibited a significant positive relationship between pCH_4(water) and water temperature. This observation indicates that higher water temperature promotes methanogenesis and methane production in the water column, which eventually escapes towards the atmosphere. This relationship also warrants that given the present rate of increase in ambient temperature due to the ongoing climate change, these ponds are about to emit more CH_4 in the near future. All of the ponds also exhibited a significant negative relationship between dissolved oxygen and pCH_4(water). The oxygen saturation levels also showed a significant negative relationship. This observation testifies that CH_4 production best took place under low DO conditions, and on the other hand, high DO levels inhibited the methanogenesis by promoting methanotrophy, i.e., methane

Table 7.4 The relationship between pCH_4(water) and other biogeochemical parameters depicted through the Pearson correlation coefficient

Parameters	P1	P2	P3	P4
Electrical conductivity	$r = -0.047$ $p > 0.05$	$r = -0.059$ $p > 0.05$	$r = -0.004$ $p > 0.05$	$r = -0.481$ $p > 0.05$
Water temperature	**$r = 0.906$ $p < 0.01$**	**$r = 0.918$ $p < 0.01$**	**$r = 0.851$ $p < 0.01$**	**$r = 0.843$ $p < 0.01$**
Dissolved oxygen	**$r = -0.810$ $p < 0.01$**	**$r = -0.596$ $p < 0.05$**	**$r = -0.758$ $p < 0.01$**	**$r = -0.885$ $p < 0.01$**
Oxygen saturation	**$r = -0.884$ $p < 0.01$**	**$r = -0.891$ $p < 0.01$**	**$r = -0.825$ $p < 0.01$**	**$r = -0.824$ $p < 0.01$**
Turbidity	$r = -0.047$ $p > 0.05$	$r = -0.043$ $p > 0.05$	$r = -0.056$ $p > 0.05$	$r = 0.178$ $p > 0.05$
UWPAR	$r = 0.294$ $p > 0.05$	**$r = 0.733$ $p < 0.01$**	$r = 0.460$ $p > 0.05$	$r = 0.255$ $p > 0.05$
Chl-a	$r = -0.186$ $p > 0.05$	$r = -0.425$ $p > 0.05$	$r = -0.393$ $p > 0.05$	$r = 0.156$ $p > 0.05$
GPP	$r = 0.390$ $p > 0.05$	**$r = 0.599$ $p < 0.05$**	$r = 0.288$ $p > 0.05$	$r = 0.371$ $p > 0.05$
CR	**$r = 0.699$ $p < 0.05$**	**$r = 0.800$ $p < 0.01$**	**$r = 0.672$ $p < 0.05$**	**$r = 0.767$ $p < 0.05$**
NPP	**$r = -0.570$ $p < 0.05$**	**$r = -0.647$ $p < 0.05$**	**$r = -0.678$ $p < 0.05$**	**$r = -0.694$ $p < 0.05$**
BOD	**$r = 0.978$ $p < 0.01$**	**$r = 0.762$ $p < 0.01$**	**$r = 0.920$ $p < 0.01$**	$r = 0.266$ $p > 0.05$

The respective significance levels are mentioned in bold below the respective r values

oxidation was prevalent under high DO levels. This observation shows us that if the photosynthetic potential of these ponds is enhanced and the DO levels are maintained at an optimum level, it can help us reduce the CH_4 emissions from these aquatic bodies. Turbidity, UWPAR, and Chl-a did not show any statistically significant relationship with pCH_4(water). The only exception was P2, which showed a significant positive relationship between UWPAR and pCH_4(water). P2 was also the only pond to show a significant positive relation between GPP and pCH_4(water). Usually higher UWPAR and GPP are signatures of higher photosynthetic activities; however, in the case of P2, which happens to be in a hypereutrophic state, higher GPP is leading to higher detritus production, and hence, providing more substrate for methanogenesis. CR showed a negative correlation with pCH_4(water) in all four ponds. This observation indicates that the higher the net heterotrophy of the system, the higher would be the rate of CH_4 emission. The negative correlation with NPP further testified to this hypothesis. Lastly, BOD exhibited a strong positive relationship with pCH_4(water). BOD gives us an idea about the abundance of organic matter present in the water column. The higher abundance of organic matter provides suitable substrates for the methanogens to act upon. However, such a relationship between BOD and pCH4(water) was not observed in P4, the most well-maintained pond, and it recorded the lowest concentration of BOD as well.

7.9 Conclusion

The present study showed that the ponds of the Indian Sundarbans acted as a net source of CH_4 towards the atmosphere. However, the ponds exhibited significant spatial variability in the CH_4 fluxes, which primarily depended on the state of the pond and the type and degree of the anthropogenic intervention. This study took into account the diffusive fluxes only. It was inferred that the gradient between pCH_4(water) and pCH_4(air) was mainly governed by the variability of pCH_4(water) as the latter varied over a comparatively smaller range than the former. The ponds where daily household activities are carried out or are anthropogenically disturbed, emitted more CH_4 than the ones that experience lesser anthropogenic intervention. The CH_4 emission from the well-maintained pond without any human intervention like bathing, washing utensils, and clothes, was the least among all four ponds studied. Water temperature, community respiration rate, and BOD showed a significant positive correlation with pCH_4(water). This observation indicates that higher temperature promotes methanogenesis and in the future, if the ambient temperature continues to rise in the rake of the ongoing climate change, these ponds can emit more CH_4 towards the atmosphere, and provide positive feedback towards the climate change. A higher degree of heterotrophy and the presence of organic matter fueled the methanogens and enhanced the dissolved CH_4 in the water column. However, a negative relationship was also observed between the dissolved oxygen levels and the pCH_4(water). This observation implied that under oxygenated conditions methanotrophy supersedes methanogenesis and reduces the methane emission rate to a large extent. Thus, promoting net autotrophy by proper maintenance of these ponds can not reduce the CO_2 emission but also the CH_4 production and its subsequent escape towards the atmosphere.

References

Aben RC, Barros N, Van Donk E et al (2017) Cross continental increase in methane ebullition under climate change. Nat Commun 8(1):1–8. https://doi.org/10.1038/s41467-017-01535-y

Adam PS, Borrel G, Brochier-Armanet C, Gribaldo S (2017) The growing tree of Archaea: new perspectives on their diversity, evolution, and ecology. ISME J 11(11):2407–2425. https://doi.org/10.1038/ismej.2017.122

Adrian R, Hessen DO, Blenckner T et al (2016) Environmental impacts—lake ecosystems. In: Quante M, Colijn F (eds) North sea region climate change assessment. Springer, Cham, pp 315–340

Allen G (2016) Biogeochemistry: rebalancing the global methane budget. Nature 538:46–48. https://doi.org/10.1038/538046a

AMAP (2017) Arctic monitoring and assessment programme. snow, water, ice and permafrost in the Arctic (SWIPA), Tromsø

APHA (2005) Standard method for the examination of water and waste water, 20th edn. American Public Health Association, Washington DC, p 541

Audet J, Neif ÉM, Cao Y et al (2017) Heat-wave effects on greenhouse gas emissions from shallow lake mesocosms. Freshw Biol 62(7):1130–1142. https://doi.org/10.1111/fwb.12930

Azar C, Johansson DJA (2012) On the relationship between metrics to compare greenhouse gases—the case of IGTP, GWP, and SGTP. Earth Syst Dynam 3:139–147. https://doi.org/10.5194/esd-3-139-2012

Barbosa PM, Farjalla VF, Melack JM, Amaral JHF, da Silva JS, Forsberg BR (2018) High rates of methane oxidation in an Amazon floodplain lake. Biogeochemistry 137(3):351–365. https://doi.org/10.1007/s10533-018-0425-2

Bastviken D, Tranvik LJ, Downing JA, Crill PM, Enrich-Prast A (2011) Freshwater methane emissions offset the continental carbon sink. Science 331(6013):50–50. https://doi.org/10.1126/science.1196808

Bastviken D, Cole J, Pace M, Tranvik L (2004) Methane emissions from lakes: dependence of lake characteristics, two regional assessments, and a global estimate. Glob Biogeochem Cycles 18:GB4009. https://doi.org/10.1029/2004GB002238

Benstead J, King GM (2001) The effect of soil acidification on atmospheric methane uptake by a Maine forest soil. FEMS Microbiol Ecol 34(3):207–212. https://doi.org/10.1111/j.1574-6941.2001.tb00771.x

Bertolet BL, West WE, Armitage DW, Jones SE (2019) Organic matter supply and bacterial community composition predict methanogenesis rates in temperate lake sediments. Limnol Oceanogr Lett 4(5):164–172. https://doi.org/10.1002/lol2.10114

Boehrer B, Dietz S, von Rohden C et al (2009) Doublediffusive deep water circulation in an iron-meromictic lake. Geochem Geophys Geosyst 10:Q06006. https://doi.org/10.1029/2009GC002389

Boehrer B, Schultze M (2008) Stratification of lakes. Rev Geophys 46(2): RG2005. https://doi.org/10.1029/2006RG000210

Borken W, Davidson EA, Savage K, Sundquist ET, Steudler P (2006) Effect of summer throughfall exclusion, summer drought, and winter snow cover on methane fluxes in a temperate forest soil. Soil Biol Biochem 38(6):1388–1395. https://doi.org/10.1016/j.soilbio.2005.10.011

Boyd CE (ed) (2012) Bottom soils, sediment, and pond aquaculture. Springer Science & Business Media

Bykova S, Boeckx P, Kravchenko I, Galchenko V, Van Cleemput O (2007) Response of CH_4 oxidation and methanotrophic diversity to NH_4^+ and CH_4 mixing ratios. Biol Fertil Soils 43(3):341–348. https://doi.org/10.1007/s00374-006-0114-5

Caraco NF, Cole JJ (1999) Human impact on nitrate export: an analysis using major world rivers. Ambio 28:167–170

Casamayor EO, Borrego CM (2009) Archaea. In: Likens GE (ed) Encyclopedia of inland waters, vol 3. Elsevier Science, p 2250

Castaldi S, Ermice A, Strumia S (2006) Fluxes of N_2O and CH_4 from soils of savannas and seasonally dry ecosystems. J Biogeogr 33(3):401–415. https://doi.org/10.1111/j.1365-2699.2005.01447.x

Castro MS, Steudler PA, Melillo JM, Aber JD, Bowden RD (1995) Factors controlling atmospheric methane consumption by temperate forest soils. Glob Biogeochem Cycles 9(1):1–10. https://doi.org/10.1029/94GB02651

Chambers LG, Osborne TZ, Reddy KR (2013) Effect of salinity-altering pulsing events on soil organic carbon loss along an intertidal wetland gradient: a laboratory experiment. Biogeochemistry 115(1–3):363–383. https://doi.org/10.1007/s10533-013-9841-5

Chistoserdova L, Laukel M, Portais JC, Vorholt JA, Lidstrom ME (2004) Multiple formate dehydrogenase enzymes in the facultative methylotroph Methylobacterium extorquens AM1 are dispensable for growth on methanol. J Bacteriol 186(1):22–28. https://doi.org/10.1128/JB.186.1.22-28.2004

Cole JJ, Caraco NF (1998) Atmospheric exchange of carbon dioxide in a low-wind oligotrophic lake measured by the addition of SF_6. Limnol Oceanogr 43:647–656. https://doi.org/10.4319/lo.1998.43.4.0647

Conrad R (2009) The global methane cycle: recent advances in understanding the microbial processes involved. Env Microbiol Rep 1:285–292. https://doi.org/10.1111/j.1758-2229.2009.00038.x

Conrad R, Klose M, Noll M, Kemnitz D, Bodelier PLE (2008) Soil type links microbial colonization of rice roots to methane emission. Glob Chang Biol 14:657–669. https://doi.org/10.1111/j.1365-2486.2007.01516.x

Cordell D, Drangert JO, White S (2009) The story of phosphorus: global food security and food for thought. Glob Environ Chang 19:292–305. https://doi.org/10.1016/j.gloenvcha.2008.10.009

Crawford GB, Collier RW (2007) Long-term observations of deepwater renewal in Crater Lake, Oregon. Hydrobiologia 574:47–68. https://doi.org/10.1007/s10750-006-0345-3

Davidson TA, Audet J, Jeppesen E, Landkildehus F, Lauridsen TL, Søndergaard M, Syväranta J (2018) Synergy between nutrients and warming enhances methane ebullition from experimental lakes. Nat Clim Chang 8(2):156–160. https://doi.org/10.1038/s41558-017-0063-z

Delaune RD, Smith CJ, Patrick WH (1983) Methane release from Gulf Coast wetlands. Tellus Ser B Chem Phys Meteorol 35:8–15. https://doi.org/10.3402/tellusb.v35i1.14581

DelSontro T, McGinnis DF, Wehrli B, Ostrovsky I (2015) Size does matter: importance of large bubbles and small-scale hot spots for methane transport. Environ Sci Technol 49(3):1268–1276. https://doi.org/10.1021/es5054286

DelSontro T, Boutet L, St-Pierre A, del Giorgio PA, Prairie YT (2016) Methane ebullition and diffusion from northern ponds and lakes regulated by the interaction between temperature and system productivity. Limnol Oceanogr 61:S62–S77. https://doi.org/10.1002/lno.10335

Downing JA, Prairie YT, Cole JJ et al (2006) The global abundance and size distribution of lakes, ponds, and impoundments. Limnol Oceanogr 51(5):2388–2397. https://doi.org/10.4319/lo.2006.51.5.2388

Dubbs LL, Whalen SC (2010) Reduced net atmospheric CH_4 consumption is a sustained response to elevated CO_2 in a temperate forest. Biol Fertil Soils 46(6):597–606. https://doi.org/10.1007/s00374-010-0467-7

Dubey SK, Trivedi RK, Chand BK, Mandal B, Rout SK (2017) Farmers' perceptions of climate change impacts on freshwater aquaculture and adaptation strategies in climatic change hotspots: a case of the Indian Sundarban delta. Environ Dev 21:38–51. https://doi.org/10.1016/j.envdev.2016.12.002

Encinas Fernandez J, Peeters F, Hofmann H (2014) Importance of the autumn overturn and anoxic conditions in the hypolimnion for the annual methane emissions from a temperate lake. Environ Sci Technol 48:7297–7304. https://doi.org/10.1021/es4056164

Enzmann F, Mayer F, Rother M, Holtmann D (2018) Methanogens: biochemical background and biotechnological applications. AMB Expr 8(1):1–22. https://doi.org/10.1186/s13568-017-0531-x

Ettwig KF, Butler MK, Le Paslier D et al (2010) Nitrite-driven anaerobic methane oxidation by oxygenic bacteria. Nature 464(7288):543–548. https://doi.org/10.1038/nature08883

Ferry JG, Kastead KA (2007) Methanogenesis. In: Cavicchioli R (ed) Archaea: molecular and cellular biology, Wiley Online, pp 288–314. https://doi.org/10.1128/9781555815516.ch13

Gålfalk M, Bastviken D, Fredriksson S, Arneborg L (2013) Determination of the piston velocity for water-air interfaces using flux chambers, acoustic Doppler velocimetry, and IR imaging of the water surface. J Geophys Res Biogeosci 118:770–782. https://doi.org/10.1002/jgrg.20064

Greening C, Geier R, Wang C et al (2019) Diverse hydrogen production and consumption pathways influence methane production in ruminants. ISME J 13(10):2617–2632. https://doi.org/10.1038/s41396-019-0464-2

Grosso SD, Parton WJ, Mosier AR, Ojima DS, Potter CS, Borken W, Brumme R, Butterbach-Bahl K, Crill PM, Dobbie K, Smith KA (2000) General CH_4 oxidation model and comparisons of CH_4 oxidation in natural and managed systems. Glob Biogeochem Cycles 14(4):999–1019. https://doi.org/10.1029/1999GB001226

Hanson RS, Hanson TE (1996) Methanotrophic bacteria. Microbiol Rev 60(2):439–471

Holgerson MA, Raymond PA (2016) Large contribution to inland water CO_2 and CH_4 emissions from very small ponds. Nat Geosci 9(3):222–226. https://doi.org/10.1038/ngeo2654

Hopcroft PO, Valdes PJ, O'Connor FM, Kaplan JO, Beerling DJ (2017) Understanding the glacial methane cycle. Nat Commun 8(1):1–10. https://doi.org/10.1038/ncomms14383

Horz HP, Rich V, Avrahami S, Bohannan BJ (2005) Methane-oxidizing bacteria in a California upland grassland soil: diversity and response to simulated global change. Appl Environ Microbiol 71(5):2642–2652. https://doi.org/10.1128/AEM.71.5.2642-2652.2005

IPCC (2014) Climate change 2014: synthesis report. In: Pachauri RK, Meyer LA (eds) Contribution of working groups I, II and III to the fifth assessment report of the intergovernmental panel on climate change.

Jacinthe PA, Lal R (2006) Methane oxidation potential of reclaimed grassland soils as affected by management. Soil Sci 171(10):772–783. https://doi.org/10.1097/01.ss.0000209357.53536.43

Jäckel U, Thummes K, Kämpfer P (2005) Thermophilic methane production and oxidation in compost. FEMS Microbiol Ecol 52(2):175–184. https://doi.org/10.1016/j.femsec.2004.11.003

Jenny JP, Francus P, Normandeau A et al (2016) Global spread of hypoxia in freshwater ecosystems during the last three centuries is caused by rising local human pressure. Glob Chang Biol 22(4):1481–1489. https://doi.org/10.1111/gcb.13193

Jeppesen E, Kronvang B, Meerhoff M et al (2009) Climate change effects on runoff, catchment phosphorus loading and lake ecological state, and potential adaptations. J Environ Qual 38:1930–1941. https://doi.org/10.2134/jeq2008.0113

Jin Q, Kirk MF (2018) pH as a primary control in environmental microbiology: 1. Thermodynamic perspective. Front Environ Sci 6:21. https://doi.org/10.3389/fenvs.2018.00021

Joyni MJ, Kurup BM, Avnimelech Y (2011) Bioturbation as a possible means for increasing production and improving pond soil characteristics in shrimp-fish brackish water ponds. Aquaculture 318:464–470. https://doi.org/10.1016/j.aquaculture.2011.05.019

Jugnia LB, Cabral AR, Greer CW (2008) Biotic methane oxidation within an instrumented experimental landfill cover. Ecol Eng 33(2):102–109. https://doi.org/10.1016/j.ecoleng.2008.02.003

Kai FM, Tyler SC, Randerson JT, Blake DR (2011) Reduced methane growth rate explained by decreased Northern Hemisphere microbial sources. Nature 476:194–197. https://doi.org/10.1038/nature10259

Kallistova AY, Kevbrina MV, Nekrasova VK, Glagolev MV, Serebryanaya MI, Nozhevnikova AN (2005) Methane oxidation in landfill cover soil. Microbiology 74(5):608–614. https://doi.org/10.1007/s11021-005-0110-z

Katsev S, Crowe SA, Mucci A, Sundby B, Nomosatryo S, Douglas Haffner G, Fowle DA (2010) Mixing and its effects on biogeochemistry in the persistently stratified, deep, tropical Lake Matano, Indonesia. Limnol Oceanogr 55(2):763–776. https://doi.org/10.4319/lo.2010.55.2.0763

Khalil MI, Baggs EM (2005) CH_4 oxidation and N_2O emissions at varied soil water-filled pore spaces and headspace CH_4 concentrations. Soil Biol Biochem 37(10):1785–1794. https://doi.org/10.1016/j.soilbio.2005.02.012

Kirschke S, Bousquet P, Ciais P, Saunois M, Canadell JG, Dlugokencky EJ et al (2013) Three decades of global methane sources and sinks. Nat Geosci 6:813–822. https://doi.org/10.1038/NGEO1955

Kosten S, Almeida RM, Barbosa I, Mendonça R, Muzitano IS, Oliveira-Junior ES, Vroom RJ, Wang HJ, Barros N (2020) Better assessments of greenhouse gas emissions from global fish ponds needed to adequately evaluate aquaculture footprint. Sci Total Environ 748:141247. https://doi.org/10.1016/j.scitotenv.2020.141247

Laanbroek HJ (2010) Methane emission from natural wetlands: Interplay between emergent macrophytes and soil microbial processes: a mini-review. Ann Bot 105:141–153. https://doi.org/10.1093/aob/mcp201

Lau E, Ahmad A, Steudler PA, Cavanaugh CM (2007) Molecular characterization of methanotrophic communities in forest soils that consume atmospheric methane. FEMS Microbiol Ecol 60(3):490–500. https://doi.org/10.1111/j.1574-6941.2007.00308.x

Le Mer J, Roger P (2001) Production, oxidation, emission and consumption of methane by soils: a review. Eur J Soil Biol 37(1):25–50. https://doi.org/10.1016/S1164-5563(01)01067-6

Li M, Peng C, Zhu Q, Zhou X, Yang G, Song X, Zhang K (2020) The significant contribution of lake depth in regulating global lake diffusive methane emissions. Water Res 172:115465. https://doi.org/10.1016/j.watres.2020.115465

Lide DR (2007) CRC handbook of chemistry and physics, 88th edn. CRC, New York, p 2660

López Bellido J, Tulonen T, Kankaala P, Ojala A (2009) CO_2 and CH_4 fluxes during spring and autumn mixing periods in a boreal lake (Pääjärvi, southern Finland). J Geophys Res 114:G04007. https://doi.org/10.1029/2009JG000923

Lyu Z, Liu Y (2018) Diversity and taxonomy of methanogens. In: Stams AJM, Sousa DZ (eds) Biogenesis of hydrocarbons. Springer, New York, pp 19–77

Lyu Z, Shao N, Akinyemi T, Whitman WB (2018) Methanogenesis. Curr Biol 28(13):R727–R732. https://doi.org/10.1016/j.cub.2018.05.021

MacIntyre S, Wanninkhof R, Chanton JP (1995) Trace gas exchange across the air-water interface in freshwater and coastal marine environments. In: Matson PA, Harriss RC (eds) Biogenic trace gases: measuring emissions from soil and water. Blackwell Science, Oxford, pp 52–97

Mancinelli RL (1995) The regulation of methane oxidation in soil. Ann Rev Microbiol 49(1):581–605. https://doi.org/10.1146/annurev.mi.49.100195.003053

Mandal S, Burman D, Sarangi SK, Bandyopadhyay BK, Maji B (2015) Homestead production systems in Sundarbans region of West Bengal, India—current status and opportunities. CGIAR-CPWF

Martinez-Cruz K, Sepulveda-Jauregui A, Walter Anthony K, Thalasso F (2015) Geographic and seasonal variation of dissolved methane and aerobic methane oxidation in Alaskan lakes. Biogeosciences 12:4595–4606. https://doi.org/10.5194/bg-12-4595-2015

Matthews E, Johnson MS, Genovese V, Du J, Bastviken D (2020) Methane emission from high latitude lakes: methane-centric lake classification and satellite-driven annual cycle of emissions. Sci Rep 10:12465. https://doi.org/10.1038/s41598-020-68246-1

Mayumi D, Mochimaru H, Tamaki H, Yamamoto K, Yoshioka H, Suzuki Y, Kamagata Y, Sakata S (2016) Methane production from coal by a single methanogen. Science 354(6309):222–225. https://doi.org/10.1126/science.aaf8821

McGinnis DF, Kirillin G, Tang KW, Flury S, Bodmer P, Engelhardt C, Casper C, Grossart HP (2015) Enhancing surface methane fluxes from an Oligotrophic lake: exploring the microbubble hypothesis. Environ Sci Technol 49:873–880. https://doi.org/10.1021/es503385d

Megonigal JP, Hines ME, Visscher PT (2004) Anaerobic metabolism: linkages to trace gases and aerobic processes. In: Schlesinger WH (ed) Biogeochemistry. Elsevier-Pergamon, Oxford, pp 317–424

Michmerhuizen CM, Striegl RG, Mcdonald ME (1996) Potential methane emission from north-temperate lakes following ice melt. Limnol Oceanogr 41:985–991. https://doi.org/10.4319/lo.1996.41.5.0985

Middelburg JJ, Nieuwenhuize J, Iversen N, Høgh N, De Wilde H, Helder W, Seifert R, Christof O (2002) Methane distribution in European tidal estuaries. Biogeochemistry 59(1–2):95–119. https://doi.org/10.1023/A:1015515130419

Mohanty SR, Bodelier PL, Floris V, Conrad R (2006) Differential effects of nitrogenous fertilizers on methane-consuming microbes in rice field and forest soils. Appl Environ Microbiol 72(2):1346–1354. https://doi.org/10.1128/AEM.72.2.1346-1354.2006

Morel FM (1983) Energetics and kinetics: principles of aquatic chemistry. Wiley, New York, p 446

Mosier AR, Delgado JA (1997) Methane and nitrous oxide fluxes in grasslands in western Puerto Rico. Chemosphere 35(9):2059–2082. https://doi.org/10.1016/S0045-6535(97)00231-2

Moss B, Kosten S, Meerhoff M et al (2011) Allied attack: climate change and eutrophication. Inland Waters 1(2):101–105. https://doi.org/10.5268/IW-1.2.359

Myhre G, Shindell D, Breon F-M, Collins W, Fuglestvedt J, Huang J et al (2013) Anthropogenic and natural radiative forcing. In: Stocker T et al (eds) Climate change 2013: the physical science basis. Contribution of working group I to the fifth assessment report of the intergovernmental panel on climate change. Cambridge, UK, and New York, Cambridge University Press

Nayak DR, Babu YJ, Datta A, Adhya TK (2007) Methane oxidation in an intensively cropped tropical rice field soil under long-term application of organic and mineral fertilizers. J Environ Qual 36(6):1577–1584. https://doi.org/10.2134/jeq2006.0501

Negandhi K, Laurion I, Lovejoy C (2016) Temperature effects on net greenhouse gas production and bacterial communities in arctic thaw ponds. FEMS Microbiol Ecol 92(8):fiw117. https://doi.org/10.1093/femsec/fiw117

Neubauer SC, Franklin RB, Berrier DJ (2013) Saltwater intrusion into tidal freshwater marshes alters the biogeochemical processing of organic carbon. Biogeosciences 10(12):8171–8183. https://doi.org/10.5194/bg-10-8171-2013

Nisbet EG, Dlugokencky EJ, Bousquet P (2014) Methane on the rise-again. Science 343:493–495. https://doi.org/10.1126/science.1247828

Oliveira Junior ES, Temmink RJM, Buhler BF, Souza RM, Resende N, Spanings T, Muniz CC, Lamers LPM, Kosten S (2019) Benthivorous fish bioturbation reduces methane emissions but increases total greenhouse gas emissions. Freshw Biol 64:197–207. https://doi.org/10.1111/fwb.13209

Pacheco FS, Roland F, Downing JA (2014) Eutrophication reverses whole-lake carbon budgets. Inland Waters 4:41–48. https://doi.org/10.5268/IW-4.1.614

Peters GP, Aamaas B, Berntsen T, Fuglestvedt JS (2011) The integrated global temperature change potential (iGTP) and relationships between emission metrics. Environ Res Lett 6:044021. https://doi.org/10.1088/1748-9326/6/4/044021

Phelps AR, Peterson KM, Jeffries MO (1998) Methane efflux from high-latitude lakes during spring ice melt. J Geophys Res Atmos 103(D22):29029–29036. https://doi.org/10.1029/98JD00044

Phillips RL, Whalen SC, Schlesinger WH (2001) Influence of atmospheric CO_2 enrichment on methane consumption in a temperate forest soil. Glob Chang Biol 7(5):557–563. https://doi.org/10.1046/j.1354-1013.2001.00432.x

Poffenbarger HJ, Needelman BA, Megonigal JP (2011) Salinity influence on methane emissions from tidal marshes. Wetlands 31(5):831–842. https://doi.org/10.1007/s13157-011-0197-0

Prather M, Holmes C, Hsu J (2012) Reactive greenhouse gas scenarios: systematic exploration of uncertainties and the role of atmospheric chemistry. Geophys Res Lett 39:L09803. https://doi.org/10.1029/2012GL051440

Price SJ, Kelliher FM, Sherlock RR, Tate KR, Condron LM (2004) Environmental and chemical factors regulating methane oxidation in a New Zealand forest soil. Soil Res 42(7):767–776. https://doi.org/10.1071/SR04026

Priemé A, Ekelund F (2001) Five pesticides decreased oxidation of atmospheric methane in a forest soil. Soil Biol Biochem 33(6):831–835. https://doi.org/10.1016/S0038-0717(00)00246-7

Reay DS (2003) Sinking methane. Biologist 50(1):15–19

Riera JL, Schindler JE, Kratz TK (1999) Seasonal dynamics of carbon dioxide and methane in two clear-water lakes and two bog lakes in northern Wisconsin, USA. Can J Fish Aquat Sci 56(2):265–274. https://doi.org/10.1139/f98-182

Rigby M, Montzka S, Prinn R, White JWC, Young D, OŠDoherty S et al (2017) Role of atmospheric oxidation in recent methane growth. Proc Natl Acad Sci USA 114(21):5373–5377. https://doi.org/10.1073/pnas.1616426114

Saari A, Rinnan R, Martikainen PJ (2004) Methane oxidation in boreal forest soils: kinetics and sensitivity to pH and ammonium. Soil Biol Biochem 36(7):1037–1046. https://doi.org/10.1016/j.soilbio.2004.01.018

Saarnio S, Winiwarter W, Leitao J (2009) Methane release from wetlands and watercourses in Europe. Atmos Environ 43(7):1421–1429. https://doi.org/10.1016/j.atmosenv.2008.04.007

Samir KC, Lutz W (2017) The human core of the shared socioeconomic pathways: population scenarios by age, sex, and level of education for all countries to 2100. Glob Environ Chang 42:181–192. https://doi.org/10.1016/j.gloenvcha.2014.06.004

Sanches LF, Guenet B, Marinho CC, Barros N, de Assis EF (2019) Global regulation of methane emission from natural lakes. Sci Rep 9:255. https://doi.org/10.1038/s41598-018-36519-5

Saunois M, Bousquet P, Poulter B et al (2016) The global methane budget: 2000–2012. Earth Syst Sci Data 8:697–751. https://doi.org/10.5194/essd-8-697-2016

Schaefer H, Mikaloff Fletcher S, Veidt C, Lassey K, Brailsford G, Bromley T et al (2016) A 21st-century shift from fossil fuel to biogenic methane emissions indicated by $^{13}CH_4$. Science 352:80–84. https://doi.org/10.1126/science.aad2705

Schilder J, van Hardenbroek M, Bodelier P, Kirilova EP, Leuenberger M, Lotter AF, Heiri O (2017) Trophic state changes can affect the importance of methane-derived carbon in aquatic food webs. Proc R Soc B Biol Sci 284(1857):20170278. https://doi.org/10.1098/rspb.2017.0278

Schmid M, Halbwachs M, Wehrli B, Wüest A (2005) Weak mixing in Lake Kivu: new insights indicate increasing risk of uncontrolled gas eruption. Geochem Geophys Geosyst 6:7. https://doi.org/10.1029/2004GC000892

Schwietzke S, Sherwood OA, Bruhwiler LMP et al (2016) Upward revision of global fossil-fuel methane emissions based on isotope database. Nature 538:88–91. https://doi.org/10.1038/nature19797

Semrau JD, DiSpirito AA, Vuilleumier S (2011) Facultative methanotrophy: false leads, true results, and suggestions for future research. FEMS Microbiol Lett 323(1):1–12. https://doi.org/10.1111/j.1574-6968.2011.02315.x

Sepulveda-Jauregui A, Hoyos-Santillan J, Martinez-Cruz K, Anthony KMW, Casper P, Belmonte-Izquierdo Y, Thalasso F (2018) Eutrophication exacerbates the impact of climate warming on lake methane emission. Sci Total Environ 636:411–419. https://doi.org/10.1016/j.scitotenv.2018.04.283

Serrano-Silva N, Sarria-Guzmán Y, Dendooven L, Luna-Guido M (2014) Methanogenesis and methanotrophy in soil: a review. Pedosphere 24(3):291–307. https://doi.org/10.1016/S1002-0160(14)60016-3

Shukla AK, Vishwakarma P, Upadhyay SN, Tripathi AK, Prasana HC, Dubey SK (2009) Biodegradation of trichloroethylene (TCE) by methanotrophic community. Bioresour Technol 100(9):2469–2474. https://doi.org/10.1016/j.biortech.2008.12.022

Sinha E, Michalak AM, Balaji V (2017) Eutrophication will increase during the 21st century as a result of precipitation changes. Science 357:405–408. https://doi.org/10.1126/science.aan2409

Stamp I, Baird AJ, Heppell CM (2013) The importance of ebullition as a mechanism of methane (CH4) loss to the atmosphere in a northern peatland. Geophys Res Lett 40:2087–2090. https://doi.org/10.1002/grl.50501

Sun Z, Wang L, Tian H, Jiang H, Mou X, Sun W (2013) Fluxes of nitrous oxide and methane in different coastal Suaeda salsa marshes of the Yellow River estuary, China. Chemosphere 90(2):856–865. https://doi.org/10.1016/j.chemosphere.2012.10.004

Tang KW, McGinnis DF, Ionescu D, Grossart HP (2016) Methane production in oxic lake waters potentially increases aquatic methane flux to air. Environ Sci Technol Lett 3(6):227–233. https://doi.org/10.1021/acs.estlett.6b00150

Templeton AS, Chu KH, Alvarez-Cohen L, Conrad ME (2006) Variable carbon isotope fractionation expressed by aerobic CH_4-oxidizing bacteria. Geochim Cosmochim Acta 70(7):1739–1752. https://doi.org/10.1016/j.gca.2005.12.002

Treude T, Niggemann J, Kallmeyer J, Wintersteller P, Schubert CJ, Boetius A, Jørgensen BB (2005) Anaerobic oxidation of methane and sulfate reduction along the Chilean continental margin. Geochim Cosmochim Acta 69(11):2767–2779. https://doi.org/10.1016/j.gca.2005.01.002

Turner AJ, Frankenberg C, Wennberg PO, Jacob DJ (2017) Ambiguity in the causes for decadal trends in atmospheric methane and hydroxyl. Proc Natl Acad Sci USA 114:5367–5372. https://doi.org/10.1073/pnas.1616020114

van Bergen TJ, Barros N, Mendonça R, Aben RC, Althuizen IH, Huszar V, Lamers LP, Lürling M, Roland F, Kosten S (2019) Seasonal and diel variation in greenhouse gas emissions from an urban pond and its major drivers. Limnol Oceanogr 64:2129–2139. https://doi.org/10.1002/lno.11173

Van der Nat F, Middelburg JJ (1998) Effects of two common macrophytes on methane dynamics in freshwater sediments. Biogeochemistry 43:79–104. https://doi.org/10.1023/A:1006076527187

Verpoorter C, Kutser T, Seekell DA, Tranvik LJ (2014) A global inventory of lakes based on high-resolution satellite imagery. Geophys Res Lett 41(18):6396–6402. https://doi.org/10.1002/2014GL060641

Wang ZP, Zeng D, Patrick WH (1996) Methane emissions from natural wetlands. Environ Monit Assess 42:143–161. https://doi.org/10.1007/BF00394047

Wanninkhof R (1992) Relationship between gas exchange and wind speed over the ocean. J Geophys Res 97:7373–7381

Wendlandt KD, Stottmeister U, Helm J, Soltmann B, Jechorek M, Beck M (2010) The potential of methane-oxidizing bacteria for applications in environmental biotechnology. Eng Life Sci 10(2):87–102

Werner C, Zheng X, Tang J, Xie B, Liu C, Kiese R, Butterbach-Bahl K (2006) N_2O, CH_4, and CO_2 emissions from seasonal tropical rainforests and a rubber plantation in Southwest China. Plant Soil 289(1–2):335–353. https://doi.org/10.1007/s11104-006-9143-y

West WE, Creamer KP, Jones SE (2016) Productivity and depth regulate lake contributions to atmospheric methane. Limnol Oceanogr 61(S1):S51–S61. https://doi.org/10.1002/lno.10247

Weston NB, Vile MA, Neubauer SC, Velinsky DJ (2011) Accelerated microbial organic matter mineralization following salt-water intrusion into tidal freshwater marsh soils. Biogeochemistry 102(1):135–151. https://doi.org/10.1007/s10533-010-9427-4

Wik M, Thornton BF, Bastviken D, Uhlbäck J, Crill PM (2016) Biased sampling of methane release from northern lakes: a problem for extrapolation. Geophys Res Lett 43:1256–1262. https://doi.org/10.1002/2015GL066501

Worden J, Bloom A, Pandey S, Jiang Z, Worden HM, Walker TW et al (2018) Reduced biomass burning emissions reconcile conflicting estimates of the post-2006 atmospheric methane budget. Nat Commun 8:2227. https://doi.org/10.1038/s41467-017-02246-0

Xiong J, Liu Y, Lin X, Zhang H, Zeng J, Hou J et al (2012) Geographic distance and pH drive bacterial distribution in alkaline lake sediments across Tibetan Plateau. Environ Microbiol 14(9):2457–2466. https://doi.org/10.1111/j.1462-2920.2012.02799.x

Yang P, Zhang Y, Lai DY, Tan L, Jin B, Tong C (2018) Fluxes of carbon dioxide and methane across the water–atmosphere interface of aquaculture shrimp ponds in two subtropical estuaries: The effect of temperature, substrate, salinity, and nitrate. Sci Total Environ 635:1025–1035. https://doi.org/10.1016/j.scitotenv.2018.04.102

Yvon-Durocher G, Allen AP, Bastviken D, Conrad R, Gudasz C, St-Pierre A et al (2014) Methane fluxes show consistent temperature dependence across microbial to ecosystem scales. Nature 507(7493):488–491. https://doi.org/10.1038/nature13164

Zhang G, Xiao X, Dong J, Xin F, Zhang Y, Qin Y et al (2020) Fingerprint of rice paddies in spatial-temporal dynamics of atmospheric methane concentration in monsoon Asia. Nat Commun 11:554. https://doi.org/10.1038/s41467-019-14155-5

Chapter 8
Characterizing the Drivers of the Productivity and Greenhouse Gas Fluxes from the Aquaculture Ponds of Indian Sundarbans

Abhra Chanda, Sourav Das, and Niloy Pramanik

Abstract The aquaculture ponds have been recognized as significant sources of greenhouse gases like CO_2 and CH_4 as a substantial amount of organic matter remains unspent during fish culture, which eventually mineralizes into inorganic form and escapes the water bodies as CO_2 or the methanogens act upon these organic substrates to form CH_4. The Indian Sundarban Biosphere Reserve shelters a population as high as 4.4 million people, and in the recent past, a trend in livelihood switch from traditional agriculture to aquaculture has been noticed and reported by many eminent researchers. However, endeavors of characterizing the rate of greenhouse gas emissions and their regulators in these ponds are yet to be reported. The present chapter takes this opportunity to monitor and report the variability in the partial pressure of CO_2 and CH_4 in water [pCO_2(water) and pCH_4(water), respectively] and the air–water CO_2 and CH_4 flux from two aquaculture ponds situated in the north and south of the Indian Sundarbans that culture *Penaeus monodon* (tiger prawn). Both of the ponds emitted CO_2 and CH_4 of varying magnitudes throughout the year. However, the pond in the north which utilized the adjacent estuarine water full of anthropogenic load emitted more CO_2 and CH_4 compared to the one situated down south, which had higher dominance of seawater. A significant positive relationship of pCO_2(water) and pCH_4(water) with water temperature warrants that in the future the ongoing climate change would lead to an increase in the emission rates. Enhanced lime treatment was found to increase the pH levels which favored reducing pCO_2(water). An oxygen-rich environment was found to dampen both the CO_2 and CH_4 fluxes, which indicates that if the autotrophic potential of these ponds could be increased, a reduction in CO_2 and CH_4 fluxes is possible.

Keywords pCO_2(water) · pCH_4(water) · Air–water CO_2 flux · Air–water CH_4 flux · Primary productivity · Methanogens · Methanotrophs · Anaerobic conditions

---

A. Chanda (✉) · S. Das · N. Pramanik
School of Oceanographic Studies, Jadavpur University, 188 Raja S. C. Mullick Road, Kolkata 700032, India

S. Das et al. (eds.), *Pond Ecosystems of the Indian Sundarbans*, Water Science and Technology Library 112, https://doi.org/10.1007/978-3-030-86786-7_8

8.1 Introduction

In the primitive days, man used to hunt an array of wild animals to quench the protein requirement. As we became more and more civilized, the concept of the market came into action, and instead of hunting, purchasing food became an essential aspect of our daily life routine. Non-vegetarians have always relied on various animal-derived food products, which principally act as the source of protein. This includes a wide variety of meat. However, in all corners of the world, fishing emerged to be an important avenue to serve protein in the daily diet. According to the various surveys and analyses furnished by the Food and Agriculture Organization (FAO), the global population extracts 17% of the total animal protein from the fishes (Duan et al. 2020). Even at the beginning of the last century, the term fishery indicated open fishing in natural water bodies, like rivers, lakes, and oceans. Marine fishing has played a significant role in providing a substantial amount of fishes throughout the globe. This practice is still on and it continues to provide a bulksome quantity of fishes, which acts as the backbone of several fishing industries. However, as the steady demand for fishes kept on increasing with the rise in global population, marine reserves appeared to be not sufficient to meet this pressure. The preceding sentence might reflect that the fish reserves in the open marine sector are steadily decreasing. However, the actual scenario is not that simple. To meet the growing demand for fishes, endeavors of fishing in the marine sector have increased manifold throughout the world since the 1950s (Bell et al. 2017). Prolonged practice of indiscriminate fishing in the open waters led to overexploitation of fishes and such observations were widespread in many places (Ye and Gutierrez 2017). Moreover, harvesting fishes from a natural system, which is vast, remains associated with a suite of uncertainties and unpredictability (Chen 2020). Thus, even if there is an infinite abundance of several fish species in the open waters, catching these fishes behind the nets depends on several complicated factors. Ever since this crisis came into the forefront, ecologists and policy managers have joined hands to cultivate and implement several options to rebuild the global capture fisheries (Worm et al. 2009; Gutierrez et al. 2011). However, the irregular yield from the capture fisheries sector compelled many who are involved in this business to switch to aquaculture fishing.

The history of aquaculture fishing dates back to 2000 B. C. China played a pioneering role in coming up with this method of fishing (Nash 2010). Even at present, China stands out tall as the largest mariculture industry of the world producing approximately 60% of the world's farmed fishes (FAO 2018). This mode of fishing has become one of the fastest-growing food sectors in the entire world (Ren et al. 2019). Several experts envisaged that aquaculture fishing would play a significant role in maintaining food security worldwide (FAO 2011). The year 2016 touched the landmark of almost 171 million tons of fish production throughout the world, out of which aquaculture contributed 47% (FAO 2018). The lowlands near the coastal periphery are hotspots for developing aquaculture (Primavera 2006). These regions encompass some of the most fertile and productive lands and have the advantage of accessing both marine and freshwater resources (Renaud et al. 2013). However, the rapid flourishment of the aquaculture sector along the coastlines has already left an irreversible signature on the fragile land use classes of these regions. Loss of biodiversity, land conversion of valuable coastal ecosystems, soil, and aquatic pollution are some of the ill effects that the aquaculture industry has incurred in the last few decades (Virdis 2014; Yao et al. 2016). It is true that to meet the food security, aquaculture ponds have become necessary in the present date. However, the coastal fringes are delicate and harbours several crucial ecosystems, like the mangroves, tidal flats, seagrasses, which furnish a suite of ecosystem services to humankind. The booming of aquaculture ponds would inevitably disturb the coastal setting and in the end, it would make these regions vulnerable (Murray et al. 2014; Spalding et al. 2014). Already the world has witnessed the expansion of aquaculture at the expense of the destruction of substantial areas covered by diverse marine ecosystems (Wu et al. 2020). Maintenance of aquaculture ponds requires the use of external nutrients and antibiotics to prevent fish diseases from spreading and increase the yield. Indiscriminate and uncontrolled use of these materials leads to severe environmental problems, like eutrophication, red tides, antibiotic residues, and several other types of aquatic pollution (Wang et al. 2018; Neofitou et al. 2019). Most of the aquaculture industries utilize earthen ponds, which are easy to construct and maintain. However, the use of such ponds enables all the inputs required to carry out aquaculture to encounter the sediment surface, and this pollutes the sediment for a long period. Despite knowing about all these harmful effects, aquaculture ponds have become indispensable, given the fish demand in the present date, and there is no viable alternative. However, sustainable management of aquaculture ponds and the development of scientific technologies to minimize the ill effects of aquaculture fishing have drawn attention.

Among the various threats that aquaculture ponds pose to the environment, greenhouse gas emissions from these systems require special mention. The efficient functioning of aquaculture ponds includes the use of external fish feeds rich in both carbon and nitrogen (Naylor et al. 2000; Cao et al. 2015). The heavy uses of the aquafeeds enhance the nutrient loadings and introduce substantial quantities of carbon, which are subject to microbial transformation to greenhouse gases like carbon dioxide (CO_2), methane (CH_4), and nitrous oxide (N_2O) (Hargreaves 1998; Hu et al. 2012; Boyd and Tucker 2014; Yuan et al. 2019). A substantial share of the carbon and nitrogen loadings in the form of aquafeeds remains unutilized by the fishes, and at the same time, the fish-produced excreta re-introduces both these elements in the labile form. The prime intention behind such excessive nutrient and carbon loading is to accelerate the rate of primary production and stimulate phytoplankton growth. However, this excessive use of fertilizers renders the aquaculture systems act as significant sources of greenhouse gases (Yuan et al. 2019). The aquaculture ponds of China have received substantial attention in this regard, in the last decade (Yang et al. 2017, 2020a, b); however, studies on the aquaculture ponds of other parts of the world are scarce. On the present date, India ranks second after China, in terms of the areal coverage of aquaculture ponds (Prasad et al. 2019). Thousands of people in the coastal belt of India have adopted aquaculture fish farming as an alternative livelihood by converting agricultural plots and mangrove forests (Jayanthi et al. 2018). In the last two decades, the number of aquaculture ponds throughout the coastline of India increased exponentially (Prasad et al. 2019). The Sundarbans are also no exception, in this regard. Within India, West Bengal ranks second after Andhra Pradesh in terms of total fish production. The sea-facing districts of North and South 24 Parganas, which shelter a part of the world's largest mangrove forest of Sundarbans, contribute a lion share of the state's total fish production (Dubey et al. 2017). The Sundarbans is home to almost 4.4 million people. Aquaculture ponds enable thousands of people to earn bread and butter (Government of West Bengal 2010). Recently the Sundarbans have witnessed considerable growth in aquaculture farming, especially after the cyclone Aila, which took place in the year 2009. The earthen embankments in the Sundarbans enable the separation of sweet water (read freshwater) from the saline water, and this, in turn, facilitates the practice of both freshwater and brackish water aquaculture. Despite the boom of aquaculture in the Sundarbans sectors, endeavors of characterizing greenhouse gas emissions are virtually absent as of now. The present chapter has reported the nature and magnitude of greenhouse gas fluxes from some typical aquaculture ponds of this region. The present study also tried to understand the role of climate change and some basic pond management practices on regulating the fluxes.

8.2 Types of Aquaculture Practices

There are several categories of aquaculture in terms of water used for the culture, treatment types, and species farmed. Rearing of commercially important fishes including the cultivation of ornamental fishes, algae culture (such as the farming of seaweeds consumed in many parts of the world), shrimp culture, oyster culture, and varied types of mariculture are some of the major categories of aquaculture. Depending on the type of water used for farming, aquacultures can be of two types: freshwater and saline water. Based on the type of species farmed and techniques adopted to carry out the cultivation, aquaculture types include fish farming, mariculture, alga-culture, Integrated Multi-Trophic Aquaculture (IMTA), Inland Pond Culture, Recirculating Systems, Open-net pen, and Cage Systems, and Flow-through/Raceway (https://www.conserve-energy-future.com/aquaculture-types-benefits-importance.php).

The most common type of aquaculture practiced all over the world on a small to large scale is fish farming. This technique mostly involves the production of fishes that are commercially important and consumed as a source of protein. This technique has become popular in the past few decades owing to the ease and convenience of practicing this type of farming. Compared to many other types of aquaculture, fish farming requires much less labor and instrumentation. The land requirement to develop a profitable yield of fishes is also much smaller than producing an equivalent amount of protein through rearing other life forms. A basic infrastructure and local traditional knowledge base in maintaining the pond conditions and feed ratio usually leads to substantial and most importantly guaranteed fish production. As of now, freshwater fishes have been the principal target in most of the fish farming sectors spread throughout the world. However, to reduce the exclusive dependency on capture fisheries for the wild marine fishes, marine fish farming is emerging as an excellent alternative to supplement open fishing (Chou and Lee 1997; Goldburg and Naylor 2005). Mariculture refers to all the forms of aquaculture that involve the use of seawater. This mode of fishing often utilizes the near-shore open seas. However, the pond system utilizing seawater also comes under this domain. This technique involves the breeding of a suite of marine organisms ranging from mollusks, seaweeds, prawns, shrimps, crabs, and lobsters. Recently Oyinlola et al. (2018) estimated that globally 72,000,000 km^2 of open ocean area is suitable for open mariculture to grow finfishes, mollusks, and crustaceans. According to the latest estimates, almost 112 countries produce seafood from open mariculture worth 65.4 billion USD (Campbell and Pauly 2013). Though inland mariculture is also suitable for most of the farmed organisms, availability of land has always remained a point of concern, as in most cases crucial coastal ecosystems undergo widespread deterioration to facilitate such mode of fish production. Even then, the share of inland mariculture to the global mariculture production is substantial and increasing with time (Shobika 2020). Besides the consumption of fish protein, mariculture also includes the production of several floral and faunal species, which act as a source of materials required for the preparation of medicines, cosmetics, and jewelry. One specialized type of mariculture that involves the use of cages submerged within the open marine water column is the Open-net pen and Cage Systems. In this model, a high density of fish population grows and thrives within cages. These cages need a specific architectural design to withstand the high-energy environment in the open marine water (Chu et al. 2020). This type of model has broadened the scope of marine fish farming in locations, which otherwise remain unutilized.

The growing of algae in enclosed systems as ponds has come up as an excellent avenue to produce biofuels. The world at present is in dire need to reduce its dependency on non-renewable energy. Under such circumstances, biofuels have emerged as a fruitful alternative (Hannon et al. 2010). Moreover, several microalgae are capable of growing in waters, which are not usable in any other sectors. This quality of microalgal strains facilitates us to make effective use of wastewater to produce biofuels (Leite et al. 2013). These microalgae can grow at a very fast rate and fixes atmospheric CO_2 in the form of algal biomass. Thus, the farming of microalgae can provide negative feedback to climate change (Agarwal et al. 2017; Reese et al. 2019). Like microalgal culture, another innovation that has lately emerged is the IMTA (Buck and Langan 2017). In this system, instead of rearing only one target species, a group of ecologically linked species grows, which belong to different trophic levels (Buck et al. 2018). This farming model simulates the natural aquatic environment largely (Kite-Powell 2017). The efficacy of this method lies in the trophic transfer of resources among the various target species. This practice further ensures efficient recycling of nutrients leading to lesser waste generation accompanied by optimum yields (Milhazes-Cunha and Otero 2017).

Inland pond culture is of the common fish farming techniques, which refers to the artificial excavation of ponds and rearing fish under ambient conditions. Usually, these ponds are shallow to facilitate efficient penetration of sunlight, which in turn enables the total water column to maintain a required temperature. Various types of aeration system remain associated with these ponds, which includes both mechanized pump-driven oxygenation and wind shaft system (Mishra 2020; Shaher et al. 2020). In China and the USA, most of the farmed fish exploits this mode of aquaculture practice. Unlike the traditional inland pond culture, there are some other practices, which require a substantial power supply, like the recirculating system. The operation of this model depends on the co-functioning of two separate chambers, one where the fishes thrive and the other acts as a treatment chamber, which continually maintains the water flow and its quality in the fish chamber. This model has an advantage as it can operate with minimum land and water dependence (Verdegem et al. 2006; Eding et al. 2009). Recirculating systems can also mitigate aquaculture-driven eutrophication (Piedrahita 2003). However, the power-intensive nature of this mode is a disadvantage, as it relies exclusively on fossil fuel combustion, and hence indirectly provides positive feedback to climate change (Aubin et al. 2006; Colt et al. 2008). Despite this feature, the number of farms operating in the mode of recirculating systems is increasing throughout the world (Dalsgaard et al. 2013; Badiola et al. 2018). Another less common aquaculture type is the flow-through or raceway model. In this system, long channels with flowing water rear the fishes. Trout culture in many regions of the world follows this model in particular (Welker et al. 2019). This model can effectively utilize the same water mass from two to six times depending on the setup (MacMillan et al. 2003). However, maintaining dissolved oxygen remains a challenge in this model. Even then, the less land and water-intensive nature of this mode are advantageous for some fishes to culture. Thus, we can see that there are several types of aquaculture practiced all over the world. Given, the demand for fish at the present date and its expected rise shortly, all these systems require exhaustive study concerning their role in providing climate change feedbacks and environmental degradation.

8.3 Aquaculture Pond Biogeochemistry

We observed in the preceding sub-section that there are various types of aquaculture ponds. However, from the perspective of this book, and keeping in mind the prevalent types in Sundarbans, we have concentrated on the typical aquaculture fishponds in this chapter. These ponds have a sole purpose and that is to produce fish. However, like any other aquatic system full of life system, these ponds also undergo several biogeochemical processes. Indeed the biogeochemistry of these ponds is largely regulated by human intervention to fulfill the principal agenda of maintaining a profitable yield of fishes. However, some of these practices significantly alter aquatic biogeochemistry and lead to long-term consequences.

The physicochemical status of these aquaculture ponds plays a pivotal role in carrying out fish farming. Several types of fishes require specific physicochemical conditions for their growth. Marine organisms require higher salinity, whereas, freshwater organisms require water of negligible salinity. Similarly, other basic physicochemical parameters like dissolved oxygen, temperature, and pH require species-specific ranges (Boyd and Tucker 1998). The water quality and its characteristic physicochemical conditions depend on the soil profile (Saraswathy et al. 2019). New aquaculture ponds store water just above the unearthed soils. The soil profile of a particular region plays a crucial role in altering the chemistry of the above lying water column. The exchange of materials from the soil to water at the sediment–water interface principally regulates this process. However, as the ponds grow old with consistent operation, a suite of biological and chemical products settle down at the pond bottom, derived from the dead plankton, fish excreta, and external fish feed. This steady deposition leads to the formation of a new sediment layer and cuts of the direct contact between the native soil profile and the water (Munsiri et al. 1995). This new stratified layer of deposits acts as the sediment–water interface and regulates the aquatic chemistry in a completely different way than the original soil profile on which the pond stands. However, most aquaculture ponds practice complete drying of these ponds and cleaning the pond bottom occasionally (Chanda et al. 2019). This approach partially enables the contact between the water and the original soil profile. Thus, the type of pond management can also play a direct role in regulating the aquatic chemistry of these ecosystems.

Temperature plays a fundamental role in rearing fish. Since fishes are ectothermic, temperature regulates their survival rate, overall health, spawning cycle, and most importantly, their yield (Galtsoff 1964; Chamberlain et al. 1980; Avault 1996; Lang et al. 2003). The ambient temperature mainly regulates the temperature of the water column of these ponds. However, practices of indoor aquaculture ponds under manipulated temperature conditions are also prevalent in various places (Lamoureux et al. 2006). The tropical shallow aquaculture ponds experience substantially high temperatures all-round the year (Yang et al. 2018). Temperature plays a critical role in regulating the dissolved gases in the water column and such enhanced temperature often reduces the dissolved oxygen content (Lawson 2013). Maintaining dissolved oxygen above a threshold of 5–6 mg l^{-1} is a challenge to sustain aquaculture farming for most fishes. Low dissolved oxygen affects the fish by reducing the appetite and promoting diseases (Boyd and Hanson 2010). During daytime photosynthetic activities enhance the dissolved oxygen; however, during nighttime, especially in tropical regions, where the temperature remains substantially high, dissolved oxygen concentration falls below the minimum requirement. These situations require artificial aeration to uplift the dissolved oxygen levels.

pH is also a crucial parameter that regulates the wellbeing of fishes. Fishes irrespective of species type, have a mean blood pH of ~7.4 (Wurts and Durborow 1992). Pieces of research unanimously indicated that fishes thrive better in waters having pH close to their blood pH levels (7.0–8.0). However, a pH range of 6.0–9.0 is acceptable for various fishes. pH values beyond this range both in the upper and lower end incurs deleterious effects on fish health that could eventually lead to mortality (Boyd 1979). Due to daytime photosynthesis and night-time respiration, pH exhibits significant diurnal variability. Usually, in low alkalinity ponds, this fluctuation shows a large amplitude (Tucker 1984). To reduce the large-scale alteration of pH, increasing the alkalinity of the ponds by adding lime is one of the viable options. Despite such measures, regular monitoring of pH is essential to ensure the quality of water. Turbidity is another factor that indirectly regulates pond biogeochemistry. Turbidity often arises in these ponds from the wear and tear of the sediment profile coupled with bottom churning driven by wind shear. Some fishes prefer to thrive near the bottom of the ponds and their locomotion disturbs the sediment–water interface, which often leads to a turbid water column. Turbidity directly does not hamper the life cycle of fishes; however, it reduces the dissolved oxygen levels by obstructing sunlight to penetrate. Thus, it often becomes imperative to scale down the turbidity levels. The use of alum, calcium sulfate, and many other flocculants is common in many fish farms (Wu and Boyd 1990).

8.4 Nutrient Dynamics and Pond Productivity

Inorganic nutrients play a critical role in regulating the fish dynamics in the aquaculture ponds. Nutrients stimulate the growth of phytoplankton, which acts as food for the fishes. However, intensive aquaculture practice often leads to over-accumulation of nutrients that lead to deleterious effects (Castillo-Soriano et al. 2013; Hu et al. 2014). Fish and shrimp farming requires external aquafeed input and most of these feeds remain unutilized and undergoes mineralization to give rise to chemical species like nitrate (NO_3^-), nitrite (NO_2^-), ammonia (NH_4^+), and phosphate (PO_4^{3-}) ions (Chen et al. 2016). Excessive nutrient concentrations consequences in uncontrolled net primary production and occasionally, harmful algal blooms (Huang et al. 2016). High nutrient concentration deteriorates the health of several fish and shrimp species. Hu et al. (2014) reported that elevated nutrient concentration leads to the release of corticosteroid hormones in shrimps, which in turn, reduces the growth and yield. Various aquaculture ponds make use of wastewaters, which remain concentrated in nutrients, and fish farming further elevates the concentration. Nearby aquatic bodies like lakes, rivers, estuaries, and oceans often receive the discharge of used water from such aquaculture ponds. Thus, the enhanced nutrient level not only poses a threat to the aquaculture ponds but also to the other adjacent ecosystems (Nora'aini et al. 2005). Under anaerobic conditions, ammonia production takes place in these ponds. Chemoautotrophic bacteria converts the ammonium ions to nitrate and nitrite ions (Camargo et al. 2005). Nitrate ions, when remaining in excess in these ponds interact with the blood hemoglobin of aquatic animals and lead to the formation of methemoglobin. This conversion of hemoglobin leads to acute scarcity of oxygen in the animal body and hence mortality (Camargo et al. 2005). Nitrite ions also form nitrosamine complexes and it is responsible for various forms of cancer (Zhou et al. 2009). Realizing all these factors, minimization of aquaculture waste output became imperative (Bureau and Hua 2010). To accomplish this, research on improving the digestive capability of fishes and enhancing the efficiency of nutrients gained impetus in the last decade (NRC 2011). Proper selection of protein contents, ingredients of aquafeeds, and use of additives like, enzymes are some of the steps taken to mitigate nutrient pollution (Bouwman et al. 2013). However, case-specific generation of aqua-wastes and its release to the immediate ambiance needs meticulous monitoring to make the aquaculture sectors sustainable in the end. The policy managers associated with aquaculture pond operation must rigorously strive to generate a nitrogen and phosphorus budget for their respective farming systems. This approach can enable us to devise proper strategies to curb the release of nutrients and reduce the ill effects of excessive nutrients within the aquaculture system (Leung et al. 1999).

A crucial aspect that is directly associated with the nutrient levels of the aquaculture ponds is the primary productivity. It is a crucial ecological parameter, which gives us an idea about the new organic matter produced through photosynthesis and chemosynthesis by the existing autotrophic communities (Uddin et al. 2018). This parameter not only acts as a proxy of the overall well-being of these aquatic ecosystems but also serves as the primary source of food for the upper trophic chain organism, in this case, the fishes (Bhattacharyya et al. 2020). Maintaining a healthy primary production is a challenging task; however, upon achieving this in the aquaculture ponds, the reliance on the external fish feeds reduces significantly. This can significantly reduce the cost of fish production and at the same time, mitigate the chances of eutrophication (Datta et al. 2019). In this way, the rate of primary production can straightaway reflect a higher fish yield by improving the secondary and tertiary productivity. For extensive and semi-intensive aquaculture practice, primary productivity plays a crucial role. However, the same is not true in the case of intensive farming (Li et al. 2019). IMTA ponds particularly strive hard to utilize this natural process of new organic matter production (Kibria and Haque 2018). Several scientists went a step forward to manipulate the natural process of primary production and utilize this to make the aquaculture ponds a long-term sink of carbon (Ahmed et al. 2017; Zhang et al. 2020). However, excessive primary production often leads to the bloom of harmful algae and introduces toxins into the system (Vasas et al. 2007). Hence, these ponds should be treated like a fragile ecosystem and the management practices should concentrate on maintaining a delicate balance in the rate of primary productivity.

8.5 Pond Management Practices

Unlike the very many natural lakes and ponds, aquaculture ponds have a sole objective and that is to rear and produce fish. Maintaining an optimum fish yield is one of the prime objectives of these ponds. To meet the increasing demand and the expected outcome, the implementation of management practices is essential for these ponds. The management practices vary across the world and it depends on the species farmed (Mishra et al. 2017; Salunke et al. 2020). Aquaculture fish production has received due attention of the scientific community in the last few decades; however, traditional practices still overrule the scientific strategies in various places of the world (Beveridge and Little 2002; Rose et al. 2016), and these practices often deliver a substantial yield. In most cases, these traditional practices were put under the lens and the cruxes of these approaches have been reframed scientifically. According to the Food and Agricultural Organization, pond management includes the following typical steps irrespective of the type of fish farmed. These steps are conditioning of the pond, stocking of the fish species, fertilizing or feeding the ponds, monitoring the water quality, maintenance of the ponds, and finally harvesting the cultured fish.

Before the initiation of the aquafarming, all the water from the ponds is usually drained out and the pond bottom is allowed to dry completely (Boyd et al. 2002). This practice is essential to ensure that all the impurities are washed out of the ponds before the beginning of fresh culture. This step is followed by the application of pesticides. The pesticide application varies from pond to pond depending on the desired output. The role of pesticides is to eliminate the growth of undesirable species, which can outcompete the target species (Ali et al. 2016). The application of tea seed cake and rotenone are common in various parts of the world (Boyd and Massaut 1999). Acid sulfate pond bottoms need special attention. These ponds require repeated water filling and flushing to prevent the increase in disease and fish mortality (Gosavi et al. 2004). Water column pH plays a crucial role in the sustenance of fish within the ponds. To maintain a suitable pH, lime treatment in the pond bottom is practiced all over the world. Agricultural lime and dolomite are two of the preferred compounds used for liming the pond (Emmanuel et al. 2014). The final step of pond preparation includes the use of fertilizers. A mixture of organic as well as inorganic fertilizer is usually applied to the pond bottom. Extensive aquaculture relies heavily on the use of fertilizers, as the growth of phytoplankton, which acts as the natural food for the fishes is exclusively dependent on the fertilizer input (De Silva and Hasan 2007). Intensive aquacultures, on the contrary, do not essentially need the use of fertilizers, as these ponds are dependent on the supply of external fish feeds (Goddard 1996). Once the pond attains a suitable state after accomplishing these steps, the water is gradually allowed to enter until it reaches the required volume. Once the pond is ready, the fish larvae are introduced to the system, and this step is referred to as the stocking of fish. The stocking density is one of the crucial parameters that determine the final yield of fishes. The density of fish, in turn, depends on the size of the larvae, the pond size and volume, and many other nitty–gritty ponds- and species-specific details (Cuenco et al. 1985).

For the growth and sustenance of the fish population, feeding is a must, and its rate varies from species to species. Maize gluten and rice bran are some of the most fish feeds (Hussain et al. 2011). These food products are locally grown within the aquaculture pond sites utilizing a piece of land. Trash fish and fish meal are usually used as a protein source (Munguti et al. 2014). Commercial fish feeds are also available in the market, which is composed of a healthy mixture of mashed fish, blood meal, bones, and shrimp heads, and rice-corn and vitamins. Various food wastes are effectively recycled to prepare this type of commercial aquafeeds (Wong et al. 2016). The composition of the fish feeds in terms of carbohydrate to protein ratio varies among the different food types. Depending on the stage of the aquaculture protein requirement changes in the fish species and accordingly food composition alteration is preferable. The feeding rate also varies at different stages of aquaculture depending on the biomass of the fish (Allan et al. 1995). Besides feeding, the pond water where the fishes thrive also plays a crucial role in maintaining fish health. Firstly, a minimum depth needs to be maintained throughout the fish rearing stage, depending upon the types of fishes reared. The water quality undergoes deterioration throughout all the fish rearing stages. New water from the water source should be introduced to the pond replacing the old stagnant water mass (Yang et al. 2010). Among the various water quality parameters, dissolved oxygen is the crucial most. Especially, in the case of intensive aquafarming, dissolved oxygen levels deplete at a much faster rate and give rise to anaerobic conditions, and stimulates the decomposition of organic matter in the pond bottom (Li et al. 2019). Electricity-operated motor pumps, aerators, and paddle wheels are some of the mechanized operations used to enhance the dissolved oxygen levels. These apparatus are regularly used in those ponds where fertilization is kept at a minimum and the fish rearing does not depend on the photosynthetic activities. The overall pond maintenance requires meticulous and regular monitoring of all the essential aspects like fertilization, liming, pest control, and stock monitoring (Ntsama et al. 2018). The last step of aquafarming that follows these steps is the harvesting of fish. For most of the species, the one-time harvest technique is followed; however, under certain circumstances, periodic harvesting is also preferred. Aside from the critical points discussed above, which come under the purview of pond management practices, there are a lot of other practices carried out in different parts of the world. However, all these techniques and their implementation have a role to play in regulating the biochemistry and the greenhouse gas emission from these ponds.

8.6 Aquaculture Ponds and Greenhouse Gas Emission

To meet the ever-increasing fish demand, various countries are switching over to intensive aquaculture practice, which makes these ponds an active site of anthropogenic greenhouse gas emission (Zacharia et al. 2016). High stocking density coupled with the enhanced application of feeds and fertilizers leads to the emission of greenhouse gases like CO_2, CH_4, and N_2O (Naskar et al. 2020). Improper nutrient management and non-scientific method of feeding can lead to substantial emission of all three greenhouse gases (Singh and Tyagi 2009). The organic matter that generates within the system and the external input undergoes biodegradation and depletes the dissolved oxygen levels, which facilitates methanogenesis and leads to CH4 emission (Krüger et al. 2001). Almost all aquaculture activities tend to decrease the pH levels of these ponds and the mineralization of organic matter to CO_2 is dependent on these processes (Weston et al. 2011). Excessive use of nitrogen-based fertilizer leads to the formation and subsequent emission of N_2O (Marton et al. 2012).

According to the best possible estimates, the amount of fish farmed through aquaculture practices was 101 million tons in the year 2014 (FAO 2014). By the year 2030, this amount is expected to reach 230 Mt (FAO 2016). This increase in fish volume is inevitably going to increase the greenhouse gas emission, especially that of CH_4 and N_2O, which have a global warming potential of 34 and 298 times higher than that of CO_2 (Myhre et al. 2013). However, some of the studies indicated that aquaculture ponds can act as sinks for CO_2, where the rate of primary productivity supersedes the production of CO_2 through heterotrophic respiration (Flickinger et al. 2020). Yuan et al. (2019) carried out an exhaustive metal analysis compiling the CH_4 and N_2O emissions from 21 top fish-producing countries. They estimated a present annual emission of 11 Tg CO_2-eq (in the form of N2O) and 218 Tg CO_2-eq (in the form of CH_4). If the predicted rate of increase in aquaculture pond area holds by the year 2030, the N_2O emissions from these ponds can account for ~5% of the global total N_2O emissions (Paudel et al. 2015). Most of the aquaculture ponds have earthen soil matter in the pond bottoms and their shallow nature facilitates more than 80% of the greenhouse gas emissions that take place from this type of fish farming practice (FAO 2018; Yuan et al. 2019). MacLeod et al. (2020) estimated that in the year 2017, the aquaculture ponds throughout the world emitted 263 million tons of CO_2-equivalent greenhouse gases. UNEP (2018) estimated the global total anthropogenic greenhouse gas emission of 53.5 gigatons of CO_2-equivalent. Thus, it is evident that aquaculture ponds account for almost 0.5% of the total anthropogenic greenhouse gas emissions.

8.7 Characteristic Features of the Aquaculture Ponds of the Sundarbans

The Sundarbans is a part of the world's largest delta, the Ganges–Brahmaputra-Meghna delta. Situated in the coastal periphery and crisscrossed by a network of creeks and estuaries, this region has an abundance of brackish water. The fish farmers of this region practiced small-scale aquaculture for a long time (Dubey et al. 2017). Shrimps, prawns, crabs are some of the species that are cultured under brackish water conditions. Despite being a saline water-dominated region, the Indian carps and few other freshwater fishes are also farmed in some of the ponds. The freshwater ponds are mostly situated inland and away from the coastal periphery. Similarly, most of the brackish water ponds exist near the embankments (Dubey et al. 2017). Sundarbans is known for producing salt-tolerant rice paddies. Integrated rice-fish-prawn farming is also a common practice in Sundarbans. Besides dedicated ponds used exclusively for aquaculture purposes, local people of Sundarbans often rear carps in the homestead ponds. A section of the entire rural population of Sundarbans owns small lands, where they carry out agriculture, mostly rice. Many such low-lying rice paddy fields are ideal for carrying out simultaneous fish farming. The majority of the aquaculture ponds are perennial and these ponds rely on monsoonal rain. Few of these ponds are dependent on both rainwater and groundwater. Besides the Indian Major Carps (*Cirrhinus mrigala*, *Catla catla*, and *Labeo rohita*), some exotic species, like *Ctenopharyngodon idella*, *Barbonymus gonionotus*, *Hypophthalmichthys nobilis*, *Cyprinus carpio*, and

Hypophthalmychthys molitrix are cultured throughout the Indian Sundarbans (Dubey et al. 2017).

8.8 Materials and Methods

Aquaculture ponds emit substantial amounts of greenhouse gases. However, characterizing these emissions from Indian aquaculture ponds is seldom reported. To the best of our knowledge, such studies are absent from the Indian Sundarbans Biosphere Reserve to date. Aquafarming of fishes is emerging as an alternate livelihood in the Sundarbans (Dubey et al. 2016). The local agricultural farmers are steadily opting for aquaculture due to a better profit margin (Philcox et al. 2010; Sundaray et al. 2019). Most of these aquaculture practices are carried out in intensive mode (Ghoshal et al. 2019), and this type of practice can lead to more nutrient pollution and greenhouse gas emission (da Silva et al. 2018). Agricultural fields are also known as hotspots of CH_4 emission. However, a steady switch over from agricultural fields to full-time aquaculture can bring about a net change in the emission potential of this region. Moreover, the ongoing climate change is affecting the Sundarbans from various aspects. The increase in ambient temperature, erratic monsoon pattern, and increased intensities of cyclonic storms can potentially alter the biogeochemistry of the lentic ecosystems, like the aquaculture ponds. Thus, characterizing the greenhouse gas fluxes is of utmost importance, given the present scenario of the Indian Sundarbans. In the present chapter, we took into account the CO_2 and the CH_4 fluxes. N_2O fluxes could not be studied due to instrumental deficiency. Measurement of fluxes through deploying eddy-covariance setups has become common; however, this was beyond the ambit of our study. We measured the fluxes through the gradient technique. We measured the partial pressure of CO_2 and CH_4 in the surface water [pCO_2(water) and pCH_4(water), respectively]. We also measured the partial pressure of CO_2 and CH_4 in the ambient air [pCO_2(air) and pCH_4(air), respectively]. The difference in partial pressure between the two phases for the respective gases enabled us to compute the fluxes. The gas transfer velocities based on wind speed were implemented to derive the fluxes. Besides these fluxes, a suite of biogeochemical parameters was also measured, and the relationship between these parameters and the fluxes was analyzed. The detailed methodology is given in the following sub-sections.

8.8.1 Sampling Stations

We selected two aquaculture ponds situated at the northern and the southern part of the Indian Sundarbans, at two different administrative blocks of the South 24 Paraganas district of the Sundarban Biosphere Reserve, namely Minakhan and Namkhana (hereafter referred to as P1 and P2, respectively) (Fig. 8.1). The Minakhan community development block is situated at the northern end of the Indian Sundarban Biosphere

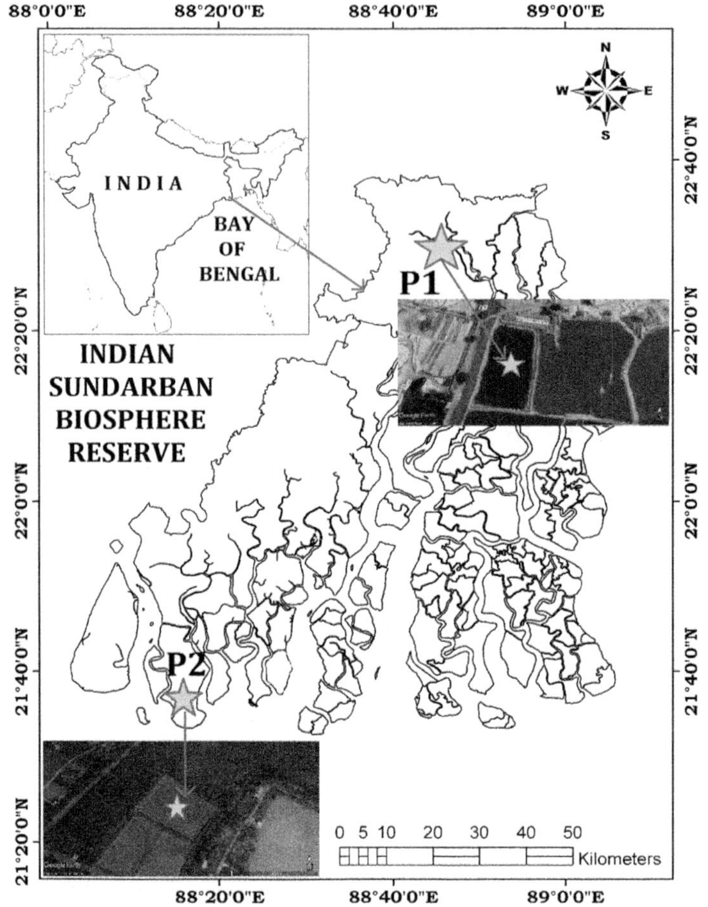

Fig. 8.1 The study area map showing the location of the two ponds selected for the present study in the Indian Sundarban Biosphere Reserve

Reserve near the bank of the Bidhyadhari River. From the 1970s, the residents of this block were mainly engaged with agriculture (Naskar 1985). However, the Department of Irrigation, Govt. of West Bengal constructed dikes and embankments to prevent flooding.

Since then the local people started switching over to aquaculture by utilizing the brackish water of Bidhyadhari River for this purpose (Bunting et al. 2017). Rice-shrimp farming is also common in this region and at present, a substantial number of aquaculture ponds are operating in Minakhan Block throughout the year (Mondal and Bandyopadhyay 2015). The Namkhana community development block is situated at the southern end of the Biosphere Reserve and it faces the Bay of Bengal in the south. A substantial number of people of this block are engaged with marine capture fisheries. Despite that, the number of aquaculture ponds has increased in this block over the last few decades (Som and Gupta 2010). Contrary to the northern part's aquaculture ponds, the aquaculture ponds of this block implement newer scientific technologies; whereas, in the northern part, people still rely on the traditional practices (Goswami et al. 2006). Compared to the two different ponds, the pond of Namkhana is much more prone to climate anomalies, especially the tropical cyclones, owing to its proximity to the Bay of Bengal. *Penaeus monodon* (Tiger prawn) was cultured in both of these ponds. We deliberately chose the same species to ascertain the spatial variability in fluxes within the Sundarban Biosphere Reserve. Ponds of the Minakhan area are mostly controlled by operating sluice gates which allow the brackish water of Bidhyadhari River to enter the ponds. In the Minakhan Block ponds, traditionally fishermen do not maintain any specific stocking density and utilize natural food, and periodic harvesting is carried out during full moon and new moon (Alagarswamy 1995; De Roy 2012). The ponds of Namkhana also allow brackish water to enter the ponds; however, these ponds were often treated with external fish feeds, and practices comparatively more intensive farming than the ponds of Minakhan.

8.8.2 Sampling Strategy

The sampling in both of these ponds was carried out during the years 2016, 2017, and 2018. This part of the world has broadly three seasons, namely the pre-monsoon season (February to May), the monsoon season (June to September), and the post-monsoon season (October to January). We sampled in each of the three seasons over the years. However, we could not cover all the months of the same calendar year. Instead, we sampled in different months of a particular season in the three respective years to characterize the short-term temporal variability in the physicochemical parameters as well as the CO_2 and CH_4 fluxes. In this way, we sampled in all the twelve months, but of different calendar years. We presented the seasonal mean data taking into account the measurements carried out in all the months of that particular season. We assumed that there was no significant inter-annual variability in the measured parameters.

We sampled in both of the ponds once a month. The samples were collected twice; once during the daytime (between 10:00 a.m. and 12:00 p.m.) and once during the nighttime (between 06:00 p.m. and 08:00 p.m.). We collected both the day and night samples to analyze the role of the presence and absence of sunlight on the CO_2 and CH_4 flux dynamics. In the year 2016, we sampled in March, May, September, and November. Similarly, in the year 2017, we sampled in January, April, August, and October. Finally, in the year 2018, we sampled in February, June, July, and December. We collected three replicate samples from each of the ponds with the help of a precleaned polypropylene glass bottle from the surface water. We collected the water samples from at least a meter distance from the edge of the ponds.

8.8.3 Biogeochemical Measurements

Most of the physicochemical parameters were measured in situ by deploying standard probes. However, some of the parameters require laboratory analysis. For such parameters, samples were collected in specific containers and were subsequently preserved by adding necessary preservatives for each parameter. The sample bottles were transferred in freezing conditions to the laboratory. Adequate care was taken to calibrate and standardize the sensors and probes used in this study. It was ensured that all the reagents used for the different laboratory analyses were of analytic grade.

8.8.3.1 In-Situ Measurements

The aquaculture ponds sampled in this study had substantial salinity throughout the year, hence we did not report electrical conductivity. The salinity of the surface water of these ponds was measured with the help of a Multikit (Multi 340 i set, WTW, Germany) fitted with a standard probe (Tetracon 325, WTW, Germany). The in-situ water temperature was measured with the same sensor. The sensors had a measurement resolution of 0.1 and 0.1 °C, for salinity and water temperature, respectively. The analytical precision of the sensors was estimated by measuring a few of the same samples repeatedly for ten times. The precision of measurements for salinity was ±0.2 and for water temperature, it was ±0.1 °C. The dissolved oxygen (DO) concentration was measured with the help of a standard probe (FiveGo portable F4 Dissolved Oxygen meter, Mettler Toledo, Switzerland) having a measurement resolution of 0.01 mg l^{-1}. The DO was measured once during each sampling endeavor following the modified Winkler's titrimetric method. The offset of the sensor's DO measurement readings were checked and calibrated with the help of the results obtained from Winkler's method. The DO sensor was regularly calibrated with the help of boiled water (having zero DO) and fully oxygenated water (having a hundred percent oxygen saturation). The DO sensor exhibited an accuracy of ±1%. The saturation state of DO was also measured using the empirical formula of Weiss (1970), and from the difference between the in-situ oxygen concentrations and their corresponding saturation state, the apparent oxygen utilization (AOU) was computed.

The pH was measured in-situ with the help of a micro-pH meter equipped with standard glass electrodes (Orion PerpHecT ROSS Combination pH Microelectrode) fitted to a data logger [Thermo Scientific, U.S.A.]. The analytical resolution of the pH meter was 0.001. The glass electrodes for pH measurements were calibrated (once before the sampling endeavor begun for each month and once after the completion) using the NBS scale technical buffers (Merck, Germany) at a controlled temperature of 25 °C. A nephelometer was used (Eutech TN-100, Singapore) to measure the turbidity of the water samples. With the help of a standard underwater light sensor (LI-192SA, Li-Cor, USA; analytical resolution 0.1 μ mol m^{-2} s^{-1}) fitted to a data logger (Li-250A, Li-Cor, USA), the underwater photosynthetically active radiation

(UWPAR) was measured. A portable weather station (WS-2350, La Crosse Technology, Wisconsin, USA) was carried and installed in the sampling spots to measure the ambient temperature, atmospheric pressure, and wind velocity. The standard light–dark bottle method was implemented by incubating the samples for 12 h from dawn to dusk to measure the gross primary productivity (GPP) and community respiration (CR) of the surface water by measuring the changes in the DO concentrations with standard probes (mentioned above). We deployed three replicate sets of bottles for GPP and CR measurements.

8.8.3.2 PCO_2 and Associated Measurements

100 ml water samples were quickly filtered through GF/C filter papers and the samples were preserved for total alkalinity (TAlk) analysis (Frankignoulle et al. 1996) by poisoning each sample with 20 μl saturated $HgCl_2$ solution (7.2 g $HgCl_2$ in 100 ml distilled water) (Kattner 1999). Usually, it is not advisable to filter the water samples for TAlk analysis; however, turbid waters often change the carbonate chemistry of the water with time. Thus, we filtered the samples and took special care to complete the filtration quickly to avoid contact with ambient air's CO_2. An automated titrator (905 Titrando, Metrohm, Switzerland) was used to measure the TAlk following a closed chamber titration. pCO_2(water), dissolved inorganic carbon (DIC), and hydroxyl ion (OH^-) concentration were estimated from the water temperature, atmospheric pressure, TAlk, and pH data with the help of the software CO_2SYS.EXE (Lewis and Wallace 1998). Oceanographers and marine scientists mostly use this software to estimate pCO_2 in seawater. However, Chanda et al. (2020) have successfully used this software to estimate the pCO_2(water) in the urban tidal river of Hooghly, which flows by the twin cities of Kolkata and Howrah, situated in the north of the Sundarban Biosphere Reserve. Chanda et al. (2019) estimated the pCO_2(water) in the aquaculture ponds of the East Kolkata Wetlands by implementing this software. Since the ponds were of brackish water, we used the dissociation constants, K_1 and K_2, of Roy et al. (1993). We used the corrections of Khoo et al. (1977) for sulfate concentrations. We used a nondispersive infrared (NDIR) sensor (Li-840A; Li-COR, USA) to measure the CO_2 concentration in the ambient air. We calibrated the instrument before each sampling campaign with certified reference standard gases of known concentrations of CO_2 (0, 300, and 600 ppm CO_2 concentration) in N_2 as base gas (Chemtron Science Laboratories, India). We converted the mole fraction of CO_2 in ambient air to the partial pressure of air [pCO_2(air)] by using the ambient temperature and atmospheric pressure, and the virial equation of state (Weiss 1974). 1 l of a water sample was collected from the pond surface, and it was stored in an amber-colored container. The samples were preserved under ice-cold conditions and were sent back to the laboratory within 24 h for chlorophyll-a (chl-a) analysis. We kept the samples in the dark in a freezer until further analysis. We implemented standard spectrophotometric protocols (Parson et al. 1992) to measure chl-a. The precision of the chl-a measurement was ± 0.02 mg m^{-3}. We used standard lyophilized chlorophyll-a

(Sigma-Aldrich, Merck, Germany) for the preparation of standard stock solutions and calibration of the spectrophotometer.

8.8.3.3 pCH_4 Measurements

Surface water samples were collected in 40 ml glass vials each fitted with a rubber septum and leaving no headspace. The water samples were poisoned by adding 20 µl of 8% $HgCl_2$ solution to stop all the biological activities. Using a pre-cleaned syringe, 20 ml water was removed from the glass vial and the remaining 20 ml water sample was purged with nitrogen gas (99.99% pure) in the laboratory. The glass vials were kept undisturbed for 2 h to equilibrate. 5 ml of the headspace gas was collected by a Hamilton syringe and CH_4 concentration was analyzed using a gas chromatograph (GC) (Systronics GC–8205) having a mean relative uncertainty of ±2.9%. Pure nitrogen gas was used as the carrier gas and the retention time for CH_4 was 0.623 min. The moisture was removed by raising the injector temperature to ~105 °C. Certified standard methane gas (0 and 20 ppm) was used for calibration of the GC. Ambient air samples above the water surface were collected in glass bulbs fitted to a battery-operated air pump. The stopcock of the glass bulbs was sealed by parafilm to prevent leakage. Air samples were also brought to the laboratory within 24 h of sampling and the CH_4 concentration was analyzed in GC by using the above-mentioned injection method.

8.8.3.4 Estimation of Air–Water CO_2 Exchange

For the sake of the CO_2 flux computation across the air–water interface, we converted the pCO_2(water) and the pCO_2(air) to concentrations of carbon dioxide in water (CO_2wc) and air (CO_2ac), respectively, following the equations given below (Weiss 1974; Anderson 2002).

$$CO_2wc = K_H \times pCO_2(water).$$

$$CO_2ac = K_H \times pCO_2(air).$$

where K_H stands for the gas partition constant of CO_2 in freshwater at the sampling temperature. We computed K_H (in mole l^{-1} atm^{-1}) following the equation below.

$$\ln K_H = -58.0931 + 90.5069 \times (100/T_K) + 22.294 \times \ln (T_K/100).$$

where T_K denotes the water temperature in the unit of Kelvin (Weiss 1974).

We computed the air–water CO_2 exchange rate (flux) according to the formula furnished in MacIntyre et al. (1995) as shown below.

$$\text{Air–water } CO_2 \text{ flux } [FCO_2] = k_x \times (CO_2wc - CO_2ac).$$

where k_x stands for the gas transfer coefficient (cm h^{-1}) that we estimated according to the formula given by Wanninkhof (1992).

$k_x = k_{600} \times (600/Sc)^x$.

where Sc denotes the Schmidt number (for CO_2). The Sc depends on the water temperature (T, in the unit of Kelvin) according to the formula given below.

We calculated k_{600} using the wind speed at a height of 10 m above the pond interface (U_{10}), according to Cole and Caraco (1998), where, the magnitude 'x' is equal to 0.5 and 0.66 for wind speeds >3 m s^{-1} and $\leq$3 m s^{-1}, respectively. We measured the wind speed with the help of a weather station at 1 m height from the pond interface and converted the speed magnitudes for 10 m height, using the empirical wind profile equations given by Kondo (2000).

8.8.3.5 Estimation of Air–Water CH_4 Exchange

pCH_4(water) and pCH_4(air) were converted to concentration of methane in water (CH_4wc) and air (CH_4ac) according to the following equations (Morel 1983; Lide 2007):

$CH_4wc = K_H \times pCH_4(water)$.

$CH_4ac = K_H \times pCH_4(air)$.

$\ln K_H = -115.6477 + 155.5756/(T_K/100) + 65.2553 \times \ln (T_K/100) - 6.1698 \times (T_K/100)$.

where K_H denotes the gas partition constant of CH_4 in the water at sampling temperature, expressed in mole l^{-1} atm^{-1}, and T_K denotes the temperature in Kelvin. The air–water CH_4 flux was calculated according to equation (MacIntyre et al. 1995):

CH_4 Flux (mg m^{-2} h^{-1}) = k_x (CH_4wc − CH_4ac).

where, k_x stands for the mass transfer coefficient (cm h^{-1}) and it is computed according to equation (Wanninkhof 1992):

$k_x = k_{600} \times (S_c/600)^{-x}$.

where S_c is the Schmidt number for CH_4 and it is dependent on water temperature according to the equation:

$S_c = 1897.8 - 114.28 \times T + 3.290 \times T^2 - 0.039061 \times T^3$.

k_{600} is computed from the wind speed (U_{10}) and 'x' is equal to 0.66 for wind speed $\leq$3 m s^{-1} and is equal to 0.5 for wind speed >3 m s^{-1} (Cole and Caraco 1998).

$k_{600} = 2.07 + (0.215 \times U_{10}^{1.7})$.

8.8.4 Statistical Analyses

We used the Statistical Product and Service Solutions software (SPSS version 16.0, Inc., USA) and Microsoft Excel for Windows 2010 to carry out all the statistical analyses and prepare the graphical illustrations for this chapter. We examined whether the arithmetic means of pCO_2(water), CO_2 fluxes, pCH_4(water), CH_4 fluxes, and other biogeochemical parameters were significantly different or not between the two ponds and across all the seasons by carrying out the Student's t-test and the one-way analysis of variance test (ANOVA; F-test), respectively. We conducted a posthoc Tukey's honest significant difference (HSD) test to specify the difference in seasonal mean of the biogeochemical parameters if any. To analyze the inter-relationship between the pCO_2(water) or pCH_4(water) and the other regulating factors, we calculated the Pearson correlation coefficient. We considered the outcomes of these tests statistically significant at $p \leq 0.05$.

8.9 Results and Discussion

8.9.1 Variability of Physicochemical Parameters

The annual mean water temperature was 31.0 ± 4.4 °C and 30.2 ± 4.4 °C in P1 and P2, respectively, with no significant difference ($p > 0.05$) amongst each other (Table 8.1). Seasonally, in both of the ponds, the highest mean temperature was observed during the monsoon season (34.3–35.2 °C), followed by the pre-monsoon season (30.0–30.1 °C), and the post-monsoon season (26.4–27.8 °C) (Fig. 8.2). The depth and the area of the ponds were almost the same. Thus, the land effect played a similar role in both of the ponds in regulating the temperature. The variability in water temperature is usually governed by the variation in ambient temperature. Though the summer months of April and May are a part of the pre-monsoon season, the ambient temperature remains consistently high throughout the monsoon months and thus, leading to the high temperature during this season. The post-monsoon season covers the winter months and hence, the lowest seasonal mean is observed during this season. There was a significant difference in salinity between the two ponds.

The annual mean salinity in P1 was 6.5 ± 1.2, whereas in P2 it was 24.2 ± 2.1. Both of these ponds used the adjacent estuarine waters for carrying out pisciculture. Pond P1 is located in the North where substantial freshwater inputs from the Bidyadhari River coupled with run-offs from the upstream reduces the salinity significantly. On the contrary, P2 is situated in the down south of the SBR and close to the Bay of Bengal. The salinity in these estuarine channels remains significantly higher than those situated in the north. Though the admixture of estuarine and freshwater is maintained artificially by the fish farmers throughout the year, a slight reduction in salinity during the monsoon season was observed in both of the ponds followed

Table 8.1 The annual mean ± standard deviation along with the range of biogeochemical parameters observed in the four ponds including the air–water CO_2 and CH_4 fluxes

Parameters	P1	P2
Water temperature (°C)	31.0 ± 4.4 (23.7–35.7)	30.2 ± 4.4 (23.5–34.8)
Salinity	6.5 ± 1.2 (4.2–8.2)	24.2 ± 2.1 (20.6–27.9)
Dissolved oxygen (mg l^{-1})	4.0 ± 0.4 (3.4–4.8)	5.8 ± 0.4 (5.0–6.7)
Oxygen saturation (mg l^{-1})	6.8 ± 0.5 (6.3–7.7)	6.9 ± 0.5 (6.4–7.7)
Turbidity (NTU)	44.8 ± 5.1 (36.7–53.4)	23.8 ± 3.0 (19.4–30.1)
BOD (mg l^{-1})	8.3 ± 0.8 (6.9–9.4)	4.7 ± 0.5 (3.9–5.2)
UWPAR (μmol m^{-2} s^{-1})	6.0 ± 1.4 (4.1–8.5)	6.9 ± 0.9 (5.7–8.7)
Chl-a (mg m^{-3})	48.3 ± 13.2 (28.0–69.0)	17.4 ± 2.7 (13.0—22.0)
GPP (gO_2 m^{-2} d^{-1})	13.0 ± 1.7 (10.1–16.1)	8.0 ± 0.7 (6.8–9.4)
CR (gO_2 m^{-2} d^{-1})	25.6 ± 5.8 (17.0 – 35.9)	13.2 ± 2.4 (10.2–17.6)
pH	7.759 ± 0.072 (7.633–7.893)	8.103 ± 0.064 (7.995–8.209)
Total alkalinity (μmol kg^{-1})	2371 ± 123 (2174–2603)	2595 ± 166 (2353–2888)
Dissolved inorganic carbon (μmol kg^{-1})	2369 ± 116 (2165–2577)	2348 ± 163 (2116–2610)
Hydroxyl ion (μmol kg^{-1})	1.7 ± 0.6 (0.8–2.5)	6.8 ± 1.7 (3.7–10.3)
pCO_2(water) (μatm)	1931 ± 297 (1342–2488)	665 ± 148 (455–962)
CO_2 flux (μmol m^{-2} h^{-1})	2991 ± 549 (1923–3785)	519 ± 318 (94–1131)
pCH_4(water) (μatm)	5.53 ± 0.61 (4.43–6.42)	4.13 ± 0.56 (3.03–5.02)
CH_4 flux (mg m^{-2} h^{-1})	13.53 ± 5.69 (5.79–22.07)	8.55 ± 4.32 (2.64–15.15)

by a steady increase in the post-monsoon and the pre-monsoon seasons. During the pre-monsoon season, the salinity was highest and this could be due to the high rate of evaporation during the summer months. Like salinity, there was a statistically significant difference in DO levels between the two ponds. P1 (4.0 ± 0.4 mg l^{-1}) exhibited a much lower annual mean DO than that observed in P2 (5.8 ± 0.4 mg l^{-1}). The oxygen saturation levels in both of the ponds, however, showed no significant difference. This observation showed that P1 was much less saturated with DO than P2. The monsoon season recorded the lowest mean DO in both of the ponds, followed by the post-monsoon season, and the highest was observed during the pre-monsoon season. Light limitation could have led to the low DO in the monsoon season, whereas during the pre-monsoon season, a high rate of insolation might have led to higher DO production in the surface layers by favoring autotrophic processes. The annual mean turbidity was almost two times in P1 (44.8 ± 5.1 NTU) compared to that observed in P2 (23.8 ± 3.0 NTU). This could be because the estuarine waters near P1 bear a significantly higher anthropogenic load that has substantially high suspended particulate matters. In the downstream stations near P2, a substantial part of these suspended matters settle down and a dilution with seawater leads to an overall reduction in turbidity. This hypothesis was justified by the variability of BOD observed between the two ponds. The annual mean BOD was as high as 8.3 ± 0.8 mg l^{-1} in

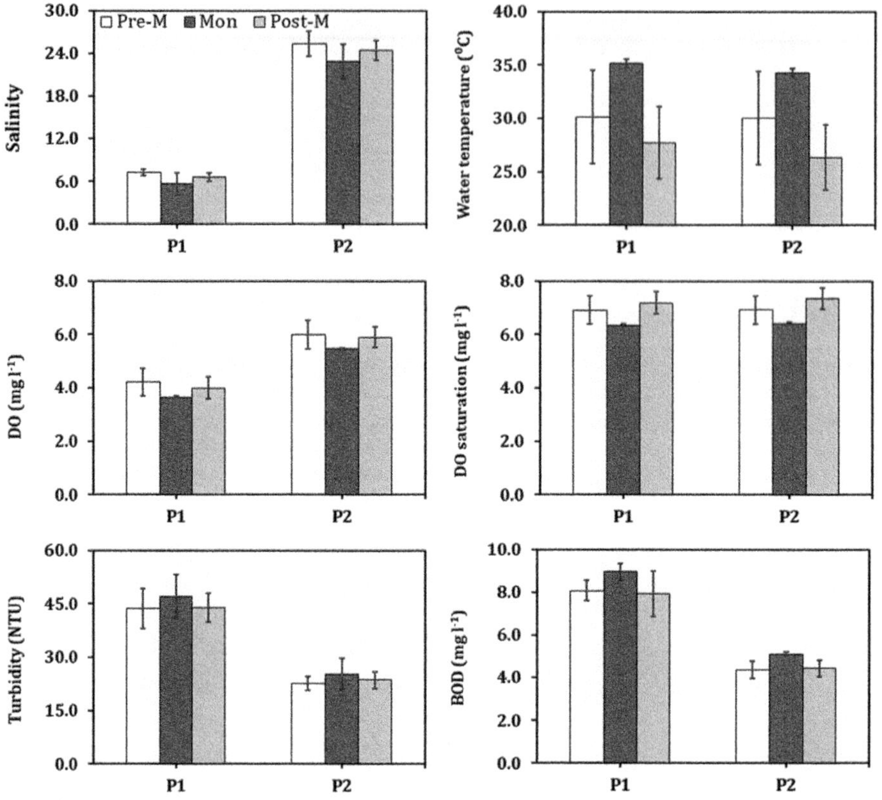

Fig. 8.2 The seasonal variability of salinity, water temperature, dissolved oxygen, dissolved oxygen saturation level, turbidity, and BOD observed in the selected two ponds

P1, whereas in P2 it was 4.7 ± 0.5 mg l^{-1}. The high BOD can be considered here as a proxy of anthropogenic load in the estuarine water. The domestic discharge from the metropolis of Kolkata ends in the Bidhyadhari channel and thus, the estuarine water near P1 remains rich in both organic matter and suspended particles, which led to the high BOD and turbidity in these waters. Near P2, the mixing with seawater reduced both the turbidity and the BOD levels.

8.9.2 Variability of Primary Productivity Parameters

The magnitudes of UWPAR help us understand the suitability of the water surface environment to carry out primary autotrophic processes, as it indicates the presence of photosynthetically active radiation underwater, which in turn depends on the clarity of the water column and the rate of insolation. In this study, P1 (6.0 ± 1.4 μmol m^{-2} s^{-1}) exhibited a slightly lower annual mean UWPAR compared to that observed in

P2 (6.9 ± 0.9 μmol m^{-2} s^{-1}); however, the difference was statistically significant ($p < 0.05$). The lower turbidity in P2 could have facilitated the higher UWPAR and vice-versa in the case of P1 since both of the ponds experience the same rate of insolation. Unlike UWPAR, there was a stark difference in chl-a concentrations between the two ponds. P1 (48.3 ± 13.2 mg m^{-3}) recorded significantly high chl-a concentrations compared to that in P2 (17.4 ± 2.7 mg m^{-3}). The adjacent estuarine water that is utilized in P1 is expected to be rich in nutrients as this region receives substantial runoffs from the northern catchments along with a steady load of anthropogenic discharge.

This excessive presence of nutrients might have facilitated the higher primary production in P1 and hence, the higher concentration of chl-a. On the contrary, in P2, the nutrient concentrations were significantly diluted with the nutrient-lean seawater, and thus the primary productivity was low. The annual mean magnitudes of GPP served as a piece of strong evidence in favor of this assumption. The annual mean GPP was substantially high in P1 (13.0 ± 1.7 gO_2 m^{-2} d^{-1}) compared to that observed in P2 (8.0 ± 0.7 gO_2 m^{-2} d^{-1}). Even though P1 exhibited a higher GPP than P2, the former was more net heterotrophic than that of the latter, as CR in P1 (25.6 ± 5.8 gO_2 m^{-2} d^{-1}) was almost twofold than that observed in P2 (13.2 ± 2.4 gO_2 m^{-2} d^{-1}). This observation is justified by the low DO and high BOD and turbidity levels in P1, which shows that P1 is comparatively more anaerobic than P2, and thus promoted net heterotrophy to a greater extent than that of P2. The seasonal variability of the primary productivity parameters is illustrated in Fig. 8.3.

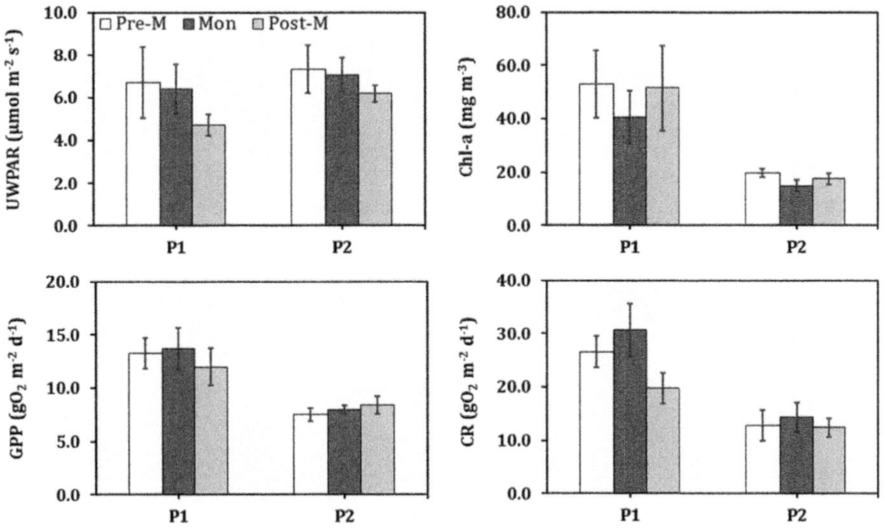

Fig. 8.3 The seasonal variability of the primary productivity parameters observed in the two selected ponds

8.9.3 Variability in Carbonate Chemistry Parameters

Throughout the annual cycle both of the ponds were alkaline (pH < 7.0); however, the pH levels in P1 (7.759 ± 0.072) were substantially lower than that observed in P2 (8.103 ± 0.064).

The low pH in P1 could be also attributed to the BOD-rich anthropogenic load in the adjacent estuaries and the high pH in P2 due to the predominance of seawater. As both of the ponds occasionally undergo lime treatment there was no trend in the seasonal differences, as the pH is often regulated by the fishermen (through lime treatment) as per their traditional knowledge. The TAlk was significantly lower in P1 (2371 ± 123 μmol kg^{-1}) compared to that observed in P2 (2595 ± 166 μmol kg^{-1}); however, there was no statistically significant difference in the estimated DIC between the two ponds (2369 ± 116 μmol kg^{-1} and 2348 ± 163 μmol kg^{-1} in P1 and P2, respectively). Unlike DIC, there was a stark difference in OH^- ion concentrations between the two ponds. The OH^- ion concentrations were significantly lower in P1 (1.7 ± 0.6 μmol kg^{-1}) than that in P2 (6.8 ± 1.7 μmol kg^{-1}). This difference in hydroxyl ion concentration could have originated due to the differential lime treatment in the two ponds. The dominance of seawater in P2 could have led to a marginal increase in pH; however, the degree of difference indicated an enhanced use of lime in P2 compared to that in P1. The seasonal variability of the carbonate chemistry parameters is illustrated in Fig. 8.4.

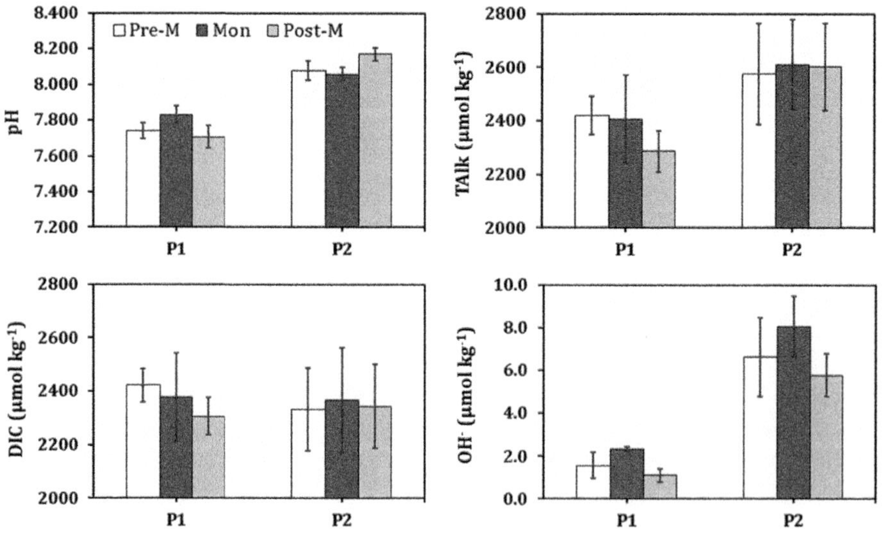

Fig. 8.4 The seasonal variability of the carbonate chemistry parameters observed in the two selected ponds for the present study

8.9.4 Variability of pCO_2(Water) and Air–Water CO_2 Flux

The present study indicated that both P1 and P2 were always supersaturated with CO_2 in comparison to the atmospheric CO_2 concentration; however, the degree of supersaturation varied significantly among the two ponds. The annual mean pCO_2(water) was 1931 ± 297 μatm and 665 ± 148 μatm in P1 and P2, respectively. The seasonal variability in pCO_2(water) followed a contrasting trend in the two ponds (Fig. 8.5). In P1, the highest pCO_2(water) was observed during the pre-monsoon season, followed by the post-monsoon, and the lowest was observed during the monsoon. This observation shows that the high pCO_2(water) was diluted by rainfall during the monsoon season in P1. However, in P2, the highest seasonal mean pCO_2(water) was observed during the monsoon, followed by the pre-monsoon, and the lowest was observed during the post-monsoon. This observation leads us to infer that during the monsoon season, the runoffs from the upstream enhanced the pCO_2(water) in the adjacent estuarine regions of P2, which in turn when entered the aquaculture ponds led to an enhanced pCO_2(water) compared to that observed in the other two seasons.

In both of the ponds, the pCO_2(air) varied over a much small range compared to that of pCO_2(water), and thus, the air–water CO_2 flux mimicked the variability of the latter. The annual mean air–water CO_2 flux was 2991 ± 549 μmol m^{-2} h^{-1} and 519 ± 318 μmol m^{-2} h^{-1} in P1 and P2, respectively.

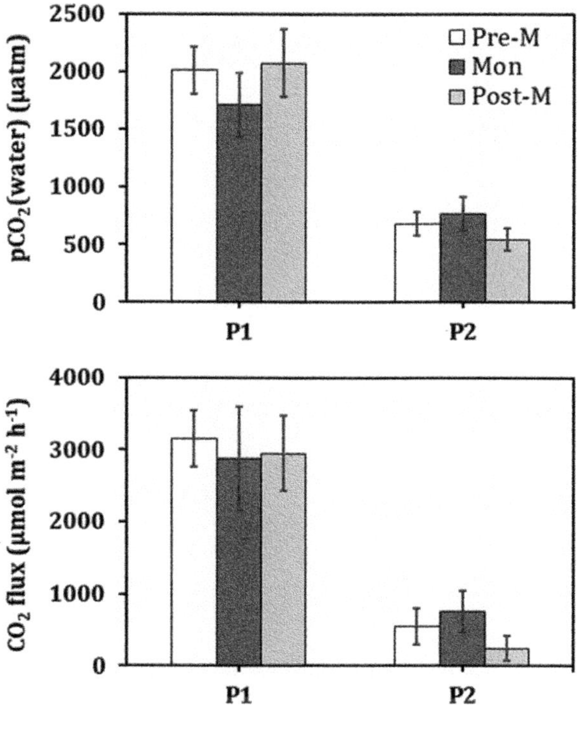

Fig. 8.5 The seasonal variability of pCO_2(water) and air–water CO_2 flux in the two selected ponds

8.9.5 Variability of pCH_4(Water) and Air–Water CH_4 Flux

Like in the case of pCO_2(water), both of the ponds were supersaturated with CH_4 in comparison to the atmospheric CH_4 concentration; however, the annual mean pCH_4(water) was comparatively higher in P1 (5.53 ± 0.61 μatm) than that of P2 (4.13 ± 0.56 μatm). In this case, also, the pCH_4(air) showed much less variability than that of pCH_4(water) and hence, the variability of air–water CH_4 fluxes mimicked the variation of pCH_4(water) in both of the ponds. P1 and P2 exhibited an annual mean air–water CH_4 flux of 13.53 ± 5.69 mg m^{-2} h^{-1} and 8.55 ± 4.32 mg m^{-2} h^{-1}, respectively. The monsoon season recorded the highest pCH_4(water) and hence, the highest air–water CH_4 fluxes in both of the ponds, whereas in the other two seasons, there was no significant difference (Fig. 8.6). This observation showed that during the monsoon season, which is the most unfavorable time for photosynthesis, net heterotrophy prevailed with low DO levels, which in turn favored the methanogens and dampened the methanotrophic activities. The increase in BOD during the monsoon season provided more organic substrates for the methane-producing microbes to act upon. Thus, it can be inferred that conditions that led to higher effluxes of CO_2 also favored higher CH_4 emission.

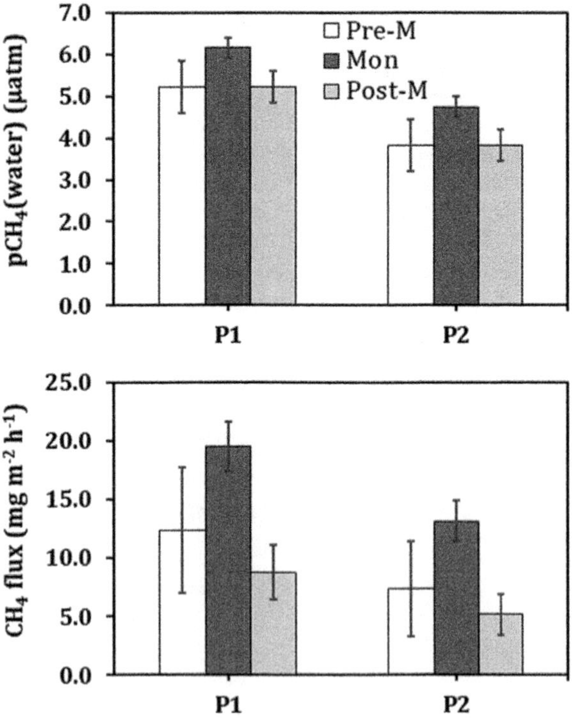

Fig. 8.6 The seasonal variability of pCH_4(water) and air–water CH_4 flux in the two selected ponds

8.9.6 Relationship Between pCO_2(Water) and Other Biogeochemical Variables

Water temperature, electrical conductivity, dissolved oxygen, oxygen saturation, turbidity, UWPAR, chl-a, GPP, CR, NPP, and BOD were correlated with pCO_2(water) to identify the potential regulators of the air–water CO_2 fluxes (Table 8.2). Both of the ponds exhibited a significant positive relationship between pCO_2(water) and water temperature. This observation showed that a higher temperature could have promoted the degradation of the organic matter, which in turn, the net heterotrophy. With the ongoing climate change, the ambient temperature in the future is expected to increase, and thus this positive relationship indicates that these ponds are going to emit more CO_2 if the temperature keeps on increasing. Salinity did not show any significant relationship with pCO_2(water) in any of the ponds, and neither did the DO levels. However, only P2 exhibited a negative relationship with oxygen saturation, which implies that a higher degree of autotrophy can reduce the dissolved CO_2 in the pond water. Turbidity showed a significant positive relationship with pCO_2(water) in P2, which indicates that a higher abundance of suspended particulate matter hampered the autotrophic processes. However, UWPAR did not show any significant relationship with pCO_2(water) in any of the ponds. However, in P1, chl-a concentration exhibited a positive relationship with pCO_2(water) which indicates that a higher chl-a instead of enhancing autotrophic CO_2 reduction, aided in the increase in net community respiration. GPP did not show any significant relationship with pCO_2(water) in any of the ponds. This observation indicates that the GPP depends on multiple factors and that the dissolved CO_2 concentration does not exclusively depend on the biological processes. However, CR and BOD exhibited a positive relationship with pCO_2(water) in P2. This observation indicates that a

Table 8.2 The relationship between pCO_2(water) and other biogeochemical parameters depicted through the Pearson correlation coefficient

Parameters	P1	P2
Water temperature	**$r = 0.606, p < 0.05$**	**$r = 0.739, p < 0.05$**
Salinity	$r = -0.410, p > 0.05$	$r = -0.450, p > 0.05$
Dissolved oxygen	$r = 0.512, p > 0.05$	$r = -0.349, p > 0.05$
Oxygen saturation	$r = 0.387, p > 0.05$	**$r = -0.731, p < 0.05$**
Turbidity	$r = -0.354, p > 0.05$	**$r = 0.529, p < 0.05$**
UWPAR	$r = -0.049, p > 0.05$	$r = 0.304, p > 0.05$
Chl-a	**$r = 0.608, p < 0.05$**	$r = -0.423, p > 0.05$
GPP	$r = -0.069, p > 0.05$	$r = 0.264, p > 0.05$
CR	$r = -0.519, p > 0.05$	**$r = 0.761, p < 0.05$**
NPP	$r = 0.519, p > 0.05$	**$r = -0.789, p < 0.05$**
BOD	$r = -0.203, p > 0.05$	**$r = 0.645, p < 0.05$**

higher BOD provided more organic substrate which upon degradation led to mineralization of organic carbon and shifted the carbonate chemistry equilibrium towards pCO_2(water). A positive relationship with CR and an equivalent negative relationship with NPP showed that net heterotrophy played a crucial role in enhancing the pCO_2(water) load. However, such biological control was absent in P1. Thus, it can be inferred that the drivers of pCO_2(water) vary from pond to pond.

8.9.7 Relationship Between pCH_4(Water) and Other Biogeochemical Variables

Water temperature, electrical conductivity, dissolved oxygen, oxygen saturation, turbidity, UWPAR, chl-a, GPP, CR, NPP, and BOD were correlated with pCH_4(water) to identify the potential regulators of the air–water CH_4 fluxes (Table 8.3). Like in the case of pCO_2(water), pCH_4(water) also showed a significant positive relationship with water temperature. This observation again shows that a higher temperature facilitates the anaerobic degradation of organic matter and favors the methanogens to produce more CH_4. Thus, keeping the ongoing climate change in mind it can be inferred that in the future, the rate of CH_4 emission would increase hand in hand along with the rate of CO_2 emission leading to positive feedback to climate change. In this case, also, salinity did not show any significant relationship. However, both dissolved oxygen levels and oxygen saturation levels exhibited a negative relationship with pCH_4(water). This observation implied that an oxygen-rich environment favored the methanotrophs, which led to a reduction in the net CH_4 production within the water column, and hence, reduced the pCH_4(water). Chl-a exhibited a negative relationship with pCH_4(water) only in P1, however, in both of the ponds, GPP did

Table 8.3 The relationship between pCH_4(water) and other biogeochemical parameters depicted through the Pearson correlation coefficient

Parameters	P1	P2
Water temperature	$\mathbf{r = 0.896, p < 0.05}$	$\mathbf{r = 0.865, p < 0.05}$
Electrical conductivity	$r = -0.309, p > 0.05$	$r = -0.159, p > 0.05$
Dissolved oxygen	$\mathbf{r = -0.746, p < 0.05}$	$\mathbf{r = -0.744, p < 0.05}$
Oxygen saturation	$\mathbf{r = -0.896, p < 0.05}$	$\mathbf{r = -0.850, p < 0.05}$
Turbidity	$r = -0.088, p > 0.05$	$r = 0.071, p > 0.05$
UWPAR	$r = 0.435, p > 0.05$	$r = 0.441, p > 0.05$
Chl-a	$\mathbf{r = -0.626, p < 0.05}$	$r = -0.425, p > 0.05$
GPP	$r = 0.301, p > 0.05$	$r = 0.456, p > 0.05$
CR	$\mathbf{r = 0.661, p < 0.05}$	$\mathbf{r = 0.767, p < 0.05}$
NPP	$\mathbf{r = -0.595, p < 0.05}$	$\mathbf{r = -0.734, p < 0.05}$
BOD	$\mathbf{r = 0.825, p < 0.05}$	$\mathbf{r = 0.978, p < 0.05}$

not show any significant relationship with pCH_4(water). Anything meaningful could not be inferred from this relationship; however, CR and BOD showed a statistically significant positive relationship with pCH_4(water) and NPP showed a negative one. This implied that net heterotrophy was associated with more CH_4 emission. In other words, conditions not favorable for photosynthesis not only led to a higher rate of CO_2 emission but also favored the methanogens to emit more CH_4.

8.10 Conclusion

Overall, both of the ponds studied in this chapter were net heterotrophic and emitted CO_2 and CH_4 throughout the annual cycle. However, the degree of net heterotrophy varied between the two ponds and so did the rate of CO_2 and CH_4 emission. Water temperature played a crucial role in governing both the dissolved CO_2 as well as CH_4 levels, which happens to be the most concerning observation, as the ambient temperature is expected to increase in the near future due to the ongoing climate change, and it can lead to enhanced rate of emissions. Thus, these emissions can imply positive feedback to climate change. Though not entirely, biological processes like photosynthesis and community respiration played a significant role in governing both the CO_2 and CH_4 fluxes. A higher rate of community respiration favored net heterotrophy, which in turn reduced the oxygen levels and favored the production of more CH_4. The study showed that despite experiencing similar climatic conditions, the drivers of greenhouse gas fluxes can vary from pond to pond. The present study indicates that from the perspective of GHG emission abatement, the autotrophic potential of these ponds should be increased to reduce both CO_2 as well as CH_4 emissions.

References

Agarwal AK, Agarwal RA, Gupta T, Gurjar BR (eds) (2017) Biofuels: technology, challenges and prospects. Springer, Singapore, p 245

Ahmed N, Bunting SW, Glaser M, Flaherty MS, Diana JS (2017) Can greening of aquaculture sequester blue carbon? Ambio 46(4):468–477. https://doi.org/10.1007/s13280-016-0849-7

Alagarswamy K (1995) Regional study and workshop on the environmental assessment and management of shrimp farming. Organized by Food and Agriculture Organisation and Network of Aquaculture Centres in Asia–Pacific (NACA), Bangkok, Thailand, pp 21–26

Ali H, Rico A, Murshed-e-Jahan K, Belton B (2016) An assessment of chemical and biological product use in aquaculture in Bangladesh. Aquaculture 454:199–209. https://doi.org/10.1016/j.aquaculture.2015.12.025

Allan GL, Moriarty DJ, Maguire GB (1995) Effects of pond preparation and feeding rate on production of Penaeus monodon Fabricius, water quality, bacteria, and benthos in model farming ponds. Aquaculture 130(4):329–349. https://doi.org/10.1016/0044-8486(94)00316-G

Anderson CB (2002) Understanding carbonate equilibria by measuring alkalinity in experimental and natural systems. J Geosci Educ 50:389–403. https://doi.org/10.5408/1089-9995-50.4.389

Aubin J, Papatryphon E, Van der Werf HMG, Petit J, Morvan YM (2006) Characterization of the environmental impact of a turbot (*Scophthalmus maximus*) re-circulating production system using life cycle assessment. Aquaculture 261(4):1259–1268. https://doi.org/10.1016/j.aquaculture.2006.09.008

Avault JW (1996) Fundamentals of aquaculture: a step-by-step guide to commercial aquaculture. AVA Publishing Company Inc., LA, USA

Badiola M, Basurko OC, Piedrahita R, Hundley P, Mendiola D (2018) Energy use in recirculating aquaculture systems (RAS): a review. Aquac Eng 81:57–70. https://doi.org/10.1016/j.aquaeng.2018.03.003

Bell J, Watson W, Ye Y (2017) Global fishing capacity and fishing effort from 1950 to 2012. Fish Fish 18:489–505. https://doi.org/10.1111/faf.12187

Beveridge MC, Little DC (2002) The history of aquaculture in traditional societies. In: Costa-Pierce BA (ed) Ecological aquaculture. The evolution of the Blue Revolution, Wiley-Blackwell, pp 3–29

Bhattacharyya S, Chanda A, Hazra S, Das S, Choudhury SB (2020) Effect of nutrient alteration on pCO_2 (water) and chlorophyll-a dynamics in a tropical aquaculture pond situated within a Ramsar site: a microcosm approach. Environ Sci Pollut Res 27(4):4353–4364. https://doi.org/10.1007/s11356-019-07106-6

Bouwman AF, Beusen AHW, Overbeek CC, Bureau DP, Pawlowski M, Glibert PM (2013) Hindcasts and future projections of global inland and coastal nitrogen and phosphorus loads due to finfish aquaculture. Rev Fish Sci 21(2):112–156. https://doi.org/10.1080/10641262.2013.790340

Boyd CE, Tucker CS (eds) (2014) Handbook for aquaculture water quality. Craftmaster Printers, Auburn, p 439

Boyd CE, Hanson T (2010) Dissolved-oxygen concentrations in pond aquaculture. Glob Aquac Advocate. http://lcwu.edu.pk/ocd/cfiles/Zoology/PhD-FA-SP703/Lecture1DissolvedOxygen.pdf

Boyd CE, Massaut L (1999) Risks associated with the use of chemicals in pond aquaculture. Aquac Eng 20(2):113–132. https://doi.org/10.1016/S0144-8609(99)00010-2

Boyd CE, Wood CW, Thunjai T (eds) (2002) Aquaculture pond bottom soil quality management. Pond dynamics/aquaculture collaborative research support program, Oregon State University, USA

Boyd CE, Tucker CS (1998) Pond aquaculture water quality management. Springer Science Business Media, New York. https://doi.org/10.1007/978-1-4615-5407-3

Boyd CE (1979) Water quality in warm water fish ponds. Auburn University Agricultural Experiment Station, USA, p 359

Buck BH, Langan R (eds) (2017) Aquaculture perspective of multi-use sites in the open ocean: the untapped potential for marine resources in the anthropocene. Springer, Cham, Switzerland

Buck BH, Troell MF, Krause G, Angel DL, Grote B, Chopin T (2018) State of the art and challenges for offshore integrated multi-trophic aquaculture (IMTA). Front Mar Sci 5:165. https://doi.org/10.3389/fmars.2018.00165

Bunting SW, Kundu N, Ahmed N (2017) Evaluating the contribution of diversified shrimp–rice agroecosystems in Bangladesh and West Bengal India to social-ecological resilience. Ocean Coast Manag 148:63–74. https://doi.org/10.1016/j.ocecoaman.2017.07.010

Bureau DP, Hua K (2010) Towards effective nutritional management of waste outputs in aquaculture, with particular reference to salmonid aquaculture operations. Aquac Res 41:777–792. https://doi.org/10.1111/j.1365-2109.2009.02431.x

Camargo JA, Alonso A, Salamanca A (2005) Nitrate toxicity to aquatic animals: a review with new data for freshwater invertebrates. Chemosphere 58(9):1255–1267. https://doi.org/10.1016/j.chemosphere.2004.10.044

Campbell B, Pauly D (2013) Mariculture: a global analysis of production trends since 1950. Mar Pol 39:94–100. https://doi.org/10.1016/j.marpol.2012.10.009

Cao L, Naylor R, Henriksson P, Leadbitter D, Metian M, Troell M, Zhang W (2015) China's aquaculture and the world's wild fisheries. Science 347(6218):133–135. https://doi.org/10.1126/science.1260149

Castillo-Soriano FA, Ibarra-Junquera V, Escalante-Minakata P, Mendoza-Cano O, Ornelas-Paz JDJ, Almanza-Ramírez JC, Meyer-Willerer AO (2013) Nitrogen dynamics model in zero water exchange, low salinity intensive ponds of white shrimp, Litopenaeus vannamei, at Colima, Mexico. Lat Am J Aquat Res 41(1):68–79

Chamberlain GW, Hutchins DL, Lawrence AL, Parker JC (1980). Winter culture of Penaeus vannamei in ponds receiving thermal effluent at different rates. In: Proceedings of the World Mariculture Society. Blackwell Publishing Ltd, Oxford, UK, 11(1–4), pp 30–43

Chanda A, Das S, Bhattacharyya S, Akhand A, Das I, Samanta S et al (2020) CO_2 effluxes from an urban tidal river flowing through two of the most populated and polluted cities of India. Environ Sci Pollut Res 27:30093–30107. https://doi.org/10.1007/s11356-020-09254-6

Chanda A, Das S, Bhattacharyya S, Das I, Giri S, Mukhopadhyay A, Samanta S, Dutta D, Akhand A, Choudhury SB, Hazra S (2019) CO_2 fluxes from aquaculture ponds of a tropical wetland: potential of multiple lime treatment in reduction of CO_2 emission. Sci Total Environ 655:1321–1333. https://doi.org/10.1016/j.scitotenv.2018.11.332

Chen R (2020) Transient inconsistency between population density and fisheries yields without bycatch species extinction. Ecol Evol 10(21):12372–12384. https://doi.org/10.1002/ece3.6868

Chen Y, Dong S, Wang F, Gao Q, Tian X (2016) Carbon dioxide and methane fluxes from feeding and no-feeding mariculture ponds. Environ Pollut 212:489–497. https://doi.org/10.1016/j.envpol.2016.02.039

Chou R, Lee HB (1997) Commercial marine fish farming in Singapore. Aquac Res 28(10):767–776. https://doi.org/10.1046/j.1365-2109.1997.00941.x

Chu YI, Wang CM, Park JC, Lader PF (2020) Review of cage and containment tank designs for offshore fish farming. Aquaculture 519:734928. https://doi.org/10.1016/j.aquaculture.2020.734928

Cole JJ, Caraco NF (1998) Atmospheric exchange of carbon dioxide in a low-wind oligotrophic lake measured by the addition of SF_6. Limnol Oceanogr 43(4):647–656. https://doi.org/10.4319/lo.1998.43.4.0647

Colt J, Summerfelt S, Pfeiffer T, Fivelstad S, Rust M (2008) Energy and resource consumption of land-based Atlantic salmon smolt hatcheries in the Pacific Northwest (USA). Aquaculture 280(1–4):94–108. https://doi.org/10.1016/j.aquaculture.2008.05.014

Cuenco ML, Stickney RR, Grant WE (1985) Fish bioenergetics and growth in aquaculture ponds: III. Effects of intraspecific competition, stocking rate, stocking size, and feeding rate on fish productivity. Ecol Model 28(1–2):73–95. https://doi.org/10.1016/0304-3800(85)90014-6

da Silva MG, Packer AP, Sampaio FG, Marani L, Mariano EV, Pazianotto RA, Ferreira WJ, Alvalá PC (2018) Impact of intensive fish farming on methane emission in a tropical hydropower reservoir. Clim Chang 150(3):195–210. https://doi.org/10.1007/s10584-018-2281-4

Dalsgaard J, Lund I, Thorarinsdottir R, Drengstig A, Arvonen K, Pedersen PB (2013) Farming different species in RAS in Nordic countries: current status and future perspectives. Aquac Eng 53:2–13. https://doi.org/10.1016/j.aquaeng.2012.11.008

Datta A, Marella TK, Tiwari A, Wani SP (2019) The diatoms: from eutrophic indicators to mitigators. In: Gupta SK, Bux F (eds) Application of microalgae in wastewater treatment. Springer, Cham, pp 19–40. https://doi.org/10.1007/978-3-030-13913-1_2

De Roy S (2012) Impact of fish farming on employment and household income. Econ Polit Wkly 47:68–74. https://www.jstor.org/stable/41720089

De Silva SS, Hasan MR (2007) Feeds and fertilizers: the key to long-term sustainability of Asian aquaculture. In: Hasan MR, Hecht T, De Silva SS, Tacon AGJ (eds) Study and analysis of feeds and fertilizers for sustainable aquaculture development. FAO Fisheries Technical Paper No. 497, Rome, FAO, pp 19–47

Duan Y, Li X, Zhang L, Chen D, Ji H (2020) Mapping national-scale aquaculture ponds based on the Google Earth Engine in the Chinese coastal zone. Aquaculture 520:734666. https://doi.org/10.1016/j.aquaculture.2019.734666

Dubey SK, Chand BK, Trivedi RK, Mandal B, Rout SK (2016) Evaluation on the prevailing aquaculture practices in the Indian Sundarban delta: an insight analysis. J Food Agric Environ 14(2):133–141

Dubey SK, Trivedi RK, Chand BK, Mandal B, Rout SK (2017) Farmers' perceptions of climate change, impacts on freshwater aquaculture and adaptation strategies in climatic change hotspots: a case of the Indian Sundarban delta. Environ Dev 21:38–51. https://doi.org/10.1016/j.envdev.2016.12.002

Eding E, Verdegem M, Martins C, Schlaman G, Heinsbroek L, Laarhoven B et al (2009) Tilapia farming using recirculating aquaculture systems (RAS)–case study in the Netherlands. Handbook for sustainable aquaculture. Project no. COLL-CT-2006-030384. http://www.sustainaqua.org

Emmanuel O, Chinenye A, Oluwatobi A, Peter K (2014) Review of aquaculture production and management in Nigeria. J Exp Agric Int 1137–1151. https://doi.org/10.9734/AJEA/2014/8082

FAO (2011). Fisheries. Aquaculture department. 2013. Global aquaculture production statistics for the year. FAO, Rome, Italy

FAO (2014) The state of world fisheries and aquaculture 2014: opportunities and challenges. FAO, Rome, Italy

FAO (2016) The state of world fisheries and aquaculture 2016: contributing to food security and nutrition for all. FAO, Rome, Italy

FAO (2018) The state of world fisheries and aquaculture 2018-meeting the sustainable development goals. FAO, Rome, Italy

Flickinger DL, Costa GA, Dantas DP, Proença DC, David FS, Durborow RM, Moraes-Valenti P, Valenti WC (2020) The budget of carbon in the farming of the Amazon river prawn and tambaqui fish in earthen pond monoculture and integrated multitrophic systems. Aquaculture Rep 17:100340. https://doi.org/10.1016/j.aqrep.2020.10034

Frankignoulle M, Bourge I, Wollast R (1996) Atmospheric CO_2 fluxes in a highly polluted estuary (the Scheldt). Limnol Oceanogr 41(2):365–369. https://doi.org/10.4319/lo.1996.41.2.0365

Galtsoff PS (1964) The American oyster, *Crassostrea virginica* Gmelin. Fish Bull 64. United States Department of the Interior, Fish and Wildlife Service, Bureaue of Commercial Fisheries, US Government Printing Office

Ghoshal TK, De D, Biswas G, Kumar P, Vijayan KK (2019) Brackishwater aquaculture: opportunities and challenges for meeting livelihood demand in Indian Sundarbans. In: Sen H (ed) The Sundarbans: a disaster-prone eco-region, coastal research library, vol 30. Springer, Cham, pp 321–349. https://doi.org/10.1007/978-3-030-00680-8_11

Goddard S (1996) Feed management in intensive aquaculture. Springer Science & Business Media, Chapman & Hall, New York, USA

Goldburg R, Naylor R (2005) Future seascapes, fishing, and fish farming. Front Ecol Environ 3(1):21–28. https://doi.org/10.1890/1540-9295(2005)003[0021:FSFAFF]2.0.CO;2

Gosavi K, Sammut J, Gifford S, Jankowski J (2004) Macroalgal biomonitors of trace metal contamination in acid sulfate soil aquaculture ponds. Sci Total Environ 324(1–3):25–39. https://doi.org/10.1016/j.scitotenv.2003.11.002

Goswami B, Mondal S, Dana SS (2006) Indigenous technological knowledge in fish farming. Indian J Tradit Knowl 5(1):60–63

Government of West Bengal (2010) Annual report 2010–11. Department of Fisheries, Aquaculture, Aquatic Resources, and Fishing Harbours. Government of West Bengal, Salt Lake, Kolkata

Gutierrez NL, Hilborn R, Defeo O (2011) Leadership, social capital, and incentives promote successful fisheries. Nature 470:386–389. https://doi.org/10.1038/nature09689

Hannon M, Gimpel J, Tran M, Rasala B, Mayfield S (2010) Biofuels from algae: challenges and potential. Biofuels 1:763–784. https://doi.org/10.4155/bfs.10.44

Hargreaves JA (1998) Nitrogen biogeochemistry of aquaculture ponds. Aquaculture 166:181–212. https://doi.org/10.1016/S0044-8486(98)00298-1

https://www.conserve-energy-future.com/aquaculture-types-benefits-importance.php. Accessed 31 Dec 2020

Hu Z, Lee JW, Chandran K, Kim S, Khanal SK (2012) Nitrous oxide (N_2O) emission from aquaculture: a review. Environ Sci Technol 46:6470–6480. https://doi.org/10.1021/es300110x

Hu Z, Lee JW, Chandran K, Kim S, Sharma K, Khanal SK (2014) Influence of carbohydrate addition on nitrogen transformations and greenhouse gas emissions of intensive aquaculture system. Sci Total Environ 470:193–200. https://doi.org/10.1016/j.scitotenv.2013.09.050

Huang S, Wu M, Zang C, Du S, Domagalski J, Gajewska M et al (2016) Dynamics of algae growth and nutrients in experimental enclosures culturing bighead carp and common carp: phosphorus dynamics. Int J Sediment Res 31(2):173–180. https://doi.org/10.1016/j.ijsrc.2016.01.003

Hussain M, Hussain SM, Afzal M, Raza SA, Hussain N, Mubarik MS (2011) Comparative study of replacement of maize gluten with rice bran (3: 1 and 1: 3) feed supplement: effect on fish growth in composite culture. Pak J Agric Sci 48(4):321–326

Jayanthi M, Thirumurthy S, Muralidhar M, Ravichandran P (2018) Impact of shrimp aquaculture development on important ecosystems in India. Glob Environ Chang 52:10–21. https://doi.org/10.1016/j.gloenvcha.2018.05.005

Kattner G (1999) Storage of dissolved inorganic nutrients in seawater: poisoning with mercuric chloride. Mar Chem 67(1–2):61–66. https://doi.org/10.1016/S0304-4203(99)00049-3

Khoo KH, Ramette RW, Culberson CH, Bates RG (1977) Determination of hydrogen ion concentrations in seawater from 5 to 40 °C: standard potentials at salinities from 20 to 45‰. Anal Chem 49(1):29–34. https://doi.org/10.1021/ac50009a016

Kibria ASM, Haque MM (2018) Potentials of integrated multi-trophic aquaculture (IMTA) in freshwater ponds in Bangladesh. Aquac Rep 11:8–16. https://doi.org/10.1016/j.aqrep.2018.05.004

Kite-Powell H (2017) Economics of multi-use and co-location. In: Buck BH, Langan R (eds) Aquaculture perspective of multi-use sites in the open ocean: the untapped potential for marine resources in the Anthropocene. Springer, Cham, Switzerland, pp 233–249. https://doi.org/10.1007/978-3-319-51159-7_10

Kondo J (2000) Atmospheric science near the ground surface. University of Tokyo, Japan

Krüger M, Frenzel P, Conrad R (2001) Microbial processes influencing methane emission from rice fields. Glob Chang Biol 7:49–63. https://doi.org/10.1046/j.1365-2486.2001.00395.x

Lamoureux J, Tiersch TR, Hall SG (2006) Pond heat and temperature regulation (PHATR): modeling temperature and energy balances in earthen outdoor aquaculture ponds. Aquac Eng 34(2):103–116. https://doi.org/10.1016/j.aquaeng.2005.06.002

Lang RP, Romaire RP, Tiersch TR (2003) Induction of early spawning of channel catfish in heated earthen ponds. North Am J Aquac 65(2):73–81. https://doi.org/10.1577/1548-8454(2003)65%3c73:IOESOC%3e2.0.CO;2

Lawson T (2013) Fundamentals of aquacultural engineering. Springer Science & Business Media

Leite GB, Abdelaziz AEM, Hallenbeck PC (2013) Algal biofuels: challenges and opportunities. Bioresour Technol 145:134–141. https://doi.org/10.1016/j.biortech.2013.02.007

Leung KMY, Chu JCW, Wu RSS (1999) Nitrogen budget for the areolated grouper Epinephelus areolatus cultured under laboratory conditions and in open-sea cages. Mar Ecol Progr Ser 186:271–281. https://doi.org/10.3354/meps186271

Lewis E, Wallace DWR (1998) Program developed for CO_2 system calculations. ORNL/ CDIAC–105. Carbon Dioxide Information Analysis Center, Oak Ridge National Laboratory, U.S. Department of Energy, Oak Ridge, Tennessee

Li F, Sun Z, Qi H, Zhou X, Xu C, Wu D, Fang F, Feng J, Zhang N (2019) Effects of rice-fish co-culture on oxygen consumption in intensive aquaculture pond. Rice Sci 26(1):50–59. https://doi.org/10.1016/j.rsci.2018.12.004

Lide DR (2007) CRC handbook of chemistry and physics, 88th edn. CRC, New York, p 2660

MacIntyre S, Wanninkhof R, Chanton JP (1995) Trace gas exchange across the air-water interface in freshwater and coastal marine environments. In: Matson PA, Harriss RC (eds) Biogenic trace gases: measuring emissions from soil and water. Blackwell Science, Oxford, pp 52–97

MacLeod MJ, Hasan MR, Robb DH, Mamun-Ur-Rashid M (2020) Quantifying greenhouse gas emissions from global aquaculture. Sci Rep 10:11679. https://doi.org/10.1038/s41598-020-68231-8
MacMillan JR, Huddleston T, Woolley M, Fothergill K (2003) Best management practice development to minimize environmental impact from large flow-through trout farms. Aquaculture 226(1–4):91–99. https://doi.org/10.1016/S0044-8486(03)00470-8
Marton JM, Herbert ER, Craft CB (2012) Effects of salinity on denitrification and greenhouse gas production from laboratory-incubated tidal forest soils. Wetlands 32:347–357. https://doi.org/10.1007/s13157-012-0270-3
Milhazes-Cunha H, Otero A (2017) Valorisation of aquaculture effluents with microalgae: the integrated multi-trophic aquaculture concept. Algal Res 24:416–424. https://doi.org/10.1016/j.algal.2016.12.011
Mishra SS, Rakesh D, Dhiman M, Choudhary P, Debbarma J, Sahoo SN, Barua A, Giri BS, Ramesh R, Ananda K, Mishra CK (2017) Present status of fish disease management in freshwater aquaculture in India: state-of-the-art-review. HSOA J Aquac Fish 1:(003). https://doi.org/10.24966/AAF-5523/100003
Mishra Y (2020) Climate change risks and mitigating measures of freshwater aquaculture in unmanaged ponds of Kaushambi district UP. GSC Biol Pharm Sci 13(2):253–259. https://doi.org/10.30574/gscbps.2020.13.2.0385
Mondal I, Bandyopadhyay J (2015) Recent trend of aquaculture land of Bidyadhari River catchment area using geospatial techniques: a case study of Haroa and Minakhan Block, North–24 Parganas, West Bengal, India. Am Res Thoughts. https://doi.org/10.6084/m9.figshare.1492986
Morel FM (1983) Energetics and kinetics: principles of aquatic chemistry. Wiley, New York, p 446
Munguti JM, Musa S, Orina PS, Kyule DN, Opiyo MA, Charo-Karisa H, Ogello EO (2014) An overview of current status of Kenyan fish feed industry and feed management practices, challenges and opportunities. Int J Fish Aquat Stud 1(6):128–137. https://repository.maseno.ac.ke/handle/123456789/2279
Munsiri P, Boyd CE, Hajek BF (1995) Physical and chemical characteristics of bottom soil profiles in ponds at Aubum, Alabama, USA, and proposed method for describing pond soil horizons. J World Aquac Soc 26:346–377. https://doi.org/10.1111/j.1749-7345.1995.tb00831.x
Murray NJ, Clemens RS, Phinn SR, Possingham HP, Fuller RA (2014) Tracking the rapid loss of tidal wetlands in the Yellow Sea. Front Ecol Environ 12(5):267–272. https://doi.org/10.1890/130260
Myhre G, Shindell D, Bréon F, Collins W, Fuglestvedt J, Huang J, Koch D, Lamarque J, Lee D, Mendoza B (2013) Anthropogenic and natural radiative forcing. Climate change 2013: the physical science basis. Contribution of working group 1 to the fifth assessment report of the intergovernmental panel on climate change, p 714
Nash C (2010) The history of aquaculture. Wiley
Naskar KR (1985) A short history and the present trends of brackish water fish culture in paddy fields at the Kulti-Minakhan areas of Sundarbans in West Bengal. J Indian Soc Coast Agric Res 3:115–124
Naskar S, Pailan GH, Datta S, Banerjee SP, Bharti VS (2020) Effect of different organic manures and salinity levels on greenhouse gas emission and growth of common carp in aquaculture systems. Aquac Res. https://doi.org/10.1111/are.15041
Naylor RL, Goldburg RJ, Primavera JH, Kautsky N, Beveridge MC, Clay J et al (2000) Effect of aquaculture on world fish supplies. Nature 405(6790):1017–1024. https://doi.org/10.1038/35016500
Neofitou N, Papadimitriou K, Domenikiotis C, Tziantziou L, Panagiotaki P (2019) GIS in environmental monitoring and assessment of fish farming impacts on nutrients of Pagasitikos Gulf, Eastern Mediterranean. Aquaculture 501:62–75. https://doi.org/10.1016/j.aquaculture.2018.11.005

Nora'aini A, Mohammad AW, Jusoh A, Hasan MR, Ghazali N, Kamaruzaman K (2005) Treatment of aquaculture wastewater using ultra-low pressure asymmetric polyethersulfone (PES) membrane. Desalination 185(1–3):317–326. https://doi.org/10.1016/j.desal.2005.03.084

NRC (2011) Nutritional requirements of fish and shrimp. The National Academies Press, Washington, DC

Ntsama ISB, Tambe BA, Takadong JJT, Nama GM, Kansci G (2018) Characteristics of fish farming practices and agrochemicals usage therein in four regions of Cameroon. Egypt J Aquat Res 44(2):145–153. https://doi.org/10.1016/j.ejar.2018.06.006

Oyinlola MA, Reygondeau G, Wabnitz CC, Troell M, Cheung WW (2018) Global estimation of areas with suitable environmental conditions for mariculture species. PLoS One 13(1):e0191086. https://doi.org/10.1371/journal.pone.0191086

Parson TR, Maita Y, Lalli CM (1992) A manual of chemical and biological methods for sea water analysis. Pergamon, New York, pp 101–103

Paudel SR, Choi O, Khanal SK, Chandran K, Kim S, Lee JW (2015) Effects of temperature on nitrous oxide (N_2O) emission from intensive aquaculture system. Sci Total Environ 518:16–23. https://doi.org/10.1016/j.scitotenv.2015.02.076

Philcox N, Knowler D, Haider W (2010) Eliciting stakeholder preferences: an application of qualitative and quantitative methods to shrimp aquaculture in the Indian Sundarbans. Ocean Coast Manag 53(3):123–134. https://doi.org/10.1016/j.ocecoaman.2010.02.001

Piedrahita RH (2003) Reducing the potential environmental impact of tank aquaculture effluents through intensification and recirculation. Aquaculture 226(1–4):35–44. https://doi.org/10.1016/S0044-8486(03)00465-4

Prasad KA, Ottinger M, Wei C, Leinenkugel P (2019) Assessment of coastal aquaculture for India from Sentinel-1 SAR time series. Remote Sens 11(3):357. https://doi.org/10.3390/rs11030357

Primavera JH (2006) Overcoming the impacts of aquaculture on the coastal zone. Ocean Coast Manag 49(9–10):531–545. https://doi.org/10.1016/j.ocecoaman.2006.06.018

Reese KL, Fisher CL, Lane PD, Jaryenneh JD, Moorman MW, Jones AD, Frank M, Lane TW (2019) Chemical profiling of volatile organic compounds in the headspace of algal cultures as early biomarkers of algal pond crashes. Sci Rep 9:13866. https://doi.org/10.1038/s41598-019-50125-z

Ren C, Wang Z, Zhang Y, Zhang B, Chen L, Xi Y, Xiao X, Doughty RB, Liu M, Jia M, Mao D, Song K (2019) Rapid expansion of coastal aquaculture ponds in China from Landsat observations during 1984–2016. Int J Appl Earth Obs Geoinfom 82:101902. https://doi.org/10.1016/j.jag.2019.101902

Renaud FG, Syvitski JP, Sebesvari Z, Werners SE, Kremer H, Kuenzer C, Ramesh R, Jeuken A, Friedrich J (2013) Tipping from the Holocene to the Anthropocene: how threatened are major world deltas? Curr Opin Environ Sustain 5(6):644–654. https://doi.org/10.1016/j.cosust.2013.11.007

Rose D, Bell D, Crook DA (2016) Restoring habitat and cultural practice in Australia's oldest and largest traditional aquaculture system. Rev Fish Biol Fish 26(3):589–600. https://doi.org/10.1007/s11160-016-9426-1

Roy RN, Roy LN, Vogel KM, Porter-Moore C, Pearson T, Good CE et al (1993) The dissociation constants of carbonic acid in seawater at salinities 5 to 45 and temperatures 0 to 45 C. Mar Chem 44(2–4):249–267. https://doi.org/10.1016/0304-4203(93)90207-5

Salunke M, Kalyankar A, Khedkar CD, Shingare M, Khedkar GD (2020) A review on shrimp aquaculture in India: historical perspective, constraints, status and future implications for impacts on aquatic ecosystem and biodiversity. Rev Fish Sci Aquac 1-20. https://doi.org/10.1080/23308249.2020.1723058

Saraswathy R, Muralidhar M, Sanjoy D, Kumararaja P, Suvana S, Lalitha N, Katneni VK, Nagavel A, Vijayan KK (2019) Changes in soil and water quality at sediment-water interface of Penaeus vannamei culture pond at varying salinities. Aquac Res 50(4):1096–1106. https://doi.org/10.1111/are.13984

Shaher S, Chanda A, Das S, Das I, Giri S, Samanta S, Hazra S, Mukherjee AD (2020) Summer methane emissions from sewagewater–fed tropical shallow aquaculture ponds characterized by different water depths. Environ Sci Pollut Res 27:18182–18195. https://doi.org/10.1007/s11356-020-08296-0

Shobika U (2020) Global status of mariculture. Biotica Res Today 2(7):618–621

Singh SN, Tyagi L (2009) Nitrous oxide: sources, sinks, and mitigation strategies. Nitrous oxide emissions research progress. Nova Science Publisher, pp 127–150

Som S, Gupta K (2010) A production model of aquaculture shrimp farming for the Namkhana Block of the Indian Sunderbans. Rabindra Bharati Univ J Econ 6:17–26

Spalding MD, Ruffo S, Lacambra C, Meliane I, Hale LZ, Shepard CC, Beck MW (2014) The role of ecosystems in coastal protection: adapting to climate change and coastal hazards. Ocean Coast Manag 90:50–57. https://doi.org/10.1016/j.ocecoaman.2013.09.007

Sundaray JK, Chakrabarti PP, Mohapatra BC, Das A, Hussan A, Ghosh A, Hoque F (2019) Freshwater aquaculture in Sundarbans India. In: Sen H (ed) The Sundarbans: a disaster-prone eco-region. Springer, Cham, pp 295–319, https://doi.org/10.1007/978-3-030-00680-8_10

Tucker CS (1984) Carbon dioxide. In: Wellborn, Jr TR, MacMillan JR (eds) For fish farmers. Mississippi Cooperative Extension Service

Uddin N, Kibria ASM, Haque MM (2018) Assessment of primary productivity of integrated multi-trophic aquaculture ponds. Int J Fish Aquat Stud 6(3):306–314

UNEP (2018) The emissions gap report 2018. United Nations Environment Programme, Nairobi

Vasas V, Lancelot C, Rousseau V, Jordán F (2007) Eutrophication and overfishing in temperate nearshore pelagic food webs: a network perspective. Mar Ecol Prog Ser 336:1–14. https://doi.org/10.3354/meps336001

Verdegem MCJ, Bosma RH, Verreth JAJ (2006) Reducing water use for animal production through aquaculture. Water Resour Dev 22(1):101–113. https://doi.org/10.1080/07900620500405544

Virdis SGP (2014) An object-based image analysis approach for aquaculture ponds precise mapping and monitoring: a case study of Tam Giang-Cau Hai Lagoon, Vietnam. Environ Monit Assess 186(1):117–133. https://doi.org/10.1007/s10661-013-3360-7

Wang B, Cao L, Micheli F, Naylor RL, Fringer OB (2018) The effects of intensive aquaculture on nutrient residence time and transport in a coastal embayment. Environ Fluid Mech 18(6):1321–1349. https://doi.org/10.1007/s10652-018-9595-7

Wanninkhof R (1992) Relationship between wind speed and gas exchange over the ocean. J Geophys Res 97(CS5):7373–7382

Weiss RF (1974) Carbon dioxide in water and seawater: the solubility of a non-ideal gas. Mar Chem 2(3):203–215. https://doi.org/10.1016/0304-4203(74)90015-2

Welker TL, Overturf K, Abernathy J (2019) Effect of aeration and oxygenation on growth and survival of rainbow trout in a commercial serial-pass, flow-through raceway system. Aquac Rep 14:100194. https://doi.org/10.1016/j.aqrep.2019.100194

Weston NB, Vile MA, Neubauer SC, Velinsky DJ (2011) Accelerated microbial organic matter mineralization following salt-water intrusion into tidal freshwater marsh soils. Biogeochemistry 102:135–151. https://doi.org/10.1007/s10533-010-9427-4

Wong MH, Mo WY, Choi WM, Cheng Z, Man YB (2016) Recycle food wastes into high-quality fish feeds for safe and quality fish production. Environ Pollut 219:631–638. https://doi.org/10.1016/j.envpol.2016.06.035

Worm B, Hilborn R, Baum JK et al (2009) Rebuilding global fisheries. Science 325:578–585. https://doi.org/10.1126/science.1173146

Wu J, Zhang H, Pan Y, Krause-Jensen D, He Z, Fan W, Xiao X, Chung I, Marbà N, Serrano O, Rivkin RB et al (2020) Opportunities for blue carbon strategies in China. Ocean Coast Manag 194:105241. https://doi.org/10.1016/j.ocecoaman.2020.105241

Wu R, Boyd CE (1990) Evaluation of calcium sulfate for use in aquaculture ponds. Prog Fish Cult 52:26–31. https://doi.org/10.1577/1548-8640(1990)052%3c0026:EOCSFU%3e2.3.CO;2

Wurts WA, Durborow RM (1992) Interactions of pH, carbon dioxide, alkalinity and hardness in fish ponds. Souther Region Aquaculutre Centre, USA

Yang HJ, Xie CX, He XG, Xian Y, Hu X, Chen BX (2010) Studies on plankton community structure and its dynamics in recycling and non-recycling aquaculture ponds. Freshw Fish 3. https://en.cnki.com.cn/Article_en/CJFDTotal-DSYY201003007.htm
Yang P, Bastviken D, Lai DYF, Jin BS, Mou XJ, Tong C, Yao YC (2017) Effects of coastal marsh conversion to shrimp aquaculture ponds on CH_4 and N_2O emissions. Estuar Coast Shelf Sci 199:125–131. https://doi.org/10.1016/j.ecss.2017.09.023
Yang P, Zhang Y, Yang H, Guo Q, Lai DY, Zhao G et al (2020a) Ebullition was a major pathway of methane emissions from the aquaculture ponds in southeast China. Water Res 184:116176. https://doi.org/10.1016/j.watres.2020.116176
Yang P, Zhang Y, Bastviken D, Lai DY, Yang H, Zhang YF et al (2020b) Large increase in diffusive greenhouse gas fluxes from subtropical shallow aquaculture ponds during the passage of typhoons. J Hydrol 583:124643. https://doi.org/10.1016/j.jhydrol.2020.124643
Yang P, Zhang Y, Lai DY, Tan L, Jin B, Tong C (2018) Fluxes of carbon dioxide and methane across the water–atmosphere interface of aquaculture shrimp ponds in two subtropical estuaries: the effect of temperature, substrate, salinity and nitrate. Sci Total Environ 635:1025–1035. https://doi.org/10.1016/j.scitotenv.2018.04.102
Yao YC, Ren CY, Wang ZM, Wang C, Deng PY (2016) Monitoring of salt ponds and aquaculture ponds in the coastal zone of China in 1985 and 2010. Wetl Sci 14:874–882
Ye Y, Gutierrez NL (2017) Ending fishery overexploitation by expanding from local successes to globalized solutions. Nat Ecol Evol 1(7):0179. https://doi.org/10.1038/s41559-017-0179
Yuan J, Xiang J, Liu D, Kang H, He T, Kim S, Lin Y, Freeman C, Ding W (2019) Rapid growth in greenhouse gas emissions from the adoption of industrial-scale aquaculture. Nat Clim Chang 9(4):318–322. https://doi.org/10.1038/s41558-019-0425-9
Zacharia PU, Gopalakrishnan A, Gopakumar G, Muralidhar M, Vijayan KK (2016) Climate change impact on coastal fisheries and aquaculture in the SAARC region: country paper–India. In: SAARC agriculture centre video conference on "Climate Change Impact on Coastal Fisheries and Aquaculture", 20 Dec 2016
Zhang D, Tian X, Dong S, Chen Y, Feng J, He RP, Zhang K (2020) Carbon budgets of two typical polyculture pond systems in coastal China and their potential roles in the global carbon cycle. Aquac Environ Interact 12:105–115. https://doi.org/10.3354/aei00349
Zhou M, Wang W, Chi M (2009) Enhancement on the simultaneous removal of nitrate and organic pollutants from groundwater by a three-dimensional bio-electrochemical reactor. Bioresour Technol 100(20):4662–4668. https://doi.org/10.1016/j.biortech.2009.05.002

Chapter 9
Valuation of Pond Ecosystem Services of Indian Sundarbans: A Methodological Approach

Somnath Hazra and Rabindra N. Bhattacharya

Abstract The State of West Bengal has an uncountable number of ponds and occupies 7.45% of the total water resources of India. Apart from small ponds, the State also has large ponds, known as *Dighi* and *Beel*. These large ponds serve as rainwater reservoirs, which find extensive use during the post-monsoon and the summer seasons. In Sundarbans, since there are no large ponds or natural lakes, the small ponds are crucial for domestic agriculture and allied activities. These water bodies are the only source of surface water in the Sundarbans. The demand for fresh water is increasing day by day due to the population growth in the region. Freshwater storage is the primary ecosystem service of the ponds because these water bodies provide substantial provisioning and Supporting services and very few cultural and regulating services. Due to changes in the agricultural activity in Sundarbans, the provisioning services (irrigation, fish farming, and domestic purposes) are declining day by day. Low awareness of ecosystem services and extreme weather events are the main reasons for the decrease in the social benefits from these ponds. So to protect the only source of freshwater ecosystem services, it is high time to assess the value of the services. Apart from ecosystem services evaluation, various stakeholders (fisher, agricultural farmer, and pond user) engagement is essential to know the importance of different ecosystem services.

Keywords Ecosystem services · Provisional · Regulating · Cultural · Supporting · Pond management · Valuation of ecosystem services

9.1 Introduction

The connection between people and nature leads to the concept of services of the ecosystem. Ecosystem services indicate the paybacks that human beings acquire from ecosystems (MEA 2005). Wetland ecosystems provide a range of services, including

S. Hazra (✉) · R. N. Bhattacharya
School of Oceanographic Studies, Jadavpur University, 188 Raja S. C. Mullick Road, Kolkata 700 032, India

S. Das et al. (eds.), *Pond Ecosystems of the Indian Sundarbans*, Water Science and Technology Library 112, https://doi.org/10.1007/978-3-030-86786-7_9

retention of water, irrigation for agriculture, flood control, conservation of biodiversity, and micro-climatic regulation, carbon sequestration (Zedler and Kercher 2005; McLaughlin and Cohen 2013). Throughout the world, wetlands cover approximately 12.8 million km^2, 8.5 percent of the earth's total land area. Continental wetlands encompass about 9.5 million km^2 (Finlayson et al. 1999). Globally, 2193 prominent wetlands are recognized (Ramsar 1971). However, the last century has witnessed a rampant loss of more than 50% of the world's wetland area (UNWWAP 2003). Small water bodies are an essential part of the ecosystem. Ecosystem of the small area also can play a critical role in ecological processes globally. These small ecosystem types show a substantial intensity of many environmental processes. These intense activities can make small ecosystems more dynamic than large water bodies.

Ponds are small wetlands with ecological roles and one of the most crucial water sources available for humans in rural society. According to Shiklomanov and Rodda (2003), globally, 90 percent of liquid water is contained in ponds and lakes. Ponds are the most crucial freshwater habitat, which plays an essential role in biodiversity maintenance. The ecosystem of the ponds directly connects with rural people, but rural ponds are vulnerable to freshwater habitat because people are not aware of their economic benefits. So, proper water management of ponds can mitigate climate change, provide recreation of water and watering of livestock and irrigation support, reduce pollutions, flooding alleviation, and capture the heavy rainfall event. Ecosystem services in small water bodies are most active and globally significant, and so are their contribution and role in the global ecosystem processes. Ponds which have a better quality of water are more economically desirable. The ecosystem of the pond is also essential for rural production and livelihoods sustainability. However, these ponds are also vulnerable and degrading very fast. Proper planning of pond restoration can generate significant benefits in biodiversity conservation, relief from flooding, reduce the impact of pollution and climate change.

The State of West Bengal has an uncountable number of ponds and occupies 7.45% of the total water resources of India. The State also has lots of large ponds, known as 'Dighi' and beel. These large ponds serve as the reservoir of rainwater, which can be used during the post-monsoon and the summer seasons. In Sundarbans, since there are no large ponds or natural lakes, the small ponds are crucial for domestic and agricultural purposes. These water bodies are the only source of surface water in the Sundarbans. The demand for freshwater is increasing every day concerning the population growth in the region. In terms of ecosystem services, freshwater storage is the primary ecosystem service of the Ponds because these water bodies provide provisioning and supporting services and very few cultural and regulating services.

Like others assets of capital, the ecosystem is a natural asset. The ecosystem is providing a flow of services by which the consumers are getting benefits. Demand for Ecosystem services' will be increasing with an increase of community awareness and environmental degradation. According to MEA (2005), economic valuation is a prerequisite to manage an ecosystem. The ecosystem plays a vital role in terms of provisioning, regulating, cultural and other supporting services in the surroundings. However, a wide range of populations is overusing many ecosystem services, and over-exploitation of these ecosystem services is happening due to low awareness about the services. Since human wellbeing is highly dependent on ecosystem services, a consequence of this over-exploitation is affecting the wellbeing of the current and the future generation.

Efforts for the valuation of ecosystem services continued throughout the last few decades. Costanza et al. (1997) assessed the worth of ecosystem services of the entire world to be $33 trillion. According to Costanza et al. (1997), ecosystem services are a bouquet of benefits that the society can retrieve from the ecosystems. These services vary from provisioning and supporting to regulating and cultural services (Costanza et al. 1997). Apart from the restoration of natural resources, enhancement of conservation and management mechanisms of these ecosystems are also necessary; otherwise, we may lose multiple benefits for society. Therefore, to restrict natural resource degradation and associated ecosystem services, the valuation of ecosystem services is highly essential. Ecosystem services evaluation can help us in decision-making by assessing the give and take relationship between human beings and the ecosystems (MEA 2005).

According to the Millennium Ecosystem Assessment, a pond can offer several benefits, like regulation, provision, and cultural services (Mace 2008). Any degradation and conservation changing of wetlands to other land use impact the biodiversity and livelihoods of people living around surrounding ponds. Due to changes in the agricultural activity in Sundarbans, the provisioning services (irrigation, fish farming, and domestic purposes) are declining day by day. Low awareness of the value of ecosystem services and extreme weather events are the main reasons that bar society from reaping benefits from this crucial lentic ecosystem. So to protect the only source of freshwater ecosystem services here, it is high time to assess the value of the services.

The Indian Sundarbans Delta consists of 19 administrative blocks. 13 blocks belong to South-24-Parganas, and 6 to North-24-Parganas. The region is under the coastal zone of the State and high salt-induced landscape. We conducted a focus group discussion and a participatory rural appraisal in February 2021involving the fishers, agricultural farmers, and pond managers (as chalked out by Chambers 1994). The primary data is essential for the valuation of Ponds Ecosystem services. A pre-tested questionnaire should generate the data on ecosystem services and socio-economic aspects, and it has to be specific for the pond ecosystem services. Some secondary data are also crucial from the different government data books like the latest District Statistical Handbook, Economic survey, and Agricultural Statistics. In the context of increasing demand for irrigation and agricultural management apart from the valuation of ecosystem services, this paper aims to know the importance of different pond ecosystem services for various stakeholders.

9.2 Challenges and Studies on Valuation of Wetland Ecosystem Services

Ponds can act as a perennial water source, which is an essential ecosystem service. However, a lack of proper assessment of wetland ecosystem services hampered the adoption approach (Crossman et al. 2013; Kull et al. 2015). Earlier pieces of research exclusively focused on the large wetland ecosystems and characterized their services from an economic point of view (Wang et al. 2004; Lai et al. 2013; Howard et al. 2016; Guimaraes and Lowe 2016). Very recently, the assessment of wetland ecosystem services, changes, and patterns of services have become a significant concern globally (Corrigan and Nieuwenhuis 2016; Li and Gao 2016; Li et al. 2016; Hu et al. 2017).

Unlike the giant wetlands, ponds occupy a much smaller surface area. These lentic water bodies are mostly not natural wetlands (constructed), and they are the crucial sources of water in agricultural landscapes (Son et al. 2014; Natsumeda et al. 2015). The pond systems provide many ecosystem services like fish production, supply of water, retention of nutrients, sequestration of carbon, biodiversity, and recreational use (EPCN 2007). However, the developmental activities and urbanization process like construction of the road have seriously degraded wetland areas, mainly ponds (Na et al. 2008; Choi and Bury 2003; Van Dam et al. 2015). Broadly wetlands are of three types- inland, coastal, and constructed. According to the Ramsar convention, 42 specialized sub-categories of wetlands exist (Ramsar 1971). In the conservation program of the U.K. in 1989, an assessment characterized the number, characteristic traits, spatial distribution, and biodiversity of the ponds (Jeffries 2010; Williams et al. 2010; Munns et al. 2016).

According to Céréghino et al. (2014), ponds provide sustainable solutions to the significant environmental problems of climate change and water management. Ponds with fresh water are multifunctional ecosystems that offer a spectrum of economic and social benefits (IUCN 1997; Bekefi and Varadi 2007; EPCN 2007; Downing 2010). They provide multiple services like water retention and diversity of vegetation (Son et al. 2014). The vegetation structure and pond size play a crucial role in regulating the biodiversity of ponds (Feng et al. 2015; Natsumeda et al. 2015; Ouyang et al. 2017). Prach and Tolvanen (2016) discussed the difficulties in the application of programs and pond conservation.

In most of the areas, the conservation and protection of ponds remain most neglected. The demise, abandonment, and pollution of ponds have enhanced continuously over the years (Cirovic et al. 2016; Mupepele et al. 2016). Scotland witnessed the disappearance of more than half of its total ponds due to urban and agricultural encroachment (SEPA 2000). The agricultural ponds of china suffer from intense cultural eutrophication (Martinez et al. 2016). The ponds are capable of providing a bundle of benefits to humankind. However, assessments primarily emphasize a handful of services, like the fish yield, neglecting the other ecological roles (Pechar 2000). Hence a holistic understanding of the ecosystem services from ponds remains poorly constrained (Son et al. 2014; Prach and Tolvanen 2016).

Previously, several endeavors emphasized designing suitable support systems to safeguard the freshwater ponds and lakes and device effective management strategies (Gutierrez Estrada et al. 2012; Gawne et al. 2012). Soto et al. (2008) devised a mechanism that can effectively identify the services, model their role in the society, and evaluate a composite benefit from several aspects of an ecosystem. According to Newton et al. (2012), cost–benefit analysis (CBA) is an outstanding example to evaluate ecosystem services. This approach calculates the monetary worth of each of the ecosystem services by comparing the costs and benefits. It enables optimizing the management decisions (Costanza et al. 1997). Several economists refer to this approach as environmental CBA (Atkinson and Mourato 2008). This approach considers the associated uncertainties that any natural ecosystem possesses on offering social benefits (Bianchini and Hewage 2012; Karmperis et al. 2012). Newton et al. (2012) calculated that these environmental benefits are highly sensitive to market finance fluctuations.

Further, in environmental management, characterizing risks is a crucial factor. However, there is a limited scope to consider uncertainties in ecological CBAs (Ticehurst et al. 2007; Barton et al. 2008). Several studies have been conducted on the multi-functionality of pond systems (Céréghino et al. 2010; Kloskowski 2011), though the integration of this multi-disciplinary knowledge into practical management suggestions is very rare. However, different studies on pond ecosystem services reveals few investigative reports on the quality of the ecology and environment and the disappearance of ponds in the case of India. In the absence of any good literature on the ecosystem services of ponds, it is hard to analyze the current status of ponds in India and West Bengal.

Discussion with the local people of Sundarbans, provides information on Regulating services (i.e., carbon sequestration, micro-climate regulations), provisioning services (i.e., water retention services, irrigation, fisheries, domestic use), and Cultural (ecotourism) ecosystem services for assessment. Supporting services are very negligible in Sundarbans, so we have not considered them here.

9.3 Evaluation Procedures of Different Services

9.3.1 Assessment of Provisioning Services

The provisioning services of ponds are essential services among all other services the ponds are giving to society. The households of the surrounding ponds are highly dependent on various goods and services of the pond. There are multiple goods like fish, molasses, and leafy vegetables obtained from the ponds.

Apart from the above products, according to our group discussion with the local households, the villagers use the pond water for irrigation purposes. The villagers irrigate their farmlands with the water from the nearest pond, but they do not pay any fees for the water. The local farmers usually deploy a diesel-operated pump

Table 9.1 Types of provisioning service available from the ponds of Indian Sundarbans

Sl No	Different uses of pond
1	Water retention
2	Irrigation
2	Fisheries
3	Duck keeping
4	Collection of small fishes, snails *(Sumuk,* etc.)
5	Collection of leafy vegetables grown in the pond *(like Kalmi,* etc.)
6	Domestic uses (bathing, washing, cattle bath, etc.)

set (5 horsepower) to irrigate farmland adjacent to the pond. Farms situated in the vicinity of the pond also occasionally use manually operated lifting devices. Fossil fuel combustion to operate a pump is a cost-intensive endeavor. The multiplicative result of the number of times of the pump operation for irrigation and the average hour of irrigation gives an idea of the tentative cost. Apart from paddy, some vegetables also grow in the pond bed and its peripheral regions. According to the villagers of Sundarbans, surrounding people are highly dependent on the different goods and services of the pond (Table 9.1).

Water retention, crop production, and fisheries are the central provisioning services of a Pond in Sundarbans, contributing to the livelihoods of the surrounding population of the villages. The main crop types grown around the ponds are paddy and some vegetables for household consumption and income. However, the vegetables are grown on a small scale. These ponds are also essential for keeping livestock.

9.3.1.1 Assessment Procedure of Water Retention Service

Odgaard et al. (2017) suggested that stock volume acts as a primary indicator while assessing wetland conservation, and it reflects the hydrological regulation service of wetland. In terms of the landscape, the farmland ponds differ from the natural wetlands. After fish farming and irrigational activities, there are significant changes in the downstream runoff of the ponds. So, the yield calculation due to the use of this water from the ecosystem is necessary (Nelson et al. 2011). In the first step, one has to divide the sub-catchment to compute the accumulation of surface runoff. In the next step, subtracting the evaporation of surface water in a pond derives the estimated water retention. The measurement of evaporation from water surface requires a proper methodology and calibration protocol with a conversion factor (Sheng et al. 2007).

$$\mathrm{WR} = \mathrm{P_i} + \mathrm{W_s} + \min(\mathrm{R_i}, \mathrm{R_c}) - \mathrm{E_i} \qquad (9.1)$$

where W.R. is the water retention potential of the pond (m^3), P_i is the annual rainfall (m^3), Ws denotes the supply of water (m^3), R_i denotes the runoff in the pond catchment (m^3), E_i is the evaporation of water surface volume (m^3), and R_c is the pond volume (m^3). To calculate the minimum value of the pond and the amount of runoff, it may be assumed that all the ponds are maintained at a steady water level.

$$R_i = \sum_{i=1}^{n} \left(\text{WaterYieldi} \times \frac{Aw}{Ap} \right) \tag{9.2}$$

where R_i is the accumulated runoff, WaterYield denotes runoff in unit area land (mm/a), A_w is the total catchment area of the pond, and A_p is the water body area.

The required parameters for this model have to be collected from field surveys and the published secondary references as benefit transfer. In this model, the other parameters will be calculated by using the FAO guide for irrigation (Allen et al. 1998). The valuation of water retention services is based on the current market price and availability of water.

9.3.1.2 Assessment of Irrigation Water

We can assume that due to irrigation, the crop yield enhances by ΔY from the rain-fed yield Yr (tons/ha) to the irrigated yield Yi, which means $Yi = Yr + \Delta Y$. We further presume that the cost of labor and materials (land, seeds, fertilizer, and use of machinery) are the same in irrigated and rain-fed conditions. Hence, the value of irrigation water from the pond in crop production is anticipated not more than the increased value of crop production.

We can adopt the Doorenbos and Kassam formula to evaluate the crop yield as a function of evapotranspiration. This method is one of the standard methods vividly used by FAO (Steduto et al. 2009), which is based on the assumption that yields are linearly related to the water consumption by different crops. Thus, the rain-fed yield, Yr, is a segment of the irrigated crop yield Yi. Therefore, the Yr can be calculated as a function of the actual evapotranspiration (ETa) in the rain-fed situation, and PET is the potential evapotranspiration of crops.

$$Y_r = Y_i(1 - k_y)(1 - \frac{ET_a}{PET})$$

Ky is the crop yield response factor that captures the linkages between production and water use by a crop. The above equation is the water production function that can be applied to all crops of Sundarbans.

Hence the water productivity is the difference between maximum irrigated crop yield and rain-fed yield:

$$\Delta Y = Yi - Yr$$

Accurate estimation of actual crop evapotranspiration is very difficult. According to FAO ETa can be estimated from data tables on evapotranspiration rate, available soil water, and wetting intervals. These tables are also complicated to get the value of ETa, and later these tables were replaced by more precise ETa estimations based on water balance estimations. Potential evapotranspiration, PET (ET. rate in the presence of adequate irrigation) can be evaluated by solving a vertical soil water balance equation. The crop-specific PET values can be estimated as ETa estimation, but we have to assume that there is no limitation of water availability to calculate PET values.

Irrigation water requirements (*IWR*) is the difference between crop water requirement (*CWR*) and rain-fed crop water consumption, CWCr. This can be calculated through the cumulative *PET* and the cumulative ET_a during the respective crop's lifetime. Thus, *IWR* can be expressed as

$$IWR = \frac{(CWR - CWC_r)}{e}, (m^3/ha)$$

where *e* implies irrigation efficiency, the ratio of irrigation water taking out from the pond, and water consumption for irrigation (i.e., the amount of water leakage at the time of irrigation). The types of irrigation systems can estimate the values of e.

Therefore the rise in production, ΔY (tons ha^{-1}), meets the expense of irrigation which is nothing but an amount of irrigation water, i.e. *IWR* (m^3 ha^{-1}).

Then the irrigation water use efficiency of a specific crop can be estimated as $IWUE = \Delta Y/IWR$ (tons m^{-3}). If the farm gate price of that particular crop is Pc (Rs/ton), then the maximum value of water (Vw) for that crop can be estimated as

$$\text{V.W.} = \text{Pc} \times \text{IWUE} = \text{Pc} \times \Delta\text{YIWR}$$

9.3.1.3 Assessment of Fish and Shellfishes

Several studies reported the aquatic, wetland-dependent, and wetland-associated fauna from ponds of Sundarbans. The freshwater wetlands are inhabited by a wide diversity of vertebrate and invertebrate faunal components. Therefore, only those genuinely aquatic or those associated with or directly dependent upon the wetlands were included in this estimation. A total of more than 20 species of fish have been identified from various ponds of Sundarbans. The Family Cyprinidae is represented by 17 species among the different families, including Indian major and minor carps and weed fishes. Some catfishes, mussels, and mud eels are quite common in occurrence in ponds of two districts of Sundarbans.

Fishes are the marketed products, and their per unit (kg) value is well defined. So overall production values of the fishes can be measured through the market transaction method. However, this is a conservative estimate as we do not take into account the costs of seeds, feed cost, cost of collection, etc., which are spent from the previous year's savings. Therefore, the estimation of beneficiaries of the pond is not exact.

Also, it is revealed that no such feed is used and the total value of fisheries is equal to {total production value—(cost of Seed + Cost of Collection)}. Some people surrounding the pond collecting small fish and shellfishes from the wetland. Most of the households collect small fish and shellfishes from ponds except during the rainy season. According to our Focus Group Discussion (FGD) on an average of over 8 months per year, the households collect mollusks which save on an average a cost of Rs. 360 per month per household. In some cases, the villagers revealed that people from lower strata of the society collect mollusks and sell them in the local market at a remunerative price. Villagers were asked to reveal the amount that they have to pay to purchase those items that they collect from the pond. The well-defined market price made it easier for them to tell the amount they save each month by collecting various small fishes and shellfishes from the pond.

9.3.1.4 Domestic Use

Ponds in Sundarbans serve as a primary source of water for the surrounding villages. Apart from the pond, other open water sources like bore wells, tube wells are mostly saline. According to the villagers, the pond water is supporting domestic uses in the surrounding houses. The weighted average cost for filtered water in a treatment plant is Rs.1.38/kiloliter. To estimate the value of domestic water at the household level, we can use this treatment cost multiplied by the average daily per capita water consumption in the household.

9.3.2 Assessment of Regulating Services

9.3.2.1 Carbon Sequestration

Globally, wetland ecosystems are recognized as important carbon sinks (Maltby and Immirzi 1993; Page et al. 2011). However, there is less information regarding the carbon dynamics of wetland types, such as coastal wetlands. Water allocations are increasing to maintain the ecological health of regulated floodplain wetlands. The wetlands store almost 20–25% of the world's soil organic carbon, out of which the terrestrial wetlands hold a larger proportion than their marine counterparts (Gorham 1991). In addition, wetlands contribute near about 40% of the global methane emissions, which have the highest carbon density among terrestrial ecosystems and have greater carbon sequestration capacities (Pant et al. 2003).

Apart from that, the greenery catchment also serves as a carbon sink. The figures for carbon sequestration were considered by Eid and Shaltout (2013), who reported that per hectare carbon sequestration benefit ranged from 14.9 and 8.6 g Carbon per m^2 per year. Considering the current land use and land cover (LULC) of Sundarbans, we can go for a conservative estimation of carbon sequestration. Since certified emission reduction (CER) markets are depressed and not a fair reflection of the importance of carbon sequestration to the human community, the European Union Voluntary Emission Reduction (EU VER) price of 10 USD/CER around 700 INR per ton of carbon was calculated here (Forest Trends Ecosystem Marketplace 2020).

9.3.2.2 Microclimate Regulation

Micro-climate regulation is an important ecosystem service of wetlands. The wetland essentially regulates the climatic conditions in its vicinity by improving the evaporation-transpiration process. Existing literature suggests that the monetary worth of micro-climate regulation service from wetlands ranged from INR 3300 to 29,040 per hectare (Costanza et al. 1997; Torras 2000; Pearce 2001). Based on the above literature, by using the mean value of the service, we can use a conservative estimation of the microclimate regulation for the ponds of Sundarbans.

9.3.3 Assessment of Cultural Services

9.3.3.1 Ecotourism

If marketed correctly, these sites have immense possibilities to appear as a significant Eco tourism support. It has all attributes that define ecotourism sites. Assuming that tourism here would be confined to West Bengal for the time being, we estimated a significant chunk of the population spending a substantial amount of money on Sundarbans tourism. Further, we calculated that once the ecotourism developed, 2000 groups of tourists would visit these sites once a year (multiple visits were ruled out, as we were making a conservative estimate). Thus, each tourist (or group) might spend 5% of their annual tourism budget to visit the ecotourism of Sundarbans. Therefore, based on these very conservative assumptions, considerable tourism earnings can be an additional benefit. This gave us a significant part of the total economic benefit from tourism. We, however, did not concern ourselves with the cost of developing institutional mechanisms and infrastructure, apart from depreciation and operations and maintenance.

9.4 Discussion

Natural resource management is one of the biggest challenges in developing countries. In the coastal area, we have a lot of sources of water but mostly saltwater. The pond is an important wetland in the coastal region that can support freshwater for domestic and other uses. So proper valuation of these wetlands is necessary. This paper aims to identify the essential services that determine wetland values and how to estimate the values of the different services. Here we have identified how the ponds influence human welfare the various benefits and characteristics of the pond and valuation methods. It has been observed that the pond size is much essential in terms of the valuation of water retention service. Sundarbans' rural ponds produce locally traded goods and services that are more valuable than other regional wetlands in West Bengal. We have identified that the ponds that provide water regulation and habitation for biodiversity are more valuable in the Sundarbans than cultural services. This means that the conversion of the pond into a supporting service for tourism development can reduce their valuation, even though the conservation may increase valuation.

According to the literature review, it has been found that there are very few studies on the valuation of pond ecosystem services. If we consider other wetland systems, then the review observed that most of the studies on wetland ecosystem services mainly focused on supporting and regulating services. If we compare different ecosystem services, it has been observed that the valuation of provisioning and cultural services is not estimated like the other two services. It is also true that wetland ecosystems are giving more supporting and regulating services than provisioning and cultural services (de Groot et al. 2012). Lack of data and the complexity in understanding services identification are the main reasons to take less initiative for calculating the provisioning and cultural services (Fish et al. 2016). Due to a lack of data and monitoring capacities, most researchers from developing countries are using the benefit-transfer method (de Groot et al. 2012; Xu et al. 2018). Though the benefit transfer is one of the cost-effective methods for policymaking in developing countries, the benefits transfer function may create a lower value than the actual one if we are not aware of the proper application process. Therefore, one must be careful about applying benefit transfer for future policy development and keep in mind space, time, and other dimensions when we are pooling values for different wetland services from different countries. It has been observed that anthropogenic factors are the major drivers to determining pond ecosystem services' valuation. The concept of ecosystem services valuation backed by human wellbeing, and if the ecosystem services are not providing any benefits to human wellbeing, then they have no value (MEA 2005). Like manufactured products and services, humans can change ecosystem services value through land-use change, ecosystem management and utilization rate, and extraction of Ecosystem services (de Groot et al. 2010; Jones et al. 2016). To estimate the valuation of pond Ecosystem services, policymakers must consider the role of anthropogenic factors for changes in economic values due to human activities.

9.5 Conclusion

Valuation of pond ecosystem services is precious because it helps to decide how to enhance the quality of life for those who are exclusively dependent on these crucial lentic water bodies. Valuation of pond ecosystem services gives us a platform to illustrate the concept of market failure and knowledge on why the sustainable use of natural resources is necessary. Market efficiency cannot be taken for granted. Valuation of pond ecosystem services has to begin with the nature of sustainability, based on justice to users. On the other hand, it should be kept in mind that the values of different services should be maximized. In conclusion, we can say that the economic valuation of pond ecosystem services should be based on normative values, should not have error of double-counting, and should follow scientific objectivity. However, the economic evaluation of the ponds of Indian Sundarbans from several standpoints is in its infancy. A vast scope of research lies for the economists to reveal the ecosystem services of these crucial ecosystems.

References

Allen RG, Pereira LS, Raes D, Smith M (1998) Crop evapotranspiration. Guidelines for computing crop water requirements. FAO irrigation and drainage paper no 56

Atkinson G, Mourato S (2008) Environmental cost-benefit analysis. Annu Rev Environ Resour 33:317–344. https://doi.org/10.1146/annurev.environ.33.020107.112927

Barton DN, Saloranta T, Moe SJ, Eggestad HO, Kuikka S (2008) Bayesian belief networks as a meta-modeling tool in integrated river basin management Pros and cons in evaluating nutrient abatement decisions under uncertainty in a Norwegian river basin. Ecol Econ 66:91–104. https://doi.org/10.1016/j.ecolecon.2008.02.012

Bekefi E, Varadi L (2007) Multifunctional pond fish farms in Hungary. Aquac Int 15:227–233

Bianchini F, Hewage K (2012) Probabilistic social cost-benefit analysis for green roofs: a lifecycle approach. Build Environ 58:152–162. https://doi.org/10.1007/s10499-007-9090-5

Céréghino R, Biggs J, Oertli B, Declerck S (2007) The ecology of European ponds: defining the characteristics of a neglected freshwater habitat. In: Oertli B, Céréghino R, Biggs J, Declerck S, Hull A, Miracle MR (eds) Pond Conservation in Europe. Developments in Hydrobiology 210, Springer, Dordrecht. https://doi.org/10.1007/978-90-481-9088-1_1

Céréghino R, Boix D, Cauchie HM, Martens K, Oertli B (2014) The ecological role of ponds in a changing world. Hydrobiologia 723(1):1–6

Céréghino R, Leroy C, Dejean A, Corbara B (2010) Ants mediate the structure of phytotelm communities in an ant-garden bromeliad. Ecology 91(5):1549–1556

Chambers R (1994) The origins and practice of participatory rural appraisal. World Dev 22(7):953–969. https://doi.org/10.1016/0305-750X(94)90141-4

Choi YD, Bury C (2003) Process of floristic degradation in urban and suburban wetlands in northwestern Indiana, USA. Nat Areas J 23:320–331

Cirovic D, Penezic A, Krofel M (2016) Jackals as cleaners: ecosystem services provided by a mesocarnivore in human-dominated landscapes. Biol Conserv 199:51–55. https://doi.org/10.1016/j.biocon.2016.04.027

Corrigan E, Nieuwenhuis M (2016) A linear programming model to biophysically assess some ecosystem service synergies and trade-offs in two Irish landscapes. Forests 7:1–21. https://doi.org/10.3390/f7070128

Costanza R, d'Arge R., De Groot R, Farber S, Grasso M, Hannon B, Limburg K, Naeem S, O'Neill RV, Paruelo J (1997) The value of the world's ecosystem services and natural capital. Nature 387:253–260. https://doi.org/10.1038/387253a0

Crossman ND, Burkhard B, Nedkov S, Willemen L, Petz K, Palomo I, Drakou EG, Martín-Lopez B, McPhearson T, Boyanova K, Alkemade R, Egoh B, Dunbar MB, Maes J (2013) A blueprint for mapping and modeling ecosystem services. Ecosyst Serv 4:4–14. https://doi.org/10.1016/j.ecoser.2013.02.001

De Groot RS, Alkemade R, Braat L, Hein L, Willemen L (2010) Challenges in integrating the concept of ecosystem services and values in landscape planning, management and decision making. Ecol Complex 7(3):260–272

de Groot R, Brander L, Van Der Ploeg S, Costanza R, Bernard F, Braat L et al (2012) Global estimates of the value of ecosystems and their services in monetary units. Ecosyst Serv 1(1): 50–61. https://doi.org/10.1016/j.ecoser.2012.07.005

Downing JA (2010) Emerging global role of small lakes and ponds: little things mean a lot. Limnetica 1:9–24

Eid EM, Shaltout KH (2013) Evaluation of carbon sequestration potentiality of Lake Burullus, Egypt to mitigate climate change. Egypt J Aquat Res 39(1):31–38. https://doi.org/10.1016/j.ejar.2013.04.002

EPCN (2007) Developing the pond Manifesto. Ann Limnol Int J Limnol 43:221–232. https://doi.org/10.1051/limn:2007001

Feng G, Ouyang Y, Read J, Adeli A, Reginelli D, Jenkins J (2015) A farm pond water irrigation management system in Mid–South United States. In: Proceedings of the American Society of Agricultural and Biological Engineers International (ASABE). New Orleans, LA, July 26–29, 2015

Finlayson CM, Davidson NC, Spiers AG, Stevenson NJ (1999) Global wetland inventory—current status and future priorities Mar Freshw Res 50(8):717–727. https://doi.org/10.1071/MF99098

Fish R, Church A, Winter M (2016) Conceptualising cultural ecosystem services: a novel framework for research and critical engagement. Ecosyst Serv 21:208–217. https://doi.org/10.1016/j.ecoser.2016.09.002

Forest Trends Ecosystem Marketplace (2020) Voluntary carbon and the post-pandemic recovery. State of voluntary carbon markets report, special climate week NYC 2020 installment. Forest Trends Association, Washington DC, 21 Sep 2020

Gawne B, Price A, Koehn JD, King AJ, Nielsen DL, Meredith S, Beesley L, Vilizzi L (2012) A Bayesian belief network decision support tool for watering Wetlands to Maximize Native fish outcomes. Wetlands 32:277–287. https://doi.org/10.1007/s13157-011-0255-7

Gorham E (1991) Northern peatlands: role in the carbon cycle and probable responses to climatic warming. Ecol Appl 1(2):182–195. https://doi.org/10.2307/1941811

Guimaraes J, Lowe S (2016) Good management of biodiversity and ecosystem services makes economic sense for farmers and agricultural supply chains. Oryx 50:389–390. https://doi.org/10.1017/S0030605316000466

Gutierrez Estrada JC, Pulido-Calvo I, de la Rosa I, Marchini B (2012) Modeling inflow rates for the water exchange management in semi-intensive aquaculture ponds. Aquac Eng 48:19–30. https://doi.org/10.1016/j.aquaeng.2011.12.009

Howard P, Parker G, Holland T, Marsh D, Jenner N, Parham E (2016) Launch of innovative guidance to protect marine biodiversity and ecosystem services. Oryx 50:389

Hu S, Niu Z, Chen Y, Li L, Zhang H (2017) Global wetlands: potential distribution, wetland loss, and status. Sci Total Environ 586:319–327. https://doi.org/10.1016/j.scitotenv.2017.02.001

IUCN (1997) Fishing for a living: the ecology and economics of fishponds in Central Europe. IUCN, Gland

Jeffries M (2010) Small ponds and big landscapes: the challenge of invertebrate spatial and temporal dynamics for European pond conservation. Aquat Conserv Mar Freshwat Ecosyst 15:541–547. https://doi.org/10.1002/aqc.753

Jones L, Norton L, Austin Z, Browne AL, Donovan D, Emmett BA et al (2016) Stocks and flows of natural and human-derived capital in ecosystem services. Land Use Pol 52:151–162. https://doi.org/10.1016/j.landusepol.2015.12.014

Karmperis AC, Sotirchos A, Aravossis K, Tatsiopoulos IP (2012) Waste management project's alternatives: a risk-based multi-criteria assessment (RBMCA) approach. Waste Manag 32:194–212. https://doi.org/10.1016/j.wasman.2011.09.001

Kloskowski J (2011) Differential effects of age-structured common carp (*Cyprinus carpio*) stocks on pond invertebrate communities: implications for recreational and wildlife use of farm ponds. Aquac Int 19:1151–1164. https://doi.org/10.1007/s10499-011-9435-y

Kull CA, Arnauld de Sartre X, Castro-Larrañaga M (2015) The political ecology of ecosystem services. Geoforum 61:122–134. https://doi.org/10.1016/j.geoforum.2015.03.004

Lai M, Wu S, Dai E, Ying Y, Zhao D (2013) The indirect value of ecosystem services in the three–river headwaters region. J Nat Resour 28:38–50

Li P, Chaubey I, Muenich RL, Wei XM (2016) Evaluation of freshwater provisioning for different ecosystem services in the Upper Mississippi River Basin: current status and drivers. Water 8:1–16. https://doi.org/10.3390/w8080333

Li T, Gao X (2016) Ecosystem services valuation of lakeside wetlands Park beside Chaohu Lake in China. Water 8(7):301. https://doi.org/10.3390/w8070301

Mace GM (2008) Millennium ecosystem assessment. Wiley, Hoboken. https://doi.org/10.1002/9780470015902.a0020478

Maltby E, Immirzi P (1993) Carbon dynamics in peatlands and other wetland soils regional and global perspectives. Chemosphere 27(6):999–1023. https://doi.org/10.1016/0045-6535(93)90065-D

Martinez OJA, Fremier AK, Gunter S, Bendana ZR, Vierling L, Galbraith SM, Bosque-Perez NA, Ordonez JC (2016) Scaling-up functional traits for ecosystem services with remote sensing: concepts and methods. Ecol Evol 6:4359–4371. https://doi.org/10.1002/ece3.2201

McLaughlin DL, Cohen MJ (2013) Realizing ecosystem services: wetland hydrologic function along a gradient of ecosystem condition. Ecol Appl 23:1619–1631. https://doi.org/10.1890/12-1489.1

MEA (2005) Millennium ecosystem assessment. Ecosystems and human well-being: synthesis. Island Press, Washington, DC

Munns WR, Rea AW, Suter GW, Martin L, Blake-Hedges L, Crk T, Davis C, Ferreira G, Jordan S, Mahoney M, Barron MG (2016) Ecosystem services as assessment endpoints for ecological risk assessment. Integr Environ Assess Manag 12:522–528. https://doi.org/10.1002/ieam.1707

Mupepele AC, Walsh JC, Sutherland WJ, Dormann CF (2016) An evidence assessment tool for ecosystem services and conservation studies. Ecol Appl 26:1295–1301. https://doi.org/10.1890/15-0595

Na AN, Yun GN, Liu CE (2008) Wetland degradation in China: causes, evaluation, and protection measures. Chin J Ecol 27:821–828

Natsumeda T, Takamura N, Nakagawa M, Kadono Y, Tanaka T, Mitsuhashi H (2015) Environmental and biotic characteristics to discriminate farm ponds with and without exotic largemouth bass and bluegill in western Japan. Limnology 16:139–148. https://doi.org/10.1007/s10201-015-0453-8

Nelson E, Ennaanay D, Wolny S, Olwero N et al (2011) VEST 2.1 Beta user's guide: integrated valuation of ecosystem services and tradeoffs a modeling suite developed by the natural capital project to support. Invest.ecoinformatics.org.

Newton AC, Hodder K, Cantarello E, Perrella L, Birch JC, Robins J, Douglas S, Moody C, Cordingley J (2012) Cost-benefit analysis of ecological networks assessed through spatial analysis of ecosystem services. J Appl Ecol 49:571–580. https://doi.org/10.1111/j.1365-2664.2012.02140.x

Odgaard MV, Turner KG, Bøcher PK, Svenning JC, Dalgaard T (2017) A multi-criteria, ecosystem–service value method used to assess catchment suitability for potential wetland reconstruction in Denmark. Ecol Indic 77:151–165. https://doi.org/10.1016/j.ecolind.2016.12.001

Ouyang Y, Paz JO, Feng G, Read JJ, Adeli A, Jenkins JN (2017) A model to estimate hydrological processes and water budget in an irrigation farm pond. Water Resour Manag 31:2225–2241. https://doi.org/10.1007/s11269-017-1639-0

Page DW, Khan SJ, Miotlinski K (2011) A systematic approach to determine herbicide removals in constructed wetlands using time-integrated passive samplers. J Water Reuse Desalin 1(1):11–17. https://doi.org/10.2166/wrd.2011.021

Pant HK, Rechcigl JE, Adjei MB (2003) Carbon sequestration in wetlands: concept and estimation. Food Agric Environ 1(2):308–313

Pearce D (2001) Valuing biological diversity: issues and overview OECD: valuation of biodiversity benefits; selected studies. OECD, Paris

Pechar L (2000) Impacts of long-term changes in fishery management on the trophic level water quality in Czech fish ponds. Fish Manag Ecol 7:23–31. https://doi.org/10.1046/j.1365-2400.2000.00193.x

Prach K, Tolvanen A (2016) How can we restore biodiversity and ecosystem services in mining and industrial sites? Environ Sci Pollut Res 23:13587–13590. https://doi.org/10.1007/s11356-016-7113-3

Ramsar (1971) Ramsar sites information service. https://rsis.ramsar.org/

SEPA (2000) Ponds, pools, and lochans: guidance on good practice in the management and creation of small waterbodies in Scotland. Scottish Environment Protection Agency

Sheng Q, Shen SH, Ze GU (2007) Conversion coefficient between small evaporation Pan and theoretically calculated water surface evaporation in China. J Nanjing Inst Meteorol 56:193–204

Shiklomanov IA, Rodda JC (2003) World water resources at the beginning of the twenty-first century, vol 13. Cambridge University Press, Cambridge

Son J, Kim M, Lee S, Kang D, Kang B (2014) The assessment of conservation value for agricultural pond wetland using the weighted function of modified RAM. J Korean Soc Rural Plan 20:13–24. https://doi.org/10.7851/ksrp.2014.20.4.013

Soto D, Aguilar-Manjarrez J, Brugere C, Angel D, Bailey C, Black K, Edwards P, Costa-Pierce B, Chopin T, Deudero S, Freeman S, Hambrey J, Hishamunda N, Knowler D, Silvert W, Marba N, Mathe S, Norambuena R, Simard F, Tett P, Troekk M, Wainberg A (2008) Applying an ecosystem-based approach to aquaculture: principles, scales, and some management measures. In: Soto D, Aguilar-Manjarrez J, Hishamunda N (eds) Building an ecosystem approach to aquaculture. FAO, Rome, pp 15–35

Steduto P, Hsiao TC, Raes D, Fereres E (2009) AquaCrop—the FAO crop model to simulate yield response to water: I. Concepts and underlying principles. Agron J 101:426–437

Ticehurst JL, Newham LTH, Rissik D, Letcher RA, Jakeman AJ (2007) A Bayesian network approach for assessing the sustainability of coastal lakes in New South Wales, Australia. Environ Model Softw 22:1129–1139. https://doi.org/10.1016/j.envsoft.2006.03.003

Torras M (2000) The total economic value of Amazonian deforestation, 1978–1993. Ecol Econ 33(2):283–297. https://doi.org/10.1016/S0921-8009(99)00149-4

UNWWAP (2003) Water for people, water for life. United Nations World Water Assessment Programme. http://unesdoc.unesco.org/images/0012/001297/129726e.pdf

Van Dam RA, Camilleri C, Finlayson CM (2015) The potential of rapid assessment techniques as early warning indicators of wetland degradation: a review. Environ Toxicol 13:297–312. https://doi.org/10.1002/(SICI)1098-2256(1998)13:4<297::AID-TOX3>3.0.CO;2-2

Wang ZM, Zhang SQ, Changchun (2004) Effects of land-use change on values of ecosystem services of Sanjiang Plain, China. China Environ Sci 24:125–128

Williams P, Biggs J, Crowe A, Murphy J, Nicolet P, Weatherby A, Dunbar M (2010) Countryside survey: ponds report from 2007. Technical Report No. 7/07 Pond Conservation and NERC/Centre for Ecology & Hydrology (CEH Project Number: C03259), U.K.

Xu X, Jiang B, Tan Y, Costanza R, Yang G (2018) Lake-wetland ecosystem services modeling and valuation: progress, gaps, and future directions. Ecosyst Serv 33:19–28. https://doi.org/10.1016/j.ecoser.2018.08.001

Zedler JB, Kercher S (2005) Wetland resources: status, trends, ecosystem services, and restorability. Annu Rev Environ Resour 15:39–74. https://doi.org/10.1146/annurev.energy.30.050504.144248

Chapter 10
The Socio-Cultural and Economic Role of Ponds in Delta Communities: Insights from Gosaba Block of the Indian Sundarbans Delta

Sumana Banerjee and Katharine Vincent

Abstract Ponds are intrinsic to deltaic life as they are important for both household-level activities and supporting livelihoods. Based on field observations from areas of the Gosaba Community Development Block (Indian Sundarbans), this chapter presents descriptions of the different kinds of ponds. These descriptions illustrate the form of these different kinds of ponds. Following this descriptive section, the chapter uses qualitative data to document the varied uses of these ponds by different people, applying a gender lens and emphasising their socio-cultural and economic importance. It also explores changes over time, showing how the form and uses are embedded within wider social and economic change, and what might be expected in the future.

Keywords Pond creation · Pond maintenance · Types of ponds · Livelihoods · Agriculture · Fishing · Rituals · Gender · Gosaba · Qualitative data

10.1 Introduction

> "Ekta pukur thaktei hobe."
>
> "One pond has to be there".

This statement by a woman in her early thirties illustrates how important it is for a household in the Indian Sundarbans Delta to have a pond. Based on many field visits to the Indian Sundarbans Delta, we have realised the veracity of her statement during our village walks. In this chapter, we outline the different forms of ponds that exist in the delta and using data from the Gosaba Community Development Block (henceforth Gosaba) with a gender lens to unpack the socio-cultural and economic importance of these ponds, and how it has changed over time.

S. Banerjee (✉)
School of Oceanographic Studies, Jadavpur University, West Bengal, Kolkata 700032, India

K. Vincent
Kulima Integrated Development Solutions Pty Ltd, Postnet Suite H79, Private Bag x9118, Pietermaritzburg 3200, South Africa

S. Das et al. (eds.), *Pond Ecosystems of the Indian Sundarbans*, Water Science and Technology Library 112, https://doi.org/10.1007/978-3-030-86786-7_10

Gosaba CD Block situated within the South 24 Parganas district in West Bengal is one of the 19 CD Blocks which form the Indian Sundarban Biosphere Reserve (Fig. 10.1). According to Census 2011, it has an area of 296.73 sq. km., 50 inhabited villages, a total population of 246,598, a population density of 831 persons per sq. km. and a sex ratio of 959 females per 1000 males. (Government of India 2011). Interspersed by rivers and creeks like Bidya, Gomdi, Gomor, Dutta, this CD Block is connected to the mainland and also has four prominent islands, namely, Gosaba, Bali, Satjelia, and Kumirmari. Gosaba is the most convenient gateway to the Sajnekhali Wildlife Sanctuary and thus a prominent spot for tourism. With a population residing so close to the mangrove forest, amidst the crisscrossing creeks, Gosaba offers a quintessential taste of the delta where land and water interact forming a unique symbiotic ecosystem.

In a terrain where water is in abundance but is mostly saline, it is the ponds that serve as a prominent freshwater source. Rainwater is stored in most of these ponds barring the brackish aquaculture and crab-fattening ponds. Pond ecosystems in the delta are intrinsically connected to the human lives and livelihoods in the delta and owing to changing lifestyles it can be imagined that their uses have also changed over the years.

This chapter relies on both quantitative and qualitative data. The quantitative data is based on Census of India 2001 and 2011 datasets, and other published datasets available on government websites. The qualitative data is derived from multiple field-work visits carried out between the years 2018 and 2021 in the course of academic research projects investigating issues of climate change, sustainability, migration, and

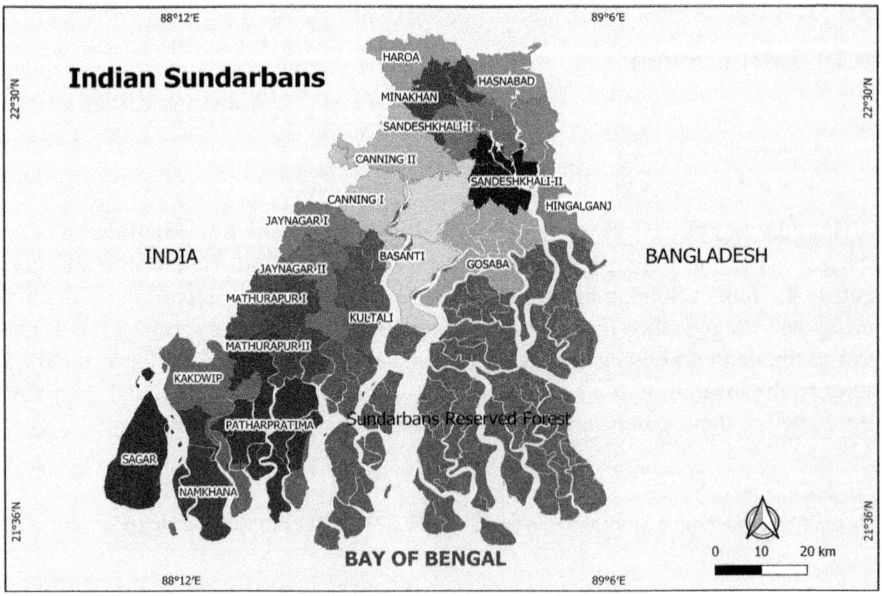

Fig. 10.1 Map showing Gosaba CD Block within the Indian Sundarbans

adaptation in the Indian Sundarbans Delta. Although ponds themselves were not the primary aim of any of this research, their critical socio-cultural and economic roles arose inductively as warranting further investigation. Drawing heavily on ethnographic methods, including participant observation, gender-disaggregated focus group discussions, life-history interviews, and key informant interviews, this chapter aims to share the form of different ponds that are found in the delta, their varied uses, and changes in their form and function over time. Section two outlines the types of ponds. Section three outlines how ponds are created and maintained. Section four looks at the socio-cultural and economic uses of ponds. Section five examines change over time, and how this might evolve in the future.

10.2 Types of Ponds

The types of the ponds are differentiated primarily based on their location, followed by their uses. The names of the types of ponds are indicative of the location or use. Homestead ponds, school ponds, tourist resort ponds, and community ponds are easily identifiable based on their location, i.e., within homestead lands, school premises, tourist resorts, and in common areas of villages. Based on their uses, we have ponds in agricultural fields for irrigation, smaller ponds to facilitate integrated fish-horticulture, ponds meant for both brackish and freshwater aquaculture, and crab-fattening ponds.

10.2.1 Homestead Ponds

Located within the homestead lands, every household has at least one pond for their personal use (Fig. 10.2). These are mostly rectangular or circular in the shape of around 6–7 feet in depth from the ground level. Bamboo, dried logs of trees, wooden planks, or bricks are structured to form the steps to access the water.

The average homestead pond has similar sights in and around them which consists of trees around the banks and other flora and fauna in them. Coconut, banana, and date palm trees are the best suited to be planted along the periphery of the ponds which help in securing the mud around the ponds. Mango and guava trees are also seen. In recent years, eucalyptus trees have seen a spurt after being distributed by government authorities. People also continued growing eucalyptus for its timber value. However, all respondents were unanimous in their opinion about the harms caused by the eucalyptus which included drying of the surrounding soil, degrading the water quality if leaves of the trees fell in the pond. People are now more aware of these cons and consciously avoid planting eucalyptus around ponds.

Some ponds also have growth of water lilies and hyacinth in them. According to all respondents, these grow naturally. The majority of respondents said that water lilies were not harmful but hyacinth adversely affects the quality of water and if

Fig. 10.2 A typical homestead pond in Gosaba

its growth is not checked it infests the entire pond. This proves detrimental for the fishes as the water's supply of sunlight and thereafter oxygen is gradually cut off with this infestation. However, one respondent added that hyacinth growth in ponds when moderated can be useful as the fish can feed on the hyacinth roots and seek temporary shelter under them during the scorching summer heat.

Ducks are also seen in ponds. There was a mixed response on the pros and cons of rearing ducks in ponds. Ducks fertilise the water and are also known as "biological aerators" (Tripathi and Sharma n.d.). Respondents shared that duck droppings help add nutrients to the pond water and also act as fish feed. The movement of ducks not only aerates the water but the disturbance caused in the water makes fishes move around more which helps in increasing their size. However, some respondents shared the cons of duck rearing in ponds but it was applicable only when the water levels were low. People mentioned that mostly during the summers when the water levels go down releasing ducks in the ponds causes the water to become murky which is detrimental to the fishes. This happens as low water levels make it easier for the ducks to stir up the sediments (Provin and Pitt 2013). The solution to this as proposed by one respondent is to barricade a portion of the pond with sticks and nets and release the ducks there during the summers. This can ensure that the pond is replenished with the droppings and the water is also aerated without disturbing the bottom sediments all over the ponds.

Besides flora and fauna, we noticed the unusual presence of extended fencing using bamboo and nets 1.5–2 feet in height around the homestead land of one respondent. The pond was excluded from the fencing. This was a preventive measure to keep toddlers away from ponds. Since the ponds do not have any walls around them, toddlers are susceptible to falling in them. A group of women shared how a toddler had fallen to her death in one of the homestead ponds. Respondents mentioned that ponds become a cause of concern if there are toddlers at home.

10.2.2 Community Ponds

These are normally located in an area intended for communal access. Some are paved with concrete on the edges and have concrete steps leading down to them, while some are not paved and have makeshift steps or a platform to access the water. Some community ponds are or were in the past, designated drinking water sources. The ones used for drinking purposes were not used for bathing, washing, or any other activities to keep the water clean and fit for drinking.

10.2.3 Tourist Resort Ponds

Tourist resorts are common in the delta, and often have a pond within their premises. These ponds too do not have concrete paving on their edges. Some have concrete steps while some have wooden platforms and steps. These ponds are mostly used to store rainwater which is then pumped into overhead tanks and finally used for bathing, washing utensils, washing laundry, and watering the gardens. Fishes are also bred in these ponds for consumption by the hotel staff. Most resorts maximise the recreational and aesthetic aspect of these ponds by encircling them with flowerbeds or adding seating. For city-dwellers who visit the delta to get away from the hubbub of cities, these spots provide relaxation for a cup of tea with friends and families.

10.2.4 Irrigation Ponds

Agricultural fields sometimes have ponds dug out in them to collect rainwater which is necessary to irrigate the crops (Ghosh and Mistri 2020). These can be circular or square in shape. Whilst these ponds are not purposefully stocked with fish, occasionally some fishes have been transported during flooding or waterlogging during monsoons, including *Channa striata* (*shol*), *Anabas cobojius* (*koi*) and *Clarias batrachus* (*magur*). After the harvest of the crops, the fishes are sold off, water from the pond is extracted and the pond bed is dredged, preparing it for the next cultivation season. Sometimes surplus water after harvesting may be used for bathing livestock (cows and goats).

10.2.5 Integrated Fish-Horticulture Ponds

Integrated fish-horticulture ponds are rare in Gosaba, but may be found elsewhere in the delta (Dubey et al. 2016). Those that are found in Gosaba are long and shallow, around 3 to 4 feet. The pond is used to breed fishes whilst dykes alongside are used for the cultivation of vegetables for sale. Vegetables that are irrigated in this way may include brinjal, ladies-finger, tomatoes, bitter gourd, pumpkin, bottle gourd, and others (Tripathi and Sharma n.d.).

10.2.6 Crab-Fattening Ponds

Crabs from the delta are exported but before they attain the "export-quality" accreditation, the crabs are released in crab-fattening ponds to fatten them up. Crabs collected by local people, mostly women, are bought by crab-farm owners. Smaller-sized crabs are kept in these crab-fattening ponds and fed till they attain a certain bodyweight that is fit for export. Not all crab-fattening ponds look the same and both mud and concrete paved ones exist. The more affluent crab-farm owners can make the concrete ones. These ponds usually have nets on the sides to prevent the crabs from crawling out.

10.2.7 Aquaculture Ponds

Although less common than a decade ago, brackish water and freshwater aquaculture are both practiced in the delta in mostly square or rectangular-shaped ponds. Brackish water aquaculture was seen close to the embankments where getting access to the saline water was easy, which is either pumped into the ponds or channels are cut in the embankments to let the water in during high tide. The ponds for freshwater fish farming are neither very deep nor very steep. Since they are relatively shallow and they take up a large area, of maybe one or two bighas,[1] and contain water of up to four to five feet in depth. One feature of these ponds which makes them easily identifiable is the presence of dried branches of trees in them, sticking out vertically from the water, which acts as a deterrent to the casting of nets to steal the fish. Additionally, in some aquaculture ponds nets are installed to prevent the fishes from escaping out of the ponds during a heavy rainfall event which causes overflowing of the ponds. Date palm (*Phoenix dactylifera*) leaves, tamarind (*Tamarindus indica*) branches, and Indian gum Arabic (*Vachellia nilotica*) branches are mostly used. Prawn aquaculture ponds often have date palm leaves to facilitate the growth of a particular kind of algae on it which serves as food for the prawns. Aquaculture ponds are exclusively

[1] Bigha is a traditional unit of measurement of land area. In West Bengal, it is equivalent to 1333 m^2.

used for fish farming until the water dries up or the quality of the water gets degraded when the pond is put to other uses such as bathing or washing.

10.2.8 Ponds in Schools

Most schools have ponds within their premises. These ponds are used by students to wash their utensils after having their mid-day meals and, in the case of residential schools, may be used for bathing and washing clothes as well. Some of these ponds of day schools served as drinking water sources until even recently.

10.2.9 Ponds for Collecting Freshwater for Wildlife

Dug out in various areas of the Reserved Forest by the Forest Department, these are man-made watering holes for the wildlife (Jalais 2014). A dearth of freshwater in the natural landscape prompts the animals to come to these ponds. Every wildlife sanctuary has a pond dug out which can be viewed from high atop the watchtowers. Tourists often see animals such as deer, wild boars, Bengal monitor lizards, monkeys, and tigers taking a drink from the ponds.

10.3 Pond Creation and Maintenance

The process of creating and maintaining ponds shows gender differences in terms of how men and women are variously involved (Fig. 10.3).

10.3.1 Digging Ponds

The dry pre-monsoon period from mid-February to mid-May is the best time for digging up ponds. This includes digging up new ones and re-excavating existing ones to increase their volume so that the dug-out pits get filled with rainwater in the following monsoon starting around June–July.

Men are tasked with digging up ponds and in areas where there exists a tribal population, tribal men are called for this task due to their perception of having greater physical strength. Normally there are groups of men who are known in the villages for digging up ponds and hired when required. Respondents recollected groups of tribal men from their villages who they have seen from their childhoods to the present day, engaging in the pond and other mud excavation jobs, like repairing embankments.

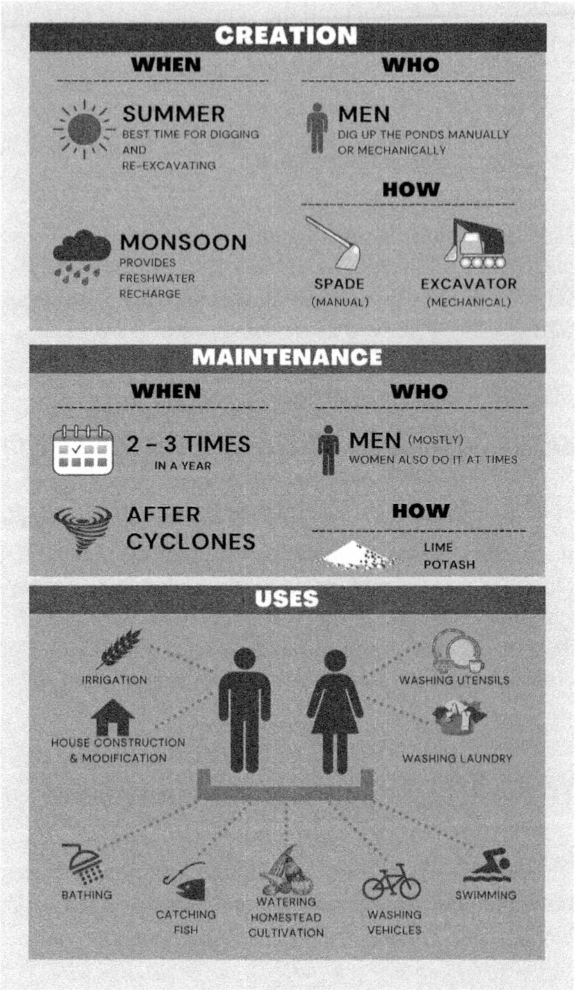

Fig. 10.3 Creation, maintenance, and uses of ponds in the Indian Sundarbans delta

For new ponds, spades are used to dig out the mud, forming V shapes on the vertical façades until eventually the pond is formed. Excavated mud is then dumped elsewhere. For existing ponds that are being re-excavated to increase their volumes, the process is slightly different. It takes place in the dry summer months when water levels are low, but care must be taken not to disturb any fish. A side of the pond is cordoned off using mud resembling a mini embankment within the pond. The water including the fishes are then transferred to this enclosure. The excavation is then carried out on the other sides beyond this enclosure. Whilst manual labour is typically used, mechanical extractors have become popular in recent years, typically for commercial operations such as tourist resorts. Excavators began appearing in 2009 when they were brought in to rapidly repair embankments damaged by Cyclone Aila.

Whilst they are more efficient than manual labour, they come with high financial costs and the potential to damage mangroves en route to excavation sites.

10.3.2 Maintaining Ponds

While digging up ponds is the first step to the creation of a pond, its maintenance is a lifelong duty. This nurturing is done mostly by the men but women are also aware of the processes and can practice those for maintaining the water quality of the ponds.

For people who are not equipped with water quality monitoring equipment, it is the assessment of the colour and turbidity of the water that are the indicators of maintenance need—and these indicators are known by men and women of all ages. A darker green to greyish or blackish colours and murkiness indicate that the water needs attention. Besides the visual indicators, people also mentioned facing skin itches or rashes when the water starts turning bad. The more alarming ones are when dead fishes float up, a film of green or red algae forms at the water surface, or leaves of trees start rotting in the water.

Lime (calcium carbonate) and potash are the go-to products to restore the water quality of the ponds. Lime is quite affordable at an approximate rate of INR 10–11 per kg and is most widely used. Both these products are administered while the fishes are in the pond. Pond maintenance using lime should be done every three to six months and only when the pond has a sufficient quantity of water. The lime is kept in a container soaked in water, after which it gradually melts and releases heat. After it cools down, the container is taken and mixed with a considerable amount of water and then thrown away in the pond thoroughly along the banks. For potash, 100 g to 150 g of it is mixed with water in a container. Then the solution is poured into the entire pond using a mug. The still water of the pond needs to be blended by either striking the surface of the water with sticks or by casting nets.

While using lime and potash are common, some locally available natural products are also used for water quality maintenance. The trunk of the banana tree is finely chopped and thrown in the ponds. According to the residents, the sap from the trunk cleans the water and then the remaining bits decompose and serve as food for the fishes. Cow dung is also used which acts as a fertiliser and food for the fish. As shared by a respondent, those who cannot afford to buy lime or potash resort to using these natural products.

10.4 Uses of Homestead Ponds

From among the wide number of forms of ponds, here we focus on the various uses of homestead ponds and use evidence from interviews, focus group discussions and ethnographic participant observation to provide insights into how this has changed over time (Fig. 10.3).

10.4.1 Drinking

Although once important sources of drinking water, water from homestead ponds are generally no longer used for drinking, having been replaced by groundwater and tap connections. In many places of Satjelia Island, even a decade ago, people used to fetch water from a specific community pond for drinking, although would rarely take it from homestead ponds due to their varied other uses.

Census data from 2001 and 2011 respectively show that only 2.11% and 0.29% of people of Gosaba had "tank, pond, lake" as their source of drinking water (Fig. 10.4)—thus it was already a minimal source, and that has declined over time (Government of India 2011; 2001). Piped water has been provided in many parts of Gosaba through many government initiatives. One such provider is the Joint Forest Management Committee (JFMC) who undertakes developmental activities in the forest fringe villages.

Where "tank, pond, lake" is the source of drinking water, the location is "away" for 73.65% and 96.49% of people in 2001 and 2011 respectively (Fig. 10.5). This indicates that most people travel to fetch water from this source and it can mean that the households with this as the sole source of drinking water either do not have any other water source nearby and/or they are aware of which sources are fit for drinking. "Within premises" as a location dropped from 3.78% in 2001 to 0% in 2011. This is corroborated by the respondents that they no longer use their homestead ponds for drinking. Women are mostly tasked to fetch water, although sometimes men are also seen fulfilling this duty. The location of this water source being away indicates more work for women who are usually tasked for fetching the water. Men and women alike recounted how fetching water from a faraway source used to be time-consuming not only because of the distance but also because of the long queues in case of tap water. A man in his early sixties recollected memories from his childhood of how

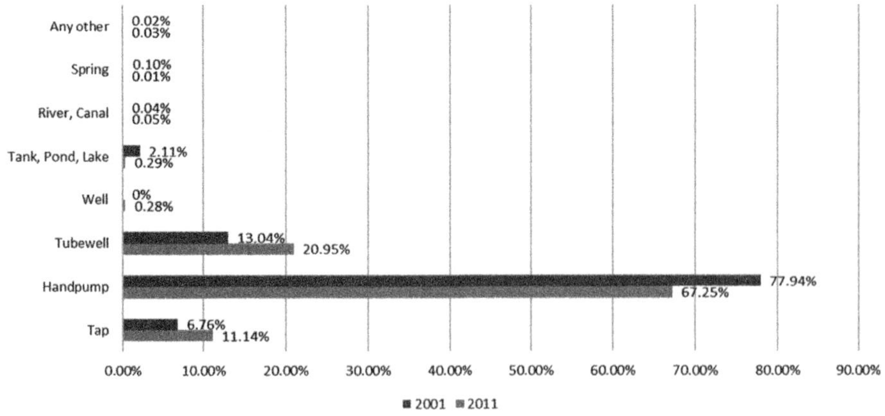

Fig. 10.4 Source of drinking water in Gosaba in 2001 and 2011 (Source Census of India 2001 and 2011)

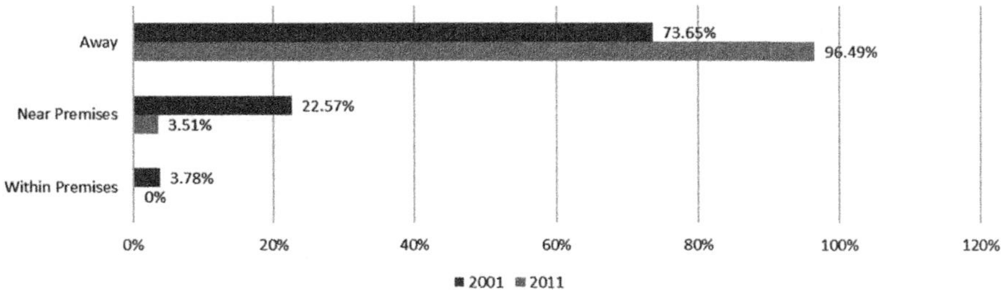

Fig. 10.5 Location of Tank, Pond, Lake as a source of drinking water in Gosaba in 2001 and 2011 (Source Census of India 2001 and 2011)

his mother used to cross a river channel during low tide to go to the other side and get safe drinking water. He recollected how she would fill up a pitcher and carry it on her head and swim back. Sixty years later, the hardship of fetching water has lessened, but the duty of fetching water is still mostly fulfilled by the women of the households.

While trying to look at 1991 Census data, "pond" was not considered as a drinking water source. However, conversations have thrown considerable light on the fact that drinking pond water after boiling was a common occurrence during that time and even earlier. Respondents mentioned how at times the poor quality of pond water would lead to illness.

10.4.2 Bathing

Men and women of all ages use the ponds for bathing. While this use has not changed much over the years, the use of cleaning products has, with implications for the water quality or chemical composition of these ponds over the years. About 25–30 years ago it was the mud from the pond beds which was used as skin and hair cleansers by both men and women. In particular mud scooped out by crabs from the tidal flats was valued for cleansing on the grounds that, since it had been extracted from the deep, it was cleaner than the surface mud. Women mentioned how using that mud used to efficiently clean their hair besides helping in retaining strength, shine, and black colour of the hair. Now that mud has largely been replaced by soap and shampoos, they complained about hair fall and losing the shine of hair. Many older people in the villages or in their families refrain from using modern soaps and shampoos and prefer using the mud from the pond beds. During menstruation, women fetch the pond water in buckets and bathe in the bathrooms rather than in the ponds to avoid contamination.

10.4.3 Washing Laundry

Ponds are the laundromats of the delta, except that all the washing is done by them manually. The steps to access the ponds or the makeshift platforms are the washing spots. The laundry is soaked in buckets using detergent powder and the pond water and then scrubbed using a detergent bar and scrubber on the steps or the platforms. Then the items are washed thoroughly in the ponds and rinsed. Detergent bars and powders have been in use for the past 60 years, with Sunlight bar being Unilever's first brand that was launched in India in 1888 (4 years after the bar's launch in the UK) (Kar 2013; Hindustan Unilever n.d.). Kolkata being the erstwhile capital of the country and a major port received packages of Sunlight. Although the product was launched with an aim to reduce women's labour and increase hygiene, laundry is still considered a "woman's job". We were fortunate to learn from one respondent who mentioned hearing from their grandmother about using the fruit of "Shonajhuri" or Earleaf Acacia (*Acacia auriculoformis*) tree for washing clothes since it had a soapy texture.

10.4.4 Washing Utensils

Like laundry, washing of household utensils is also done by the women. Earlier hay ash and cow dung cakes were used to clean the utensils. Whilst some households continue using those many have shifted to modern dishwashing bars with synthetic scrubs. A respondent mentioned how the use of these products is a result of people going to cities and bringing back urban practices.

10.4.5 Washing Vehicles (Boats, Bicycles, Motorbikes, Vans)

As of 2011, 14.92% of households of Gosaba have bicycles, 0.56% have a scooter/motorcycle/moped, and 1.94% have a car/jeep/van (Government of India 2011). At least in the case of bicycles, the proportion of ownership has increased recently as a result of government schemes, such as the West Bengal State Government's "Sabooj Sathi" scheme which provides bicycles to every student from classes 9–12. This scheme has been operational since 2015 and by February of 2021, 23,430 bicycles have been distributed in schools of Gosaba (Sabooj Sathi n.d.). There may be students in these schools who come from other CD Blocks but that number is not likely to be very high since Gosaba is largely a collection of islands and communication is not very easy. Assuming that all of these beneficiaries are from the Gosaba CD Block, the number of bicycles is around 40% of the total number of households in Gosaba as of 2011. All these assets need maintenance and washing is done using the water from the homestead ponds. With an increase in the assets, it can be deduced that the pressure on the pond water has increased over the past decade.

Another asset which many people possess is boats and pond water also has a connection to it. Boatbuilding is done in the delta. After the wooden planks are put in place using nails and before coating it in paint or tar, it is desirable to use freshwater. The source of this freshwater is the homestead pond. If freshwater cannot be obtained one uses saline water.

10.4.6 Bathing Animals

Most households have livestock, which commonly include cows and goats. While many are bathed in the fields using water from the irrigation ponds, water from the homestead ponds is fetched and the animals are bathed elsewhere.

10.4.7 Irrigation

Homestead ponds are prominent irrigation sources for the crops grown for commercial purposes and vegetables or fruits grown for personal consumption. Fetching water from the ponds and watering the homestead cultivation of a patch of vegetables or fruits for personal consumption is done by men and women alike.

As of 2011, Gosaba had 45.01% total workers, which is the highest in the South 24 Parganas district. Gosaba ranks the highest in the district for both male and female workers at 60.21% and 29.15% respectively (Government of India 2011). Out of its total working population, 73.90% were involved in agriculture with 53.76% male and 20.14% male and female workers respectively (Government of India 2011). With such a significant proportion of the population engaged in agriculture, water is clearly essential. Water usage for agriculture in the form of irrigation is done using rainwater harvested in canals and from ponds. Many homestead ponds are extensively used to irrigate agricultural fields.

Households with adjoining fields or relatively nearby fields nowadays use pumps to transfer the water from the ponds. Operating pumps need electricity and extensive areas of Gosaba have very recently been supplied with electricity. Census 2011 data reveals that only 2.68% of households had electricity as their main source of lighting and none of the villages had access to electrical connections for conducting various agricultural activities including irrigation (Government of India 2011). Not all agriculturalists have their agricultural fields in close proximity and for them, the irrigation ponds in the fields work as the source of water. Back in 2018, we met farmers who complained about the lack of irrigation support. However, in recent times canals have been dug up in areas under the West Bengal State Government's "Jal Dharo Jal Bharo" Scheme which aims at water conservation in all kinds of water bodies, viz. tanks, ponds, reservoirs, canals, etc. (Jal Dharo Jal Bharo n.d.). This should benefit the farmers and reduce the pressure on homestead ponds in the future.

Mono-cropping is widely practiced in the delta but many farmers with access to water for irrigation have been able to practice double cropping. Aman paddy is the widely cultivated crop besides the Boro paddy. A field visit in the warm, dry month of March revealed many dry homestead ponds and ones with very shallow water levels since all the water had been used up to water the paddy. During a field visit in the monsoons in June, an enthusiastic farmer who also migrated to Bengaluru seasonally showed us his big homestead pond of 15–16 years which enabled him to practice double cropping. He dug the pond mainly for irrigation purposes and now the fishes breeding in it are consumed by his family as well.

10.4.8 Breeding and Catching Fish

Most household ponds have fishes growing in them which are generally not bred for commercial purposes but personal consumption. The standard diet in the delta comprises rice, dal, and fish. Fish is a staple for the population and most of it comes from the homestead ponds. Breeds of fish such as *Labeo rohita* (rohu), *Catla catla* (catla), *Cirrhinus cirrhosis* (mrigel), *Puntius ticto* (pnuti), *Hypophthalmichthys molitrix* (Silver carp), are often bred in these ponds. Additionally, *Channa striata* (*shol*), *Anabas cobojius* (*koi)* and *Clarias batrachus* (*magur)* are also found in the ponds. *Magur* is nutritious and one tiger-widow in her late twenties reported going to the pond to collect it for her ailing daughter.

Catching fish is a skill which one learns first in the homestead ponds using fishing rod and line and also using casting nets. The duty of catching fishes is fulfilled by both men and women but more men and younger women respondents perceived it as men's task. However, intersectional contextualisation must be done to understand the variances in response. For the male respondents, most of them had more than one adult male family member besides themselves and at any given point of time at least one male member of the family was at home. For the women in their early twenties too, this was true as they had at least one adult male family member in their homes, and the women including themselves were involved in cooking, taking care of the household, and in child care duties. However, women in their 30 to 50 s said that it is the women who are tasked to catch fish from the ponds regularly and they are well trained in casting nets. They mentioned that since the men are generally toiling away in the fields during the day or doing other day jobs, it is their duty to catch the fishes, cut and clean them, and cook them, besides their other regular tasks.

From our field observations too, we have seen on multiple occasions, women of different ages and marital status catching fish from their homestead ponds. Back in 2018, when we wanted to interview a late-middle-aged widow, she was borrowing her relative's net to catch a few fish for her lunch from her own homestead pond. She mentioned that she is a single person so she does not need to catch fish regularly and thus does not want to invest in buying a net for herself. Also, as mentioned earlier was the incidence of the tiger-widow on her way to catch fish for her ailing daughter. Thus whilst adult males may have primary responsibility, women also perform the task when men are not around. With male migration on the rise in the delta, this task will increasingly be fulfilled by women. Cultural prohibition of fish consumption by widowed Hindu women is followed. For widowed Christian women, there are no such prohibitions.

10.4.9 House Construction and Modification

Ponds are a critical resource for the construction of both mud and concrete houses. The mud which is dug up from the ponds during the creation or re-excavation of a pond serves a wide range of purposes. Among the types of housing in the Gosaba CD Block, temporary houses were the highest at 75.33% according to the 2011 Census data (Government of India 2011). Temporary houses include ones constructed of mud walls, floor, and thatch or tiled roof. Most mud houses with thatch or tiled roofing in Gosaba are built using the mud dug up from the ponds.

Elevating the plinth level of mud housing is a common climate adaptation practice in the Sundarbans delta (Chakma et al. 2021). In Gosaba, people practice this using the mud dug up from the ponds. A male respondent shared how he raised the plinth level of his mud house after Cyclone Aila in 2009, using the mud by re-excavating the sides of his existing homestead pond.

Construction of concrete houses in the delta has risen in the post-Aila period owing to the housing schemes. West Bengal Government's Gitanjali Scheme provides INR 75,000 for construction of houses to beneficiaries in Sundarbans who are homeless but possess the land and have monthly income less than INR 6000 (Geetanjali n.d.). Those affected by flooding or erosion are accounted for in this. Besides this, the existing Indira Awas Yojana of the Central Government has been incorporated in the present Pradhan Mantri Awas Yojana-Gramin since 2016 (Government of India n.d.). Between 2014 and 2021, 5011 houses have been constructed under the Indira Awas Yojana and 19,309 houses have been constructed under the Pradhan Mantri Awas Yojana-Gramin. So, from the Central Government schemes, 24,320 concrete houses have been constructed in Gosaba. The mud from pond beds is also used during the construction of these concrete houses by acting as a filler between the foundation and the plinth levels. Normally construction sand is used for this filling but not everyone can afford it as its base price is not only expensive for a regular villager but also the added transportation costs of bringing it into the remote islands of the delta make it dearer.

Construction of concrete housing demands large quantities of water during every building stage. It is no surprise that this water is supplied from the homestead ponds. With the trend of building concrete houses on the rise in the delta, the homestead ponds have yet another new use.

10.4.10 Swimming

A critical skill for all residents of the delta, men, and women have all said that their first swimming lessons have been at their homestead ponds. We have seen young boys learning to swim in homestead ponds using the trunk of the banana tree as a floatation device.

10.4.11 Cultural Practices and Rituals

Hinduism is the predominant religion in Gosaba, comprising 88.06% according to the 2011 census. Many Hindu rituals require the use of water. Women mentioned how an early morning wedding ritual of fetching water by groups of married women is fulfilled from the homestead ponds. Known as "Jol Sowa", the ritual requires the bride and groom's mothers to visit the nearest water body in their respective neighbourhoods with a group of married women, where they fill up a brass pitcher, which is to be used to bathe the bride and the groom later in the day (CulturalIndia n.d.). For other rituals around the pond, we saw shallow earthen bowls, locally known as "*malsha*" dumped beside a homestead pond. Upon enquiring about those, we learnt that there had been a death in the family a few months back. The bereavement rituals include the family members cooking vegetarian food without spices and oil and eating out of those bowls (Mitra 1947).

10.5 Ponds in a Changing Environment and Climate

In looking at the uses of ponds, it is already clear that there have been changes over time, for example in the transition from traditional to modern cleaning products, and as a consequence of changing policy, including through support to non-pond sources of drinking water. Human uses of ponds within the delta also take place within the context of a changing environment and climate, which have also led to changing use.

Irrigating the second crop using the pond water affects the water levels. With a changing climate, summers have become more intense and longer, and rainfall has become erratic. This means natural recharge of the pond water is not happening at regular intervals like it used to happen earlier. This lowering of the water levels is harmful to the fishes as they struggle to survive in such low water levels and the scorching summer heat. People often construct makeshift platforms on the ponds and cover those with banana leaves to create shade for the fishes. However, these are temporary fixes and do little to overcome the situation. Some also reported buying water from ponds that have sufficient quantities of water as they have not been used for irrigation purposes. Pumps and pipes are used to transfer the water at INR 100 for

1 hour of water transfer. For a pond of around 6 kathas,[2] a rise in 1 foot of water costs around INR 200–300. But this option of buying water is also not always available if all ponds face the same fate after a prolonged summer period.

A changing climate affects the existing economic opportunities and a lack of income opportunities has acted as a driver for male migration (Safra de Campos et al. 2020). With more males migrating out of the villages, more women are left behind to carry out household and livelihood duties. This means pond uses are to be carried out more by the women. A male respondent mentioned how after a disaster, a single woman who stays with her children has to deprioritise the pond relative to other urgent matters. She would most likely have to pay neighbours to support the pond maintenance. But since the disaster has also affected these people, they will first pay attention to their own ponds and then venture out to work on others' properties. This lapse in time can be detrimental to the health of the ponds.

Most people have mentioned cyclones as the most devastating disasters. With a predicted increase in the incidence of cyclones (IPCC 2018), the delta is most vulnerable to it and the ponds are also a part of it. The impact of cyclones differs depending on their characteristics and the level of the tide with which they coincide. Cyclones Bulbul (2019) and Amphan (2020) caused many trees to be uprooted and branches to be snapped (Nag 2019; Ghosh 2020). When these leaves fell in the pond water, it degraded the water quality of the ponds. Decomposing leaves caused the water to degrade and killed the fishes. Damage to ponds is not limited to trees contaminating the water. Cyclone Yaas in 2021 caused large-scale saline water inundation in Gosaba, with particular implications for freshwater ponds (Sarkar 2021). For brackish water aquaculture ponds, the severity of the water ingress has affected the farms situated close to the embankments.

After every cyclone, pond maintenance has to be amped up, and many NGOs distribute lime to take care of their ponds. This takes secondary importance to houses, which are repaired first if they have been damaged, although the delay can worsen the water quality. If saline water enters the ponds after breaching embankments, the only way to rectify it is to pump out the water. In some cases, ponds have also provided shelters during Cyclone Aila. A woman recounted how her family stayed on a makeshift platform on one of the ponds for days as her house had washed away and the embankment had breached causing the village to flood.

Cyclones and accompanying strong winds and storm surges affect infrastructure and drinking water is one of the most important among them. Flooding, owing to saline water ingress, submerges the sources for drinking water, especially taps, and hand pumps. The uprooting of trees also causes some taps to break. This impedes access to safe drinking water, which may necessitate reliance on relief distribution or travelling further distances to the next available unaffected source. Since fetching water is mostly a woman's duty, this increases women's labour burden. Some respondents mentioned resorting to using pond water and boiling it for consumption as well.

[2] Kathas is a traditional unit of measurement. In West Bengal 1 katha is equivalent to 720 sq. ft.

Besides cyclones, some elderly respondents recollected cloud bursts as the most memorable weather events which brought continuous rain for 4–5 days, and the ponds would fill and overflow. A changing environment also includes changes in land use. Most respondents shared that they could not recollect such changes in land use. One respondent mentioned that the ponds in the agricultural fields do undergo changes. The soil from ploughing the land often gets in the ponds and after a few years, the depth of the pond decreases. Then some might leave the pond barren and fill it up with mud and dig another one elsewhere in the fields. Besides this, we have witnessed one conversion of a pond that was close to an embankment. This plot was sold and developed into a tourist resort, meaning the pond was filled in, although another was one created elsewhere on the site.

10.6 Conclusion

Ponds play a key role in life in deltas. A wide range of ponds exist, differing in their size and water composition and location, for example, homestead ponds, community ponds, tourist resort ponds, irrigation ponds, integrated fish-horticulture ponds, crab-fattening ponds, aquaculture ponds, ponds in schools and ponds for collecting freshwater for wildlife. Creating and maintaining ponds follows particular procedures, also linked to the purpose. Applying a gender lens shows that men and women have different roles in creation and maintenance, and also in the main socio-cultural and economic purposes of the ponds—with men typically playing a larger role in their creation, and women playing a larger role in their use.

Ponds have various socio-cultural and economic uses, and this shows the evolution over time. Drinking, bathing, washing laundry, washing utensils, washing vehicles, bathing animals, irrigation, breeding and catching fish, house construction and modification, swimming, and cultural values are the main uses as observed in Gosaba of the Indian Sundarbans delta. What this looks like exactly has changed over time: for example, their use for drinking water has diminished as alternative sources have become available; and increasingly traditional practices are replaced by modern ones, for example in the use of mud compared with soap and shampoo for cleaning. The evolution of uses is also linked to changing environmental and climate conditions (Nicholls et al. 2020). Ensuring freshwater for irrigation enables crop agriculture to take place, whilst aquaculture and crab fattening are also possible. Cyclones have historically caused significant problems, including causing damage to ponds; although heavy rainfall events and land-use change also have an influence. Since environmental and climate conditions are likely to change in the future, both the form and socio-cultural and economic uses of ponds in the delta are likely to continue to evolve.

Acknowledgements The authors would like to acknowledge the contribution of the countless people of Gosaba who selflessly gave them their time to talk about their ponds and what it means to them. Additionally, the authors would like to acknowledge the support of the

DECCMA (FCDO, UK & IDRC Canada) and UKRI-GCRF Living Deltas Hub (Grant Reference NE/S008926/1) projects during the multiple rounds of fieldwork in which the authors have participated.

References

Chakma S, Paul A, Rahman M, Mithun M, Sunny, A (2021) Impact of climate change and ongoing adaptation measures in the Bangladesh Sundarbans. Preprints 2021, 2021020321. https://doi.org/10.20944/preprints202102.0321.v1

CulturalIndia (n.d.) Bengali wedding, https://www.culturalindia.net/weddings/regional-weddings/bengali-wedding.html. Accessed 29 May 2021

Dubey SK, Kinkar B, Trivedi R, Mandal B, Rout SK (2016) Evaluation on the prevailing aquaculture practices in the Indian Sundarban delta: an insight analysis. J Food Agric Environ 14(2):133–141

Geetanjali (n.d.) Schemes—Geetanjali & Amar Thikana, https://wb.gov.in/government-schemes-details-geetanjali.aspx. Accessed 29 May 2021

Ghosh H (2020) Our Sundarban is unrecognisable: life after cyclone Amphan wrecked the Island. The wire, https://thewire.in/environment/sundarbans-cycle-amphan-ground-report. Accessed 29 May 2021

Ghosh S, Mistri B (2020) Coastal agriculture and its challenges: a case study in Gosaba Island, Sundarban, India. Sp Cult India 8(2):140–154. https://doi.org/10.20896/saci.vi0.747

Government of India (2001) District Census handbook south twenty four parganas, Census of India 2001, Series 20, Part A & B West Bengal, Directorate of Census Operations

Government of India (2011) District Census handbook south twenty four Parganas, Census of India 2011, Series 20, Part XII A and District Census handbook south twenty four Parganas, Census of India 2011, Series 20, Part XII B West Bengal, Directorate of Census Operations

Government of India (n.d.) Pradhan Mantri Awaas Yojana–Gramin. Framework for operations. http://www.iay.nic.in/netiay/Uploaded/English_Book_Final.pdf

Hindustan Unilever (n.d.) Sunlight, https://www.hul.co.in/brands/home-care/sunlight.html. Accessed 29 May 2021

IPCC (2018) Global warming of 1.5°C. In: Masson-Delmotte V, Zhai P, Pörtner HO et al (eds) An IPCC Special Report on the impacts of global warming of 1.5°C above pre-industrial levels and related global greenhouse gas emission pathways, in the context of strengthening the global response to the threat of climate change, sustainable development, and efforts to eradicate poverty. Geneva, WMO.

Jalais A (2014) Forest of tigers. People, politics and environment in the Sundarbans. Routledge, India. https://doi.org/10.4324/9780203085516

Kar S (2013) Sunlight had brighter days. Business standard, https://www.business-standard.com/article/management/sunlight-had-brighter-days-113092301035_1.html. Accessed 29 May 2021

Mitra DN (1947) Mourning customs and modern life in Bengal. Am J Sociol 52(4):309–311

Nag J (2019) Cyclone Bulbul: seven dead, 2.73 lakh affected in West Bengal. Mumbai Mirror, https://mumbaimirror.indiatimes.com/news/india/cyclone-bulbul-seven-dead-2-73-lakh-affected-in-west-bengal/articleshow/71996064.cms. Accessed 29 May 2021

Nicholls RJ, Adger WN, Hutton CW, Hanson SE, Lázár AN, Vincent K, Allan A, Tompkins EL, Arto I, Rahman MM, Hazra S, Codjoe SNA (2020) Sustainable deltas in the Anthropocene. In: Nicholls RJ, Adger WN, Hutton CW, Hanson SE (eds) Deltas in the Anthropocene. Palgrave Macmillan, Cham, pp 247–279. https://doi.org/10.1007/978-3-030-23517-8_11

Provin TL, Pitt JL (2013) Water management: clearing cloudy and muddy water in ponds and Lakes. Texas A&M Agrilife Extension http://publications.tamu.edu/WATER/SCS-2013-02.pdf

Sabooj Sathi (n.d.) Sabooj Sathi, https://wbsaboojsathi.gov.in/v2/. Accessed 29 May 2021

Safra de Campos R, Codjoe SNA, Adger, WN, Mortreux C, Hazra S. Siddiqui T, Das S, Atiglo DY, Bhuiyan MRA, Rocky MH and Abu M (2020) Where People Live and Move in Deltas. In: Nicholls R, Adger WN, Hutton CW, Hanson SE (eds) Deltas in the Anthropocene. Palgrave Macmillan, Cham, pp 153–177. https://doi.org/10.1007/978-3-030-23517-8_7

Sarkar P (2021) Cyclone Yaas: Sunderban rivers flow above danger level. Telegraph India, https://www.telegraphindia.com/west-bengal/calcutta/cyclone-yaas-rivers-passing-through-sunderbans-flow-above-danger-level/cid/1816726. Accessed 29 May 2021

Tripathi SD, Sharma BK (n.d.) Integrated fish-duck farming. In Integrated agriculture-aquaculture. A primer. FAO Fisheries Technical Paper 407, Rome. http://www.fao.org/3/Y1187E/y1187e14.htm

Chapter 11
Assessment of Heavy Metal Concentration in Water and Sediment of Brackish Water Ponds of the Indian Sundarbans

Sourav Das

Abstract Heavy metals in varying quantities exist in the earth's surface as well as natural aquatic bodies. However, beyond a certain threshold concentration, all these heavy metals pose toxicity to a multitude of life forms that come in contact with these metals or ingest substantial quantities through food. The Sundarbans mangrove ecosystem lies at a critical junction between the pristine mangroves in the south and the urbanized setup in the north. This ecosystem receives substantial industrial effluents and signatures of heavy metal accumulation have been observed in the estuarine water column, sediments, floras, and faunas. However, the lentic aquatic systems like that of ponds, which are integral to the daily life of the millions that reside in this zone, have received the least attention. This study was aimed to develop a preliminary idea on certain selected heavy metal accumulation levels in the brackish water ponds of Indian Sundarbans. Concentrations of Cd, Pb, Zn, and Fe were detected in two such ponds of the Indian Sundarbans mangrove ecosystem at two different locations namely Canning (north) and Frazergaunge (south). The observations indicate that immediate anthropogenic activities are increasing the heavy metal levels in the north of the Sundarbans ecosystem. Whereas, the Hooghly River Estuary, which has been carrying a substantial load of heavy metals for a long time has led to significant heavy metal accumulation in the brackish water ponds situated down south through tidal activities.

Keywords Heavy metal concentration · Indian Sundarbans · Brackish water · Sediments · Pond ecosystems

11.1 Introduction

The study was accomplished in the Indian Sundarbans delta, which is a mangrove ecosystem situated on the northeast coastal part of India. Marine water of the Bay of Bengal mixes with river water (fresh water) flowing through the Hooghly River and

S. Das (✉)
School of Oceanographic Studies, Jadavpur University, Kolkata, India

S. Das et al. (eds.), *Pond Ecosystems of the Indian Sundarbans*, Water Science and Technology Library 112, https://doi.org/10.1007/978-3-030-86786-7_11

other tidal rivers like Matla, Vidyadhari, Raimangal, Ichhamati, along with thousands of creeks make the water network of the Indian Sundarbans. The resulting brackish water submerges many low-lying areas of the Indian Sundarbans. Farmers are using such flooded lands filled with saline water for the production of shrimp with temporary embankments. Such brackish water aquacultures are used to culture Tiger shrimp and many other mullets (i.e. *Liza tade, Liza parsia, Lates calcarifer*).

Several researchers noticed that heavy metal concentrations are steadily increasing in the water and sediment of the Indian Sundarbans (Mitra and Chaudhury 1993; Guhathakurta and Kaviraj 2000). However, the fate of metal concentrations from the estuaries into these low-lying areas is very rarely documented. The goal of the present study was to estimate the metal concentration in the sediment and water of the brackish water pond ecosystem of the Indian Sundarbans.

11.2 Materials and Methods

11.2.1 Study Area

Brackish water ponds were selected at two different localities of Indian Sundarbans (Fig. 11.1). Among these two localities, Frazergaunge (B) lies closer to the Bay of Bengal than Canning (A). In these two sampling sites, tidal waters almost nourished all the ponds through narrow creeks. We have selected two brackish water ponds each from Canning (A) and Frazergaunge (B). The most important difference between these two study points is the tide timing and the salinity range of estuarine water. It is expected that the salinity range will be higher in the Frazergaunge than Canning due to the distance from the Bay of Bengal.

11.2.2 Sampling Strategy

Water and sediment samples from the brackish water pond were collected during August 2017 from at least two ponds situated in Canning and Frazergaunge, respectively. Two samples from each location were shared together to make a composite of each sample for Canning (A) and Frazergaunge (B).

The pond water samples were collected into a pre-rinsed plastic container. Water temperature and water salinity were analyzed on-site. Water samples were stored from 50 to 60 cm in depth. Pond water samples were filtered (0.45 μm filters, cellulose nitrate, Millipore) using a plastic syringe (BD Plastipak, 50 mL) for dissolved trace metals. Samples were acidified with HNO_3 (0.24 M). After that samples were kept at 4 °C (in dark environment) until analysis.

For the determination of nitrate and phosphate, filtered pond water was acidified with 1 ml HCl per litre. Acidified water was stored in glass bottles made by

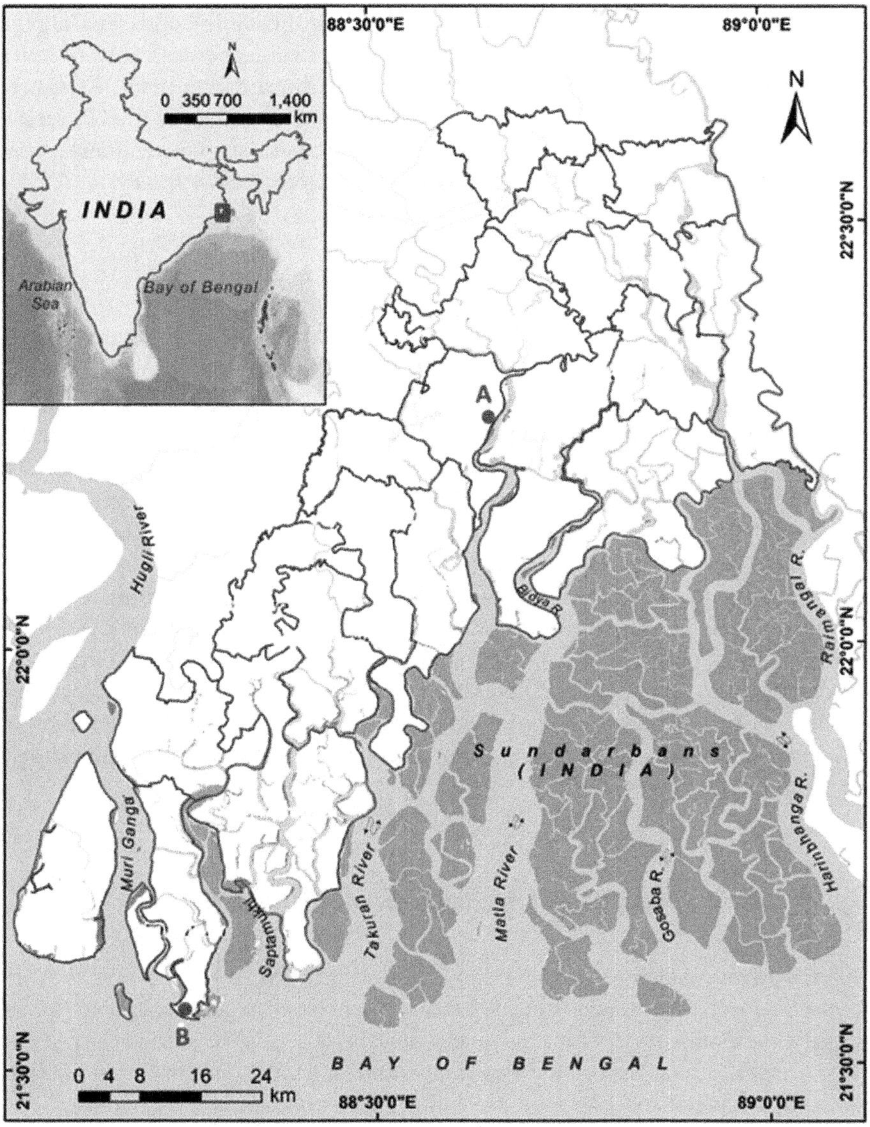

Fig. 11.1 Study area map of the present research

borosilicate. Dissolved oxygen was fixed in water samples collected in 100 ml BOD bottles by using alkaline potassium iodide and manganous sulphate. To determine pH, alkalinity, conductivity, ammonia, and nitrite; water samples were stored in airtight polyethylene bottles. Samples were brought to the laboratory as soon as possible. Water samples were preserved at −4 °C before digestion for trace metal analysis.

Sediment samples were collected randomly from the same ponds of Canning and Frazergaunge using a tubular auger. 03 different sediment cores i.e. (0–10 cm), (10–20 cm), and (20–30 cm) were collected from each sampling point. Each core was packed in clean polyethylene bags (acid-soaked) and brought to the University laboratory (store in ice buckets). In the University laboratory the sediments were dried at 105 °C, then stored in clean polyethylene packets (acid-soaked) at −20 °C.

11.2.3 Analytical Protocol

The pond water salinity was measured using a Multikit (WTW 340i Set; Merck, Germany). Water temperature was measured by micro-pH meter made of Eutech Instruments, Singapore (precision—0.001). Acid washed polyethylene bottles (Tight-capped) were used for sample storing. Whatmann filter papers were used to filter the water samples. 2–3 drops of conc. HNO_3 were added to water samples to attain a pH ~2 to stop the biological growth and precipitation of metals (Kramer 1994; Eaton et al. 1995). Thereafter, samples were stored at 4 °C until analysis.

Physico-chemical parameters of pond water were measured according to APHA (1995). Sediment sample was mixed with deionized water with a ratio of 1:20 for the measurement of conductivity and pH of sediment using digital conductivity meter and digital lectronic pH meter. Dichromate oxidation method is used to determine the total organic carbon in the sediment. Measureable Phosphate concentration of the sediment was measured according to Olsen et al. (1954), Dickman and Bray's (1940) and Jackson (1967).

11.2.4 Total Heavy Metal Measurement of Pond Water

200 ml water were digested with 5 ml of acid mixture (HNO_3: $HClO_4$ = 9: 4). Then digested sample was filtered using Whatman filter (No. 42) and made up to 50 ml using double distilled water for heavy metal analysis (i.e. Cu, Pb, Cd, and Cr) (APHA 2005). Finally, atomic absorption spectrophotometer (AAS) (Perkin Elmer-4000) is used to detect metal concentration (Scientific Research Laboratory, Kolkata).

11.2.5 Total Heavy Metal Measurement of Sediment Sample (Pond)

Dried sediment were screened through a metallic sieve (63 μm). Then one g of the screened sample was extracted with 10 ml mixture of nitric acid/hydrochloric acid (3:1 v/v) at 130 °C (Otte 1991).

AAS was used to determine the metal concentration in the acid solution. Cd, Zn, Pb and Fe were measured at the wavelength of 228.8 nm, 213.9 nm, 217.0 nm, and 248.3 nm respectively. According to Nafde et al. (1998), the accuracy and precision were checked. The minimum detection limit of Cd, Zn, Pb and Fe were 0.01, 0.01, 0.02 and 2. 50 mg l^{-1}.

11.3 Result

Cd, Zn, Pb, and Fe conc. of water and sediment samples are shortened in Table 11.1. Cd conc. in brackish pond water were very less (0.04–0.14 μg ml^{-1}) in both locations while conc. of Fe were very high (15.4–163.7 μg ml^{-1}) in both. Zn value ranged from not detectable to 0.40 μg ml^{-1}. The mean value of Pb was high in the Canning pond water than in Frazergaunge. Metal conc. showed no variation among the locations ($p > 0.05$). Metals conc. were very high in the sediment sample except Cd. Moreover, metal conc. (sediment) varied with the locations and depth (Table 11.1, Fig. 11.2).

Background conc. (at 20–30 cm depth) of Fe in sediment was very high in both study area (Fig. 11.2) while the Zn (background) was not detectable in Canning (A) and Frazergaunge (B), on the contrary, showed a high surface conc. of Zn in the sediment of Frazergaunge (B). A significant disparity among the locations was found in Zn and Fe conc. in the sediment samples. Whereas Pb concentration showed very near values among locations and depths. Background concentration of Cd was nine times higher in Canning than Frazergaunge but the surface concentration was four times higher in the Frazergaunge than Canning. Whereas the mean value of Cd was almost similar in both stations (Table 11.1). A very important observation in this study was the almost similar concentration of Fe in all the depth of both station (Fig. 11.2).

Mean (±SD) of the physicochemical parameters of sediment and water of Canning and Frazergaunge brackish water pond are shown in Tables 11.2 and 11.3, respectively. This study found a higher DO value in the brackish pond water of Frazergaunge than Canning but alkalinity showed the opposite trend (Table 11.2). The

Table 11.1 Concentrations (Mean ± SD) of cadmium (Cd), zinc (Zn), lead (Pb) and iron (Fe) in pond water (μg ml^{-1}) and sediment (μg g^{-1}) from Canning (A) and Frazergaunge (B)

Sampling station	Cd	Zn	Pb	Fe
Water				
A (Canning)	0.12 ± 0.02	<0.01	0.20 ± 0.08	157.5 ± 6.2
B (Frazergaunge)	0.05 ± 0.01	0.29 ± 0.11	0.04 ± 0.02	16.2 ± 0.8
Sediments				
A (Canning)	0.02 ± 0.01	38.2 ± 12.6	12.45 ± 6.11	7822 ± 2005
B (Frazergaunge)	0.02 ± 0.04	630.1 ± 310.1	17.12 ± 3.20	10,280 ± 580

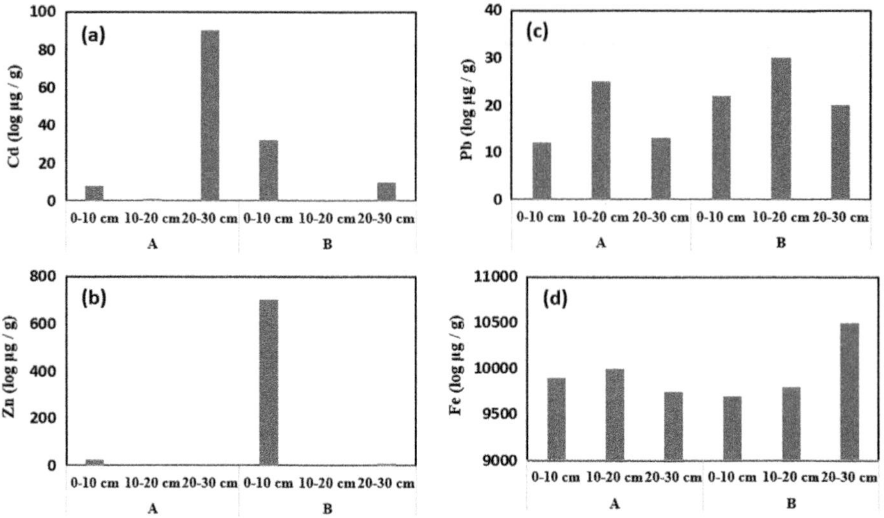

Fig. 11.2 Concentrations of heavy metals at different depths of pond sediment at each sampling station

Table 11.2 Mean values (±SD) of pond (brackish) water quality in of Canning (A) and Frazergaunge (B)

Parameters	Sampling stations	
	A (Canning)	B (Frazergaunge)
Dissolved oxygen (DO) (mg L^{-1})	10.28 ± 2.75	12.11 ± 2.15
Alkalinity (mg L^{-1} as $CaCO_3$)	325.12 ± 270.15	95.13 ± 10.11
Salinity	7.11 ± 4.05	16.11 ± 3.11
Water temp (°C)	29.5 ± 1.2	30.2 ± 0.5
pH	6.9 ± 0.2	7.4 ± 0.1
Nitrate–N (mg L^{-1})	1.94 ± 1.22	2.22 ± 2.11
Phosphate-P (mg L^{-1})	0.50 ± 0.24	0.07 ± 0.03

Table 11.3 Mean values (±SD) of sediment quality in ponds (brackish water) of Canning (A) and Frazergaunge (B)

Parameters	Sampling stations	
	A (Canning)	B (Frazergaunge)
pH at 30 °C	8.17 ± 0.03	8.11 ± 0.01
Conductivity (mmho)	0.92 ± 0.08	0.41 ± 0.19
PO_4^{-3}–P (ppm)	0.32 ± 0.11	0.30 ± 0.05
Organic carbon (%)	1.52 ± 0.06	0.96 ± 0.09

brackish pond water salinity was two and half times higher in the Frazergaunge than Canning. Pond water temperature was almost similar value during August. Whereas mean nitrate values of pond water were nearly similar in both locations but the phosphate mean value was seven times higher in the Canning than Frazergaunge (Table 11.2). The pH of the pond sediment was similar in both the station but pond water pH was more basic in Frazergaunge than Canning pond (brackish) water (Tables 11.2 and 11.3). The sediment of the Canning pond contained higher organic carbon than Frazergaunge.

11.4 Discussion

The present study reported the metal concentrations of the pond (brackish) water and sediment of the two different kind locations of Indian Sundarbans namely Canning (A) and Frazergaunge (B). The Canning area is densely populated and the anthropogenic impact is higher than the Frazergaunge area. Overall, we have found higher metal concentration in the pond water of Canning than Frazergaunge except for Zn. Higher anthropogenic input might be the reason to observe such kinds of results. On the contrary, metal concentration in the sediment of brackish ponds has shown opposite trend i.e. metal concentration of sediment was higher in the Frazergaunge than Canning except for Zn. The above-mentioned result indicated the metal source in the Frazergaunge also. Mitra (1998) and Bastidas et al. (1999) reported that riverine water supply is the leading source of metal pollution in coastal areas. Hooghly Estuary is the principal supplier of metallic pollutants in the Indian Sundarbans (Guhathakurta and Kaviraj 2000). There are also pieces of evidence of heavy metal accumulation in different biota of Hooghly estuary (Kaviraj 1989; Mitra et al. 1995). Frazergaunge is located adjacent to the mouth of the Hooghly estuary (Fig. 11.1) and incessantly obtain industrial pollution from the adjacent cities of Howrah, Haldia and Kolkata through the Hooghly estuary. Chattopadhyay and Saha (1982) reported that a significant amount of Fe and Zn also enters into the Frazergaunge areas from the Bay of Bengal through tidal waters. Whereas the Canning area (Study point A of the present study) is situated about 22 km away from the Bay of Bengal. So there is no freshwater flow from the Hooghly River. Instead of getting fresh water from the Hooghly Estuary, Canning obtains tide induced sea waters through the Matla Estuary. Thus, the heavy metal load is moderately low in the pond sediment except for Fe. From the above-mentioned discussion, we understand that there are two theories to explain the metal concentration of pond water and sediments in the Canning and Frazergaunge area. First of all, recent higher anthropogenic pressure increases the metal concentration in the pond water of Canning area than Frazergaunge. Secondly, the Frazergaunge area has been receiving industrial run-off through Hooghly estuaries for a long time, and the metal concentration is higher in the sediments of the brackish water ponds than in Canning. The Zn conc. has not increased in the different depths of the pond sediment of Canning and Frazergaunge, because of the controlled use of tidal water in aquaculture during last few years.

The present study showed that Fe conc. (sediment) is very high in the pond ecosystem of the Indian Sundarbans. The most probable reason behind this is the increasing use of shallow tube well water in the aquaculture and agriculture of Indian Sundarbans (Mitra 1998; Guhathakurta and Kaviraj 2000) and tube-well (shallow) water contains a high value of Fe leading to higher deposition of Fe (pond sediment). Bansal (1998) also reported that constant drainage of the nearby city's sewage may also lead to such a high level of Fe. Guhathakurta and Kaviraj (2000) also described the geological reason for such a higher value of Fe in the sediment of Indian Sundarbans. So that impact of both reasons (anthropogenic and geological) causing the variation of iron conc. in the water and sediment of the pond of Indian Sundarbans. Zn conc. (sediment) also differs between the A and B. In both the stations (A and B) Zn was identified only from the surface sediment (0–10 cm layer) of the pond representing the anthropogenic source of the metal.

From the literature review, we have found only two studies on the sediment Pb concentration of the Indian Sundarbans. Whereas Mitra et al. (1999) reported the value of 9–20 $\mu g\ g^{-1}$ Pb in the sediment (brackish water ponds) in the Kulti, Indian Sundarbans. Another study conducted in the different regions of the Indian Sundarbans revealed that of 7.2–25.2 $\mu g\ g^{-1}$ Pb conc. in the sediment-of the brackish water pond (Guhathakurta and Kaviraj (2000). The present study detected a little bit higher range (12–30 $\mu g\ g^{-1}$) of Pb in the sediment (brackish water pond) of the Indian Sundarbans due to the increasing anthropogenic pressure.

11.5 Conclusion

Overall, this study exhibited significant Cd, Pb, Zn, and Fe levels in the two sampled ponds from the Indian Sundarbans mangrove ecosystem at two different locations namely Canning (north) and Frazergaunge (south). The analysis revealed that the current anthropogenic pollutant load discharged from domestic and industrial sectors are enhancing the degree of pollution in the Canning region. However, in the southern part, the heavy metal load in the mighty Hooghly River Estuary accumulated substantial quantities of heavy metal in the brackish water ponds through tidal activities. Compared to the number of ponds and varying geomorphic setting, the number of studies are acutely scarce to develop a holistic understanding of the heavy metal pollution scenario in these ponds. The future researchers should emphasize on this aspect as heavy metals in the long run are capable of posing severe threat by bioaccumulating in aquatic fauna like fishes and crabs, and biomagnification to higher life forms including the human beings. Human health risk assessment from heavy metal ingestion and dermal contact has already grabbed the attention of the scientific community working in the Sundarbans. However, the scenario from the ponds of Indian Sundarbans remain poorly constrained to date.

References

APHA (1995) Standard method for examination of water wastewater, 19th edn. American Public Health Association, Washington DC

APHA (2005) Standard method for examination of water wastewater, 21st edn. American Public Health Association, Washington, DC, p 3030

Bansal OP (1998) Heavy metal pollution of soil and ponds due to sewage irrigation. Indian J Environ Health 40:51–57

Bastidas C, Bone D, Garcia EM (1999) Sedimentation rates and metal content of sediments in a Venezuelan coral reef. Mar Pollut Bull 38:16–24

Chattopadhyay GN, Saha MS (1982) Distribution of diethylenetriamine pentaacetic acid extractable iron, manganese, copper and zinc in some brackish water mullet pond sediments of West Bengal, India. J Inland Fisher Soc India 12:56–62

Dickman SR, Bray RH (1940) Colorimetric determination of phosphorus. Ind Eng Chem Anal Ed 12:665–668

Eaton AD, Clesceri LS, Greenberg AE (1995) Standard methods for the examination of water and wastewater, 19th edn. American Public Health Association, Washington, DC

Guhathakurta H, Kaviraj A (2000) Heavy metal concentration in water, sediment, shrimp (Penaeus monodon) and mullet (Liza parsia) in some brackish water ponds of Sunderban, India. Mar Pollut Bull 40(11):914–920

Jackson ML (1967) Sediment chemical analyses. Prentice-Hall, New Delhi

Kaviraj A (1989) Heavy metal concentration in mullet and prawn from the Hooghly estuary. Sci Cult 55:695–699

Kramer KJM (1994) Inorganic contaminants in the water column: sampling and sampling strategy. Int J Environ Anal Chem 57(3):179–188

Mitra A (1998) Status of coastal pollution in West Bengal with special reference to heavy metals. J Indian Ocn Std 5:135–138

Mitra A, Chaudhury A (1993) Heavy metal concentration in oyster Crassostrea cuculata of Sagar Island. Indian J Environ Health 35:119–123

Mitra A, Mandal T, Bhattacharya DP (1999) Concentrations of heavy metals in Penaeus spp. of brackish water wetland ecosystem of West Bengal, India. Indian J Environ Ecoplan 2:97–106

Mitra A, Trivedy S, Gupta A, Bag M, Ghosh I, Chaudhury A (1995) Balanus balanoides as an indicator of heavy metals. Indian J Environ Health 37:42–45

Nafde AS, Kondawar VK, Hasan MZ (1998) Precision and accuracy control in the determination of heavy metals in sediment and water by atomic absorption spectrophotometry. J Indian Assoc Environ Manage 25:83–91

Olsen SR, Cole CV, Watanbe FS, Dean LA (1954) Estimation of available phosphorus in sediments by extraction with sodium bi-carbonate. US Department of Agriculture, CIO, p 939

Otte ML (1991) Heavy metals and arsenic in vegetation of salt marshes and floodplains. Doctoral dissertation. Centrale Huisdrukkerij Vrije Universiteit

Chapter 12
Characterizing the Optical Properties of Chromophoric Dissolved Organic Matter (CDOM) of Two Different Kinds of Pond Ecosystems Situated in Indian Sundarbans

Sourav Das

Abstract Dissolved organic matter (DOM) plays a significant role in the carbon cycling of all types of ponds. Optical characteristics of chromophoric dissolved organic matter (CDOM) were examined between two different kinds of ponds ecosystem (freshwater pond and brackish water pond) in every month for one annual cycle [pre-monsoon, monsoon, and post-monsoon seasons] in Indian Sundarbans. Present study is the first research regarding the CDOM optical properties of the pond ecosystem in the Indian Sundarbans Delta. Annual data set demonstrates that there is a rise of CDOM value in both ponds during monsoon as compared to other seasons. The water salinity was found much higher in Pond B (brackish water) than Pond A (freshwater) throughout the annual period of sampling but the chl-*a* value was higher in Pond A than Pond B. The present study revealed that the CDOM of both types of ponds is allochthonous in the Indian Sundarban, except for post-monsoon season in the freshwater pond. The strong negative relationship between CDOM and salinity described the conservative nature of CDOM in the brackish water pond but non-conservative behavior in the freshwater pond.

Keywords CDOM · Freshwater pond · Brackish water pond · Allochthonous · Conservative · Indian Sundarbans

12.1 Introduction

Pond ecosystems are universally accepted as a support system of significant biodiversity of the world (Deacon et al. 2019). Moreover, ponds provide diverse ecosystem services to human beings and other life forms (Fu et al. 2018). In terms of the basic ecology, there is no difference between the different types of pond ecosystems (De Macro et al. 2014). Moreover, several studies have described that ponds have substantial potential to mitigate climate change and at the same time tackle several

S. Das (✉)
School of Oceanographic Studies, Jadavpur University, Kolkata, India

S. Das et al. (eds.), *Pond Ecosystems of the Indian Sundarbans*, Water Science and Technology Library 112, https://doi.org/10.1007/978-3-030-86786-7_12

water management issues (Quinn et al. 2007). Céréghino et al. (2014) revealed that a 500 m^2 pond can seize and store as much carbon as emitted by a car throughout one annual cycle. Downing et al. (2008) emphasized that owing to the very high numbers of these lentic bodies, they occupy a substantial portion of the land surface and if properly managed they are capable of sequestering equal amounts of carbon as the oceans do at any time. Despite many exciting research findings, the ponds have received very little attention from policy managers and stakeholders throughout the world. In the present date, ponds continue to be one of the most neglected land-use classes.

Moreover, in the pond ecosystem dissolved organic matter (DOM) used as a proxy for water quality (Hudson et al. 2008), is composed of humic like materials, protein substance and carbohydrates and it especially affects the physical transport, chemical transformation and bio-availability of heavy metals in lake or pond ecosystems (Wufuer et al. 2014). Moreover, Guenther and Valentin (2008) described that how high conc. of DOM resulting from adjacent areas of the pond system can release an unpleasant odor and may contribute to eutrophication. Chromophoric DOM (CDOM) is the colored section of DOM that strongly absorbs light in the UV and blue region. CDOM is an optically energetic constituent in the aquatic bodies and affects the remote sensing assessments of the Chl-*a* conc and total suspended matter, attenuation of UV and photosynthetically active energy, global nutrient and carbon cycling, ecosystem productivity, heavy metal transportation, and adsorption, and drinking water treatment (Coble 2007; Gonsior et al. 2014). Therefore, the sources, cycle, transformation and composition of CDOM play an important character in lentic bodies. Moreover, numerous research works have been carried out to explore how hydrological environments in the upstream branches may influence the dynamics and sources of CDOM in downstream-connected ponds or aquaculture ponds in rainy or dry season (Stedmon and Markager 2005). Because CDOM is the most vital watercolor parameters and is also linked to nutrients variability (Liu et al. 2014), certain studies have defined the lake or pond trophic state by nutrient-color paradigm with CDOM absorption values (Webster et al. 2008). Therefore, speedy development in the CDOM research has been experienced, with an upsurge of nearly 160 SCI research articles each year in the past two decades (2000–2020). But the CDOM study on the pond ecosystem is very rarely documented especially in India.

Indian Sundarbans is located in the southernmost part of West Bengal, India. The Indian Sundarbans encompasses the largest mangrove cover (~4264 km^2) in the world (Sánchez-Triana et al. 2014). The Indian Sundarbans comprise a group of 102 islands out of which 54 are populated (Mukherjee and Siddique 2019). The area is a transitional zone between freshwaters coming from the Hooghly River and the saline water from the Bay of Bengal (Akhter et al. 2018). Besides the mangrove forest and an array of diverse flora and fauna, the Indian Sundarbans hosts a highly dense population of 4.4 million people (Census of India 2011). Though agriculture is the principal means of earning a livelihood in this region, aquaculture fish farming practice is also very common among the people of Sundarbans (Dubey et al. 2016). Ponds are integral to the daily life activities of the people of Indian Sundarbans. Because the pond is the main source of water supply for agriculture in a large part of the Indian Sundarbans till data. For fish farming obviously, pond or aquaculture is needed. Natural ponds (freshwater) as well as brackish water, both types of ponds are there in the studied area. Natural ponds are nourished by rainwater only. Freshwater fishes are cultured in natural ponds. On the contrary, brackish aquaculture ponds are nourished by saline River water daily or weekly or bi-monthly or fixed time intervals according to the requirement. There are several pieces of research on the Indian Sundarbans but no research has been documented on the CDOM study of the pond water in Indian Sundarban. In the present era, it is important to know the CDOM variability of different pond waters to describe the lentic (aquatic) ecosystems. The present research aims to describe the optical properties of CDOM of two different kinds of pond ecosystems (freshwater pond and brackish water pond) situated in the Indian Sundarbans.

12.2 Materials and Methods

12.2.1 Study Region and Sampling Plan

We have selected one brackish aquaculture pond (B-Lat: 21° 34′ 37.23″ N, Long: 88° 14′ 53.50″ E) near to the Edward Creek of Frazergaunge (close to Frazergaunge Fishing Harbour), Namkhana Block, Indian Sundarbans (Fig. 12.1) and one freshwater pond (A-Lat: 21° 34′ 45.13″ N, Long: 88° 15′ 41.75″ E) in the same area. After conducting a survey, it is revealed that Pond B (brackish water) is nourished by nearby Edward creek water (saline) weekly and Pond A (freshwater pond) is nourished by

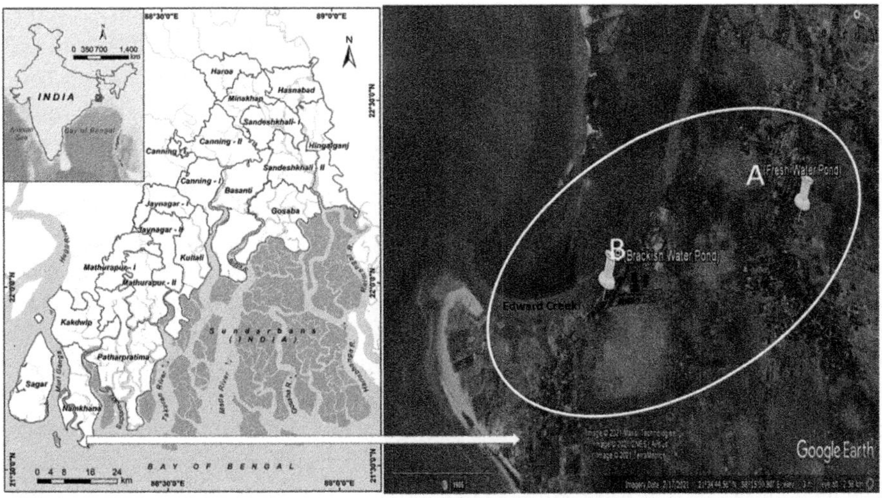

Fig. 12.1 The study area map showing the sampling locations in Indian Sundarbans

rainwater only. To examine the temporal inconsistency of the optical properties of CDOM, the study was conducted for one annual cycle (February 2016-January 2017).

During the study phase, twelve times surveys (sampling) were carried out in each pond. One survey was piloted at each month. 01 sample (mean of triplicate) was collected from each point during surveys. Hence throughout the one annual cycle, in total 24 samples were collected from both ponds. The whole survey was performed during daytime only. Pond water were collected from the surface using amber color glass bottles. Pond water temperature (PWT) and Pond water salinity (PWS) were recorded immediately on site. Samples were stored into pre-rinsed amber colored plastic containers for chlorophyll-a (chl-*a*) and total suspended matter (TSM). CDOM sample was stored in amber color glass bottle (Sasaki et al. 2005).

12.2.2 Analytical Procedure

PWS and PWT were recorded using a Multi-kit (Company:Merck, Made:Germany). To quantify chl-a content, 2 L of water sample was filtered through GF/F glass-fiber filter paper, and the filtrate were kept in a cylinder (liquid nitrogen) before analysis. The samples were extracted (90% aceton) and measured using a spectrophotometer (Shimadzu UV-visible 1600 double-beam) according to Parsons (2013) for chl-a. To quantify TSM, a thoroughly mixed water sample was filtered (by weighed glass-fiber filter, pore size: 0.45 μM), and the residue retained filter paper was dried at 103–105 °C. Electronic Balance (precision of 0.0001 g) was used to weigh the dried filter paper. The following equation is used to calculate the TSM value: TSM [g m^{-3}] = (A − B) × 1000/C (Strickland and Parsons 1972), where, A = (weight of filter + dried residue) [g], B = weight of filter [g], C = volume of water filtered [m^3].

To measure CDOM absorption, samples were stored in glass bottles (amber-colored) for 04 h to equilibrate at room temperature. At first, Whatman GF/F (47 mm) was used to filter the samples for removal of coarse particles. Secondly, 47 mm Nuclepore membrane filter (pore size: 0.2 mm) paper was used to filter the filtered samples again for the removal of fine partilces. Scanning is performed to measure the CDOM absorption from 300 to 750 nm using a spectrophotometer (Shimadzu UV-Visible 1600 double-beam). During scanning procedure a 10 cm (path-length) cuvetter was used. As a reference, Milli-Q water was used. Accorinding to Zaneveld and Pegau (1993), the measured absorbance were normalized to 0 at 600 nm to nullify the temperature dependent artifacts. The CDOM absorbance was calculated according to Sasaki et al. (2005).

The spectral slope (S) has been determined using an exponential regression (Stedmon et al. 2000). According to Para et al. (2010) S was caclculated after relating a non-linear exponential regression to original CDOM absorbance data estimated (range 400–600 nm). The R^2 values (determination coefficients) estimated from the exponential fits were always >0.97. S delivers information regarding CDOM origin (marine vs terrestrial), with commonly lower slopes in fresh and coastal waters compared to the open ocean (Blough and Del Vecchio 2002).

12.2.3 *Statistical Examination*

The regression models were verified and Pearson correlation coefficient was estimated of salinity, TSM, chl-*a*, and S with a_{CDOM} (440). Moreover, one-way ANOVA was tested to differentiate the mean of each parametes in different season. SPSS version 13.0 was used for all statistical analysis.

12.3 Results

12.3.1 *Overall Hydrography of Both Ponds*

The water temperature of the freshwater pond (Pond A) varied over a range of ~21.2 to ~31.8 °C during the course of sampling. The mean (seasonal) water temperature of Pond A was 25.1 ± 2.4, 27.3 ± 1.0, and 22.3 ± 2.2 °C during pre-monsoon, monsoon, and post-monsoon season respectively. The water temperature of the brackish water pond (Pond B) varied over a range of ~20.6 to ~30.6 °C during sampling phase. The mean (seasonal) pond (B) water temperatures were 24.6 ± 1.9 °C, 27.1 ± 1.3 °C, and 21.2 ± 2.0 °C during pre-monsoon, monsoon, and post-monsoon season respectively. The water salinity of Pond A ranged from 0.10 to 0.50 during the annual course of sampling (Table 12.1). A slight seasonal variability of salinity in Pond A was detected. The highest salinity values were observed in the pre-monsoon season (0.3 ± 0.1) followed by the post-monsoon (0.17 ± 0.1). The lowest salinity values was revealed during the monsoon period (0.1 ± 0.1). The water salinity of Pond B ranged from 11.1 to 30.1 during the sampling (Table 12.1). In Pond B, significant seasonal variability of water salinity was detected. The salinity were highest during the pre-monsoon period (28.2 ± 2.1) followed by the post-monsoon (20.2 ± 0.9). The lowest was observed during the monsoon months (13.7 ± 1.1). The water salinity was found much higher in Pond B than Pond A throughout the annual period of sampling (Fig. 12.2a). The TSM of Pond A ranged from 43.9 to 84.0 g m^{-3}. TSM concentrations (Pond A) were higher in the monsoon months (75.0 ± 9.7 g m^{-3}) followed by the post-monsoon (63.3 ± 13.7 g m^{-3}) and pre-monsoon months (61.9 ± 18.7 g m^{-3}). In Pond B, The TSM ranged from 70.1 to 135.7 g m^{-3}. TSM concentrations (Pond B) were higher

Table 12.1 Mean ± standard deviation from mean along with minimum and maximum value (within parentheses) of the annual data set for the parameters salinity, TSM, chl-*a*, a_{CDOM} (440) obtained from the respective ponds

Sampling pond	Salinity	TSM (g m^{-3})	Chl-*a* (mg m^{-3})	a_{CDOM} (440) (m^{-1})
A (Fresh water Pond)	0.19 ± 0.15 (0.1–0.5)	66.6 ± 11.4 (43.9–84.0)	21.18 ± 6.21 (13.2–32.4)	20.4244 ± 3.7159 (15.1133–27.5511)
B (Brackish water Pond)	20.75 ± 7.18 (11.1–30.1)	91.3 ± 20.4 (70.1–135.7)	3.85 ± 1.32 (2.3–6.3)	4.5866 ± 1.4975 (2.2233–6.2238)

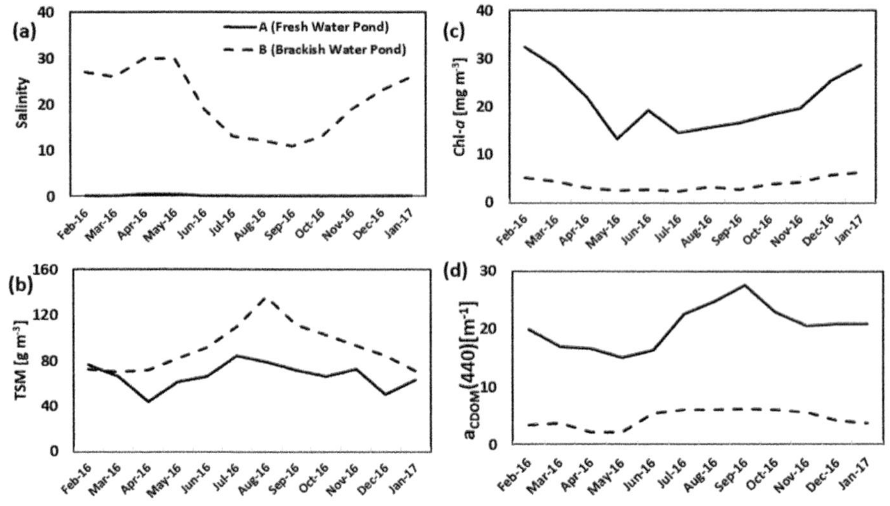

Fig. 12.2 Spatio-temporal variability of **a** Salinity, **b** TSM, **c** Chl-a and **d** $a_{CDOM}(440)$ of two different kind of pond ecosystems (Fresh water and brackish water) in Indian Sundarbans

during the monsoon period (112.1 ± 17.7 g m^{-3}) followed by the post-monsoon (87.5 ± 11.3 g m^{-3}) and pre-monsoon period (73.9 ± 21.1 g m^{-3}). Overall, TSM was higher in Pond B than Pond A in all the seasons (Fig. 12.2b). The chl-*a* of Pond A ranged from 13.2 to 32.4 mg m^{-3}. Seasonal average chl-*a* values (Pond A) as low as 16.5 ± 2.2 mg m^{-3} were detected during monsoon. The highest mean (seasonal) value of chl-*a* (Pond A) was 24.0 ± 1.8 mg m^{-3} in pre-monsoon months (Fig. 12.2c). In Pond B, The chl-*a* ranged from 2.3 to 6.3 mg m^{-3}. The chl-*a* concentrations (Pond B) were higher during the post-monsoon (5.1 ± 1.1 mg m^{-3}) followed by the pre-monsoon (3.8 ± 1.1 mg m^{-3}) and monsoon season (2.7 ± 1.1 mg m^{-3}). However, the chl-*a* value was higher in Pond A than Pond B throughout the annual cycle (Fig. 12.2c).

12.3.2 Variability of Light Absorption Characteristics of CDOM of Fresh and Brackish Water Pond

In Pond A (freshwater), the $a_{CDOM}(440)$ ranged between 15.1133 and 27.5511 m^{-1} during the sampling period and the annual mean value was 20.4244 ± 3.7159 m^{-1}. Whereas seasonal mean value was as lower as 17.1616 ± 2.1159 m^{-1} during pre-monsoon, which increased to 22.8029 ± 1.2112 m^{-1} in the monsoon period, and in the post-monsoon months, it was 21.3088 ± 3.0055 m^{-1} (Table 12.1, Fig. 12.2d). On the contrary, the $a_{CDOM}(440)$ varied between 2.2233 m^{-1} and 6.2238 m^{-1} during the annual cycle in Pond B (brackish water pond). The annual mean $a_{CDOM}(440)$ was nearly 5 times lower (i.e. 4.5866 ± 1.4975 m^{-1}) in Pond B than Pond A. The seasonal mean value was as lower as 2.9186 ± 0.5136 m^{-1} during pre-monsoon, which increased to 5.9504 ± 1.0032 m^{-1} in the monsoon months and the post-monsoon season, it was 4.8908 ± 0.8061 m^{-1} (Table 12.1, Fig. 12.2d).

In Pond A, the spectral slope values (S) varied between 0.011 and 0.029 nm^{-1} during the study period. The annual mean slope was 0.0179 ± 0.0059 nm^{-1}. The mean slope (seasonal) was lower (0.0147 ± 0.0016 nm^{-1}) during monsoon months and higher (0.024 ± 0.0033 nm^{-1}) in the pre-monsoon period. The seasonal mean slope showed a substantial difference during the study. Slope and $a_{CDOM}(440)$ showed a statistically significant exponential relationship (negative) among each other during the study period (Fig. 12.3). The coefficient of determination was found high during the annual cycle ($R^2 = 0.99$, $p < 0.05$). In Pond B, the spectral slope values (S) ranged between 0.04 and 0.049 nm^{-1} during the entire study period. The annual mean slope was 0.0436 ± 0.0032 nm^{-1}. The seasonal mean slope was 0.0472 ± 0.0065 nm^{-1}, 0.041 ± 0.0044 nm^{-1}, and 0.0425 ± 0031 nm^{-1} during pre-monsoon, monsoon, and post-monsoon season respectively. However, the seasonal slope (mean) did not display any significant difference in the entire study (one-way ANOVA: $F = 0.75$, $p = 0.48$). $a_{CDOM}(440)$ and S also revealed a statistically significant exponential relationship (negative) among each in the entire study in Pond B (Fig. 12.3). The coefficient of determination was found high during the annual cycle ($R^2 = 0.96$, $p < 0.05$). The distribution of slope showed that mean slope was considerably higher in the brackish water pond (Pond B) compared to the freshwater pond (Pond A) (one-way ANOVA: $F = 1.78$, $p < 0.05$).

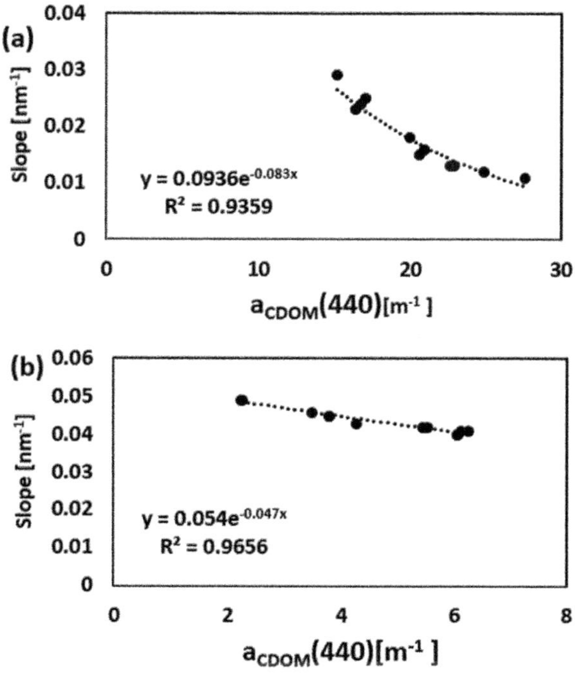

Fig. 12.3 Correlation between $a_{CDOM}(440)$ with slope of **a** fresh and **b** brackish water pond

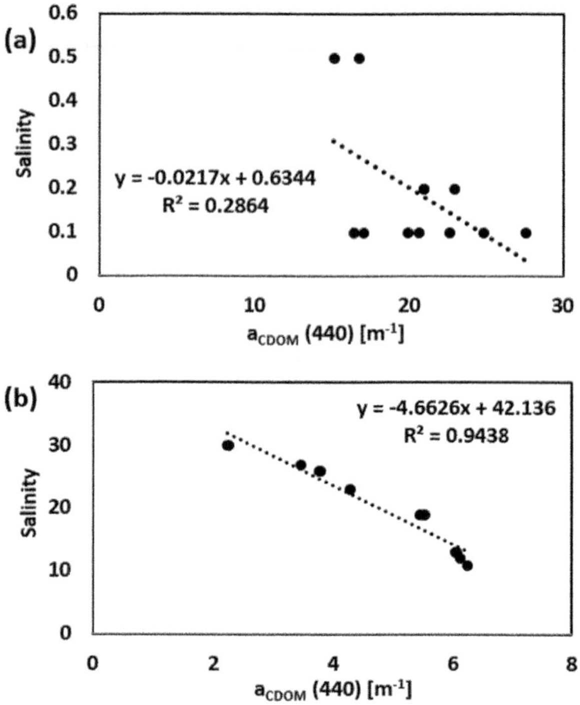

Fig. 12.4 Correlation between $a_{CDOM}(440)$ with salinity of **a** fresh and **b** brackish water pond

12.3.3 Relationship Between CDOM and Other Relevant Hydrographical Parameters of Fresh and Brackish Water Pond

The R^2 values between $a_{CDOM}(440)$ and salinity, TSM, and chl-*a* (seeing the entire dataset) of the freshwater pond (Pond A) depicted no significant linear relation between them (Figs. 12.4a, 12.5a, and 12.6a). When analyzing the data set of the brackish water pond (Pond B), we found significant linear relation of $a_{CDOM}(440)$ with salinity and TSM (Figs. 12.4b and 12.5b). $a_{CDOM}(440)$ did not reveal any such substantial relationship with chl-*a* in Pond B (Fig. 12.6b). Since the annual data of Pond B, $a_{CDOM}(440)$ was correlated with salinity according to the equation [salinity $= -\{4.6626 \times a_{CDOM}(440)\} + 42.136$] ($R^2 = 0.94$, $p < 0.05$) (Fig. 12.4b). $a_{CDOM}(440)$ was also associated with TSM according to the equation [TSM $= \{11.032 \times a_{CDOM}(440)\} + 40.702$] ($R^2 = 0.65$, $p < 0.05$) (Fig. 12.5b). According to the R^2 value, CDOM vs salinity relationship was more significant (strong) than CDOM vs TSM relationship in the brackish water pond (Pond B).

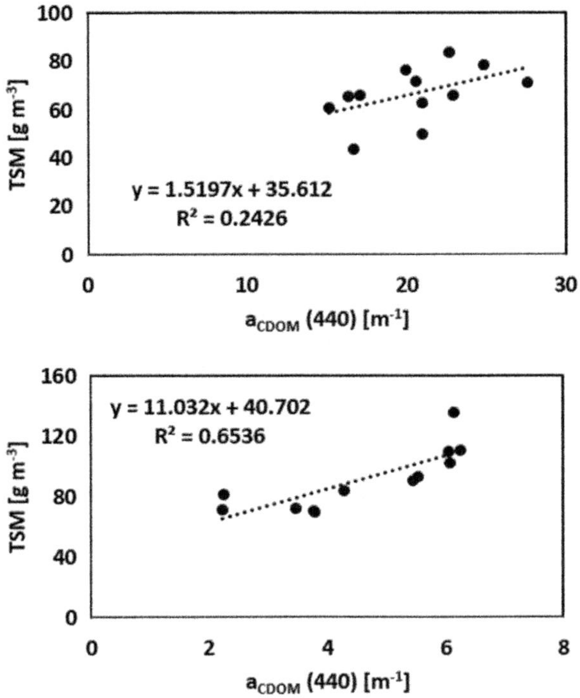

Fig. 12.5 Correlation between $a_{CDOM}(440)$ with TSM of **a** fresh and **b** brackish water pond

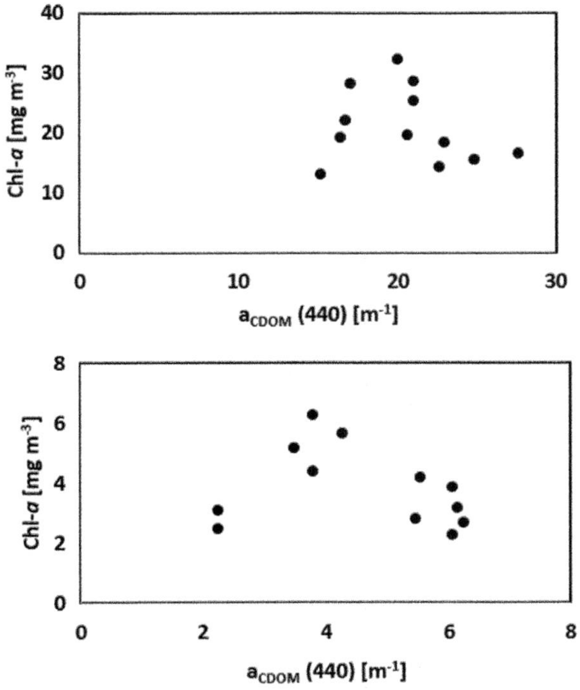

Fig. 12.6 Correlation between $a_{CDOM}(440)$ with chl-*a* of **a** fresh and **b** brackish water pond

12.4 Discussion and Final Remarks

The study region is very close to the northern Bay of Bengal (nBoB) i.e. a archetypal sea- fresh water mixing area of the world. The brackish water pond (Pond B) is nourished by nearby Edward creek water (saline) weekly and Edward creek is connected with nBoB. Therefore, it is expected that CDOM and other physicochemical parameters of Pond B would show the same trend as well nBoB. Das et al. (2017) while working in the inshore of nBoB has revealed substantial temporal variations in salinity and TSM values (particularly comapring the non-monsoon and monsoon months). Hooghly River acts as a perennial origin of freshwater input in this zone, this input is higher during the monsoon. This higher fresh-water input caused to decrease the salinity and an increase in TSM during the monsoon season compared to the non-monsoon. The brackish water pond data also showed the same trend, decreased salinity, and increased TSM values in the monsoon months. Chl-*a* values of the lentic ecosystem (in favorable conditions i.e. light, nutrients, etc.) are higher than the open water system (Bhattacharyya et al. 2020). Chl-*a* magnitudes also showed significant temporal changeability throughout the year in Pond B due to the availability of sunlight and nutrients. Though the annual mean Chl-*a* values were found five times lower in Pond B (brackish water) than Pond A (freshwater). Analyzing the annual data, it is observed that the TSM value is much higher in Pond B than Pond A. So that the light availability was higher in the freshwater pond (Pond A) and thus the chl-a values were higher in Pond A. But the water salinity was negligible in the freshwater pond (Pond A, annual mean salinity $= 0.19 \pm 0.15$) because there was no source to contribute the salinity in Pond A whereas Pond B (brackish water pond) was nourished by saline Edward creek water every week. So that Pond B showed higher water salinity values (annual mean salinity $= 20.75 \pm 7.18$). Guhathakurta and Kaviraj (2000) while working in the brackish water pond of the Indian Sundarban, revealed that the annual mean salinity was 13.01 ± 4.11 and 3.75 ± 5.07 in Sagar Island and Kakdwip respectively. Sagar Island is situated very near to the nBoB and thus showed a higher salinity value in the brackish aquaculture. The present study region is also very near to the nBoB thus showing a higher salinity value in the brackish water pond (Pond B).

a_{CDOM} (440) exhibited a substantial temporal variation in both ponds in uniformity with the dynamics of salinity and TSM, i.e. considerably higher values were detected in the monsoon months and lower values were estimated in pre-monsoon and post-monsoon. But in the freshwater pond, the a_{CDOM} (440) value of post-monsoon season (i.e. $21.3088 \pm 3.0055\ m^{-1}$) is very close to the monsoon season ($22.8029 \pm 1.2112\ m^{-1}$). During the monsoon, terrestrial organic matter runoffs enter the pond ecosystem through rainwater. This might be the reason to find higher a_{CDOM} (440) values in Pond A during monsoon. But the higher value during the post-monsoon season indicated some other issues. Seeing the temporal variations of a_{CDOM}(440) it can be showed that the a_{CDOM}(440) levels increased in the freshwater pond in post-monsoon when a alike increase in chl-*a* was also detected. This might be an sign of autochthonous CDOM production in the freshwater pond (Pond A) since chl-a magnitude designates higher phytoplankton which when degrades could act as a source of CDOM (Hong et al. 2012; Guo et al. 2011). In the brackish water pond, we have found a similar trend like Das et al. (2017) revealed while working in the nBoB. Only the magnitude of $a_{CDOM}(440)$ was higher because of the lesser dilution effect in the Edward creek water.

Spectral slopes indicate the rate at which CDOM absorption declines with increasing wavelength over various ranges. These changes have been attributed to photochemical and biological processing and differences in molecular weight (Osburn et al. 2011). Higher spectral slopes have been attributed to autochthonous material of recent production (Zhang et al. 2011). Wallace (2020) described that spectral slope (S) data decreased from dry season (low rainfall) to wet season (high rainfall) while working on the Hog's Pond, USA. In Hog's pond, in the dry season (low rainfall), spectral slopes S had an mean value was 0.018 nm^{-1}, (standard deviation of S 0.0028 nm^{-1}). In the wet season (high rainfall), spectral slopes S had an mean value was of 0.012 nm^{-1} (standard deviation of S 0.00088 nm^{-1}). In the present study also, we have found a lower value of S in the monsoon season (higher rainfall) in both ponds indicating a very rare chance of autochthonous character of CDOM. Miao et al (2019) while working in the freshwater lake (i.e. Lake Taihu and Chaohu in China) also found the same trend of S (range 0.0154–0.0291 nm^{-1}). During monsoon, terrestrial organic matter runoff to the pond ecosystem through rainwater increases the CDOM values. Hence, the CDOM of fresh water and brackish water pond is allochthonous in the Indian Sundarbans regarding the present study except for post-monsoon season in the freshwater pond.

Several studies carried out in the estuarine water of the world detected that CDOM acted conservatively and presented significant correlations (negative) with salinity (e.g., Bowers et al. 2004; Para et al. 2010; Zhang et al. 2013). The present study also observed a strong negative relationship between CDOM and salinity following the conservative behavior in the brackish water pond. In the freshwater pond, CDOM did not show any significant relation with salinity thus might indicate non-conservative behavior.

In the future, fluorometric estimations of CDOM along with the depiction of the CDOM sources and chemical configuration should be done in the pond ecosystem of Indian Sundarbans for a better understanding of CDOM dynamics. Moreover, we have cover only one block (namely Namkhana) of the Indian Sundarbans in the present study, sample number should be increased to portray the overall CDOM variability of the pond ecosystem in Indian Sundarban. Although, the present study could be a good case study for future researchers regarding CDOM dynamics of pond ecosystems in Indian Sundarban.

References

Akhter M, Reza Ms, JamilA, Uddin MN (2018) Assessment water quality and seasonal variations based on aquatic biodiversity of sundarbans Mangrove forest, Bangladesh. IOSR J Biotechnol Biochem 4(1):06–15

Bhattacharyya S, Chanda A, Hazra S, Das S, Choudhury SB (2020) Effect of nutrient alteration on pCO 2 (water) and chlorophyll-a dynamics in a tropical aquaculture pond situated within a Ramsar site: a microcosm approach. Environ Sci Pollut Res 27(4):4353–4364

Blough NV, Del Vecchio R (2002) Chromophoric DOM in the coastal environment. In: Hansell DA, Carlson CA (eds) Biogeochemistry of marine dissolved organic matter. Press, Amsterdam, Acad, pp 509–546

Bowers DG, Evans D, Thomas DN, Ellis K, Williams PLB (2004) Interpreting the colour of an estuary. Estuar Coast Shelf Sci 59(1):13–20. https://doi.org/10.1016/j.ecss.2003.06.001

Census of India (2011) Government of India

Céréghino R, Boix D, Cauchie HM, Martens K, Oertli B (2014) The ecological role of ponds in a changing world. Hydrobiologia 723(1):1–6

Coble PG (2007) Marine optical biogeochemistry: the chemistry of ocean color. Chem Rev 107(2):402–418

Das S, Das I, Giri S, Chanda A, Maity S, Lotliker AA, Hazra S (2017) Chromophoric dissolved organic matter (CDOM) variability over the continental shelf of the northern Bay of Bengal. Oceanologia 59(3):271–282

De Marco P, Nogueira DS, Correa CC, Vieira TB, Silva KD, Pinto NS, Oertli B (2014) Patterns in the organization of Cerrado pond biodiversity in Brazilian pasture landscapes. Hydrobiologia 723(1):87–101

Deacon C, Samways MJ, Pryke JS (2019) Aquatic insects decline in abundance and occupy low-quality artificial habitats to survive hydrological droughts. Freshw Biol 64(9):1643–1654

Downing JA, Cole JJ, Middelburg JJ, Striegl RG, Duarte CM, Kortelainen P, Laube KA (2008) Sediment organic carbon burial in agriculturally eutrophic impoundments over the last century. Global Biogeochem Cycles 22(1)

Dubey SK, Chand BK, Trivedi RK, Mandal B, Rout SK (2016) Evaluation on the prevailing aquaculture practices in the Indian Sundarban delta: an insight analysis. J Food Agric Environ 14(2):133–141

Fu B, Xu P, Wang Y, Yan K, Chaudhary S (2018) Assessment of the ecosystem services provided by ponds in hilly areas. Sci Total Environ 642:979–987

Gonsior M, Hertkorn N, Conte MH, Cooper WJ, Bastviken D, Druffel E, Schmitt-Kopplin P (2014) Photochemical production of polyols arising from significant photo-transformation of dissolved organic matter in the oligotrophic surface ocean. Mar Chem 163:10–18

Guenther M, Valentin JL (2008) Bacterial and phytoplankton production in two coastal systems influenced by distinct eutrophication processes. Oecol Bras 12(1):15

Guhathakurta H, Kaviraj A (2000) Heavy metal concentration in water, sediment, shrimp (Penaeus monodon) and mullet (Liza parsia) in some brackish water ponds of Sunderban. India. Mar Pollut Bull 40(11):914–920

Guo W, Yang L, Hong H, Stedmon CA, Wang F, Xu J, Xie Y (2011) Assessing the dynamics of chromophoric dissolved organic matter in a subtropical estuary using parallel factor analysis. Mar Chem 124(1–4):125–133. https://doi.org/10.1016/j.marchem.2011.01.003

Hong H, Yang L, Guo W, Wang F, Yu X (2012) Characterization of dissolved organic matter under contrasting hydrologic regimes in a subtropical watershed using PARAFAC model. Biogeochemistry 109(1):163–174. https://doi.org/10.1007/s10533-011-9617-8

Hudson N, Ayoko GA, Collman G, Gallagher E, Dunlop M, Duperouzel D (2008) Long-term assessment of efficacy of permeable pond covers for odour reduction. Biores Technol 99(14):6409–6418

Liu Y, Shen F, Li X (2014, December) Light absorption properties of colored dissolved organic matter (CDOM) in adjacent waters of the Changjiang Estuary during a flood season: implication for DOC estimation. In: Ocean remote sensing and monitoring from space, vol 9261, p 92610K. International Society for Optics and Photonics

Miao S, Lyu H, Wang Q, Li Y, Wu Z, Du C, Lei S (2019) Estimation of terrestrial humic-like substances in inland lakes based on the optical and fluorescence characteristics of chromophoric dissolved organic matter (CDOM) using OLCI images. Ecol Ind 101:399–409

Mukherjee N, Siddique G (2019) Assessment of climatic variability risks with application of livelihood vulnerability indices. Environ Dev Sustain 1–27

Osburn CL, Wigdahl CR, Fritz SC, Saros JE (2011) Dissolved organic matter composition and photoreactivity in prairie lakes of the U.S. Great Plains. Limnol Oceanogr 56:2371–2390

Para J, Coble PG, Charrière B, Tedetti M, Fontana C, Sem-péré R (2010) Fluorescence and absorption properties of chro-mophoric dissolved organic matter (CDOM) in coastal surface waters of the northwestern Mediterranean Sea, influence of the Rhône River. Biogeosciences 7(12):4083–4103. https://doi.org/10.5194/bg-7-4083-2010

Parsons TR (2013) A manual of chemical & biological methods for seawater analysis. Elsevier

Quinn PF, Hewett CJ, Jonczyk J, Glenis V (2007) The PROACTIVE approach to Farm Integrated Runoff Management (FIRM) plans: Flood storage on farms. Newcastle University, Newcastle

Sánchez-Triana E, Paul T, Leonard O (2014) Building resilience for sustainable development of the Sundarbans. The International Bank for Reconstruction and Development, The World Bank, Washington, DC

Sasaki H, Miyamura T, Saitoh SI, Ishizaka J (2005) Seasonal variation of absorption by particles and colored dissolved organic matter (CDOM) in Funka Bay, southwestern Hokkaido. Japan. Estuar Coast Shelf Sci 64(2–3):447–458

Stedmon CA, Markager S (2005) Resolving the variability in dissolved organic matter fluorescence in a temperate estuary and its catchment using PARAFAC analysis. Limnol Oceanogr 50(2):686–697

Stedmon CA, Markager S, Kaas H (2000) Optical properties and signatures of chromophoric dissolved organic matter (CDOM) in Danish coastal waters. Estuar Coast Shelf Sci 51(2):267–278

Strickland JDH, Parsons TR (1972) A practical handbook of seawater analysis. Fisheries Research Board of Canada

Wallace K (2020) Seasonal differences in the optical properties of chromophoric dissolved organic matter at Hoag's Pond, WA. WWU Honors Program Senior Projects, p 428. https://cedar.wwu.edu/wwu_honors/428

Webster KE, Soranno PA, Cheruvelil KS, Bremigan MT, Downing JA, Vaux PD, Connor J (2008) An empirical evaluation of the nutrient-color paradigm for lakes. Limnol Oceanogr 53(3):1137–1148

Wufuer R, Liu Y, Mu S, Song W, Yang X, Zhang D, Pan X (2014) Interaction of dissolved organic matter with Hg (II) along salinity gradient in Boston Lake. Geochem Int 52(12):1072–1077

Zaneveld JR, Pegau WS (1993) Temperature-dependent absorption of water in the red and near-infrared portions of the spectrum

Zhang XY, Chen X, Deng H, Du Y, Jin HY (2013) Absorption features of chromophoric dissolved organic matter (CDOM) and tracing implication for dissolved organic carbon (DOC) in Chang-jiang Estuary, China. Biogeosci Discuss 10:12217–12250. https://doi.org/10.5194/bgd-10-12217-2013

Zhang Y, Yin Y, Liu X, Shi Z, Feng L, Liu M, Zhu G, Gong Z, Qin B (2011) Spatial-seasonal dynamics of chromophoric dissolved organic matter in Lake Taihu, a large eutrophic, shallow lake in China. Org Geochem 42(5):510–519

Chapter 13
Portraying the Nutrient Variability with Relation to Variation in Partial Pressure of Carbon Dioxide in an Aquaculture Pond of Indian Sundarbans

Sourav Das

Abstract A microcosm study has been done to explain the disparities of nutrient variability vis-à-vis the partial pressure of carbon dioxide (CO_2) in water [pCO_2(water)] of brackish water pond of Frazergaunge, Namkhana block, Indian Sundarbans. The present study showed that there are a relationship between the decrease in pCO_2(water) and nutrient removal from the brackish water pond system. Although there are exceptions. The present study revealed that the brackish water pond of Indian Sundarbans acted as an important source of CO_2 during monsoon and pre-monsoon season although having a substantial amount of chlorophyll-*a* in the pond water to make the pond an autotrophic ecosystem. The present research showed that if the optimum photosynthesis can be continued in a shallow aquaculture pond, it can act as a sink of CO_2. During the post-monsoon season (water column pH is high) the ecosystem acted as a CO_2 sink. The present experiment suggests that the controlled lime treatment could be useful to reverse the character of this shallow aquaculture pond.

Keywords Aquaculture pond · Nutrient variability · Photosynthetic potential · Optimum photosysnthesis · Indian sundarbans

13.1 Introduction

Wetlands are sink in nature of carbon dioxide in the tropical areas of the World. Globally, Mitsch et al. (2013) described that tropical wetlands acted as a net carbon sink (~830 Tg C/yr) as well occupied ~5–8% of the land area. Several tropical marshlands or wetlands contained numerous inland water bodies like aquaculture ponds, swamps, riverine wetlands, etc. In India, most of the wetlands are the source of CO_2 (Panneer Selvam et al. 2014). The age and the functioning of the aquaculture pond are the main criteria to be shown the source/sink character in terms of carbon

S. Das (✉)
School of Oceanographic Studies, Jadavpur University, Kolkata, India

S. Das et al. (eds.), *Pond Ecosystems of the Indian Sundarbans*, Water Science and Technology Library 112, https://doi.org/10.1007/978-3-030-86786-7_13

dioxide (Mitsch et al. 2013). St. Louis et al. (2000) described that anthropogenic conversion of wetlands can alter the carbon dioxide flux.

The sink/source environment of the wetlands like aquaculture ponds depends upon the evenness between the community respiration and the primary production (Natchimuthu et al. 2014). In the aquaculture ponds, available resources (nutrients, etc.) and CO_2/pH coordinations play an important character in determining the trophic status (Huszar et al. 1998). Because aquaculture effluents contain high concentrations of organic matter and different nutrients. These effluents have a role to degrade the water quality as well as source/sink nature in terms of carbon dioxide. The biogeochemical processes of tropical lentic ecosystems (e.g. aquaculture ponds) are dependent on two abiotic constituents: dissolved carbon dioxide and nutrients. Dissolved carbon dioxide coexists with numerous dissociation goods like HCO_3^-, H_2CO_3, and CO_3^{2-}. Therefore, primary producers can use them during the process of photosynthesis.

However, CO_3^{2-} and HCO_3^- and are the key suppliers of the total alkalinity and the involvement of dissolved inorganic carbon (DIC) is reliant on the solubilization of particulate matter and pH of the water column (Dickson 1992). In general, tropical inland waters are supersaturated with carbon dioxide (Aufdenkampe et al. 2011). Holgerson and Raymond (2016) portrayed that several factors such as autotrophic primary production, benthopelagic respiration, and remineralization of organic materials to carbon dioxide by biological decay are responsible for variable the net biological pump of the aquaculture pond in the tropical areas. On the other hand, Song (2011) described that the dissolved nutrient pool could alter the carbon sequestration potential and phytoplankton community of aquaculture. According to Downing et al. (1993), the optimal nutrient pool is responsible to increase the utilization rate of carbon as well the increased primary production.

The present research has shown that the waters acquired from aquaculture (brackish water pond) positioned in the Frazergaunge, Namkhana Block of Indian Sundarbans. The objectives of the present research were: Change in nutrient concentration and water pCO_2 with time and the seasonal changeability of the nutrient deduction concerning pCO_2(water) dynamics.

To perform the present study, a microcosm experiment applied the water column's nutrient concentration was not changed by external nutrient addition as the chief determination of the present research was to scrutinize the rate at which pre-existent nutrients are getting drained, hence, the change in pCO_2(water) of the corresponding samples.

13.2 Materials and Method

13.2.1 Study Location and Sampling Strategy

The present study was carried out in pre-monsoon, monsoon, and post-monsoon seasons of the tropical weather. Surface (water) samples were stored from the aquaculture pond (S) of Frazergaunge, Namkhana Block, Indian Sundarbans (Fig. 13.1) around 8:00 a.m., on the 20th January 2018 (post-monsoon), 12th May 2018 (pre-monsoon), and 25th August 2018 (monsoon). Two reservoirs or tanks (made: polyvinyl chloride) was positioned on the roof of the SOS's laboratory, Jadavpur University, Kolkata. The reservoir's height was one m (the depth of the aquaculture pond is also near one meter). Both tanks were filled up with collected sample (aquaculture pond) waters. According to Biswas et al. (2015), eight microcosm bottles (5 L transparent polycarbonate sealed bottle) filled with sample (aquaculture pond surface) waters were nurtured and kept floating in the respective tanks. 01 bottle was treated as control by poisoning the microcosm bottles with $HgCl_2$ (Kattner 1999). Each bottle was mixed moderately to avoid the bottom settlement. The microcosm experiment was carried out for eight consecutive days. Samples were collected each day from one microcosm bottle (floating) respectively. The experiment was accompanied for 15 days, however, in the present research, we have conveyed the experimental data of eight days only, because most of the nutrients were used by the end of the 8th day.

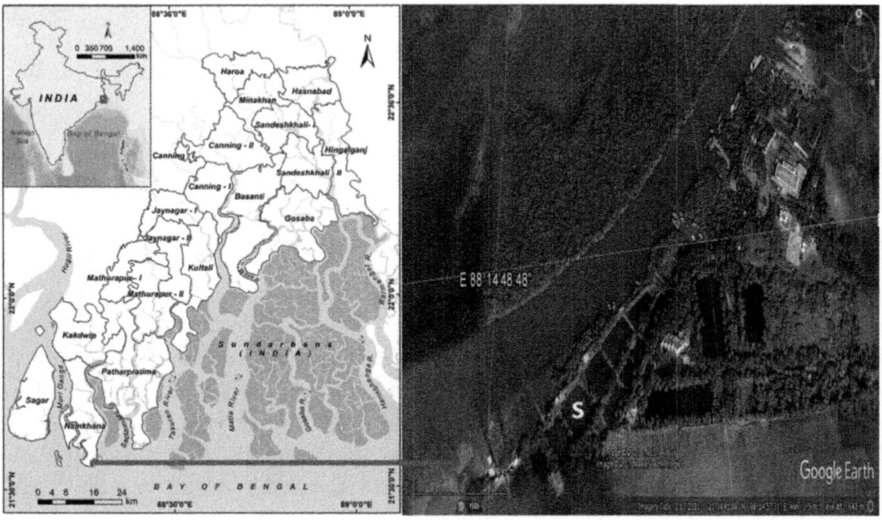

Fig. 13.1 The study area map showing the pond location in Frazergaunge, Namkhana Block, Indian Sundarbans

13.2.2 Analytical Methods

Water temperature and pH were observed with a micro-pH meter (Thermo Scientific, U.S.A.) with a precision of 0.1 °C and 0.001, respectively. The pH meter was calibrated daily (before sampling). During calibration, technical buffers of pH 4.01, pH 7.00, and pH 9.00 were used (at 25 °C). Conductivity was examined using a Multikit (Merck, Germany). A Digital DO meter (Mettler Toledo, USA) was used to monitor the dissolved oxygen (DO) of the samples. The precision of the DO meter was 0.01 mg/L. A standard sensor with a data logger (UWQ 8247, Li-Cor, NE, USA, precision 0.1 μ.mol/m^2/s) was used to observe the underwater photosynthetically active radiation (PAR) of the samples.

Sample turbidity was checked with a digital meter (Eutech Instruments, Singapore) with a precision of 0.1 NTU. A Lux digital meter (Lutron, Czech Republic) was used to monitor the incoming solar radiation on the surface water of samples. According to Parsons et al. (1992), standard spectrophotometric procedures were used to estimate the concentration of nitrite-nitrogen (NO_2–N), dissolved nitrate-nitrogen (NO_3–N), ammonium–nitrogen (NH_4–N), silicate, phosphate, iron with a precision of 0.01 μM. Chlorophyll-*a* was also measured following standard spectrophotometric procedures (Parsons et al. 1992) with a precision of 0.01 mg/m^3. Total alkalinity (TAlk) was monitored using an auto-titrator (Titrando, Switzerland). Dissolved inorganic carbon (DIC) and the pCO_2(water) were calculated through the CO2SYS.EXE software. Water temperature, pH, TAlk, silicate, and phosphate were used as primary data to calculate DIC and pCO_2 (Lewis and Wallace 1998). According to Parsons et al. (1992), the gross primary production (GPP), net primary production (NPP), and community respiration (CR) data were recorded for 08 successive days in pre-monsoon, monsoon, and post-monsoon season.

13.3 Results

13.3.1 Seasonal Changeability (In-Situ Aquatic State) of the Aquaculture Pond

The surface water temperature of the aquaculture pond ranged from ~23 to ~33 °C. Conductivity values ranging between ~389 and ~620 μS/cm (Table 13.1). D.O. was highest in pre-monsoon months (8.8 mg/L) and lowest in the monsoon (7.3 mg/L). Turbidity was comparatively high in the monsoon compared to pre-monsoon and post-monsoon. The in-situ pH was observed as the order post-monsoon > pre-monsoon > monsoon season. The in-situ pCO_2(water) surveyed the opposite order. The higher TAlk value was detected in the pre-monsoon season (5434 μmol/kg) and the lower in the post-monsoon season (2844 μmol/kg). DIC was also found to change similarly to TAlk. Minimum in-situ Chl-*a* concentration was detected during pre-monsoon months (34.2 mg/m3) followed by monsoon (61.2 mg/m3) and

Table 13.1 Initial physico-chemical settings in the start of the microcosm study (all the three seasons)

Parameters	Pre-monsoon	Monsoon	Post-monsoon
Water temperature (°C)	32.4	27.1	23.4
Conductivity (μS/cm)	620	389	512
D. O. (mg/L)	8.8	7.3	7.6
PAR (μmol/m^2/s)	6.2	1.6	2.8
Solar intensity (k lux)	75	52	28
Turbidity (NTU)	83	119	64.5
pH	8.286	7.221	8.661
Total Alkalinity (μmol/kg)	5434	3677	2844
Dissolved inorganic carbon (μmol/kg)	5298	3731	2571
pCO_2 (μ.atm)	1072	2131	334
Fe (μM)	15.3	13.1	21.2
PO_4^{3-} –P (μM)	19.3	3.6	8.1
NO_3^- –N (μM)	7.8	4.3	5.2
NH_4^+–N (μM)	21.2	11.9	11.4
SiO_3^- (μM)	80.5	32.6	67.2
Chl-*a* (mg/m^3)	34.2	61.2	69.4

post-monsoon months (69.4 mg/m3). NO_3^-–N, NH_4^+–N, PO_4^{3-}–P and SiO_3^{2-} were maximum concentrations during pre-monsoon i.e. 7.8, 21.2, 19.3, and 80.5 μM. All the above-mentioned parameters were showed the following order as pre-monsoon > post-monsoon > monsoon except ammonium. Iron was observed maximum in the post-monsoon months and minimum in the monsoon.

13.3.2 Physicochemical Factors—Temporal Trend During the Experiments days (Eight days of Microcosm)

The microcosm test was carried out for 08 consecutive days in pre-monsoon, monsoon, and post-monsoon season. In the post-monsoon and pre-monsoon, pH revealed an early increase followed by a decrease from 2nd to 4th days and started increasing again from 5 to 8th days (Fig. 13.2). During monsoon season, a steady rise of pH was found till the 8th day. The highest increase of pH happened during the pre-monsoon season (Fig. 13.2). DO exhibite an initial boost during the 1st day of the test followed by a firm decrease between day one and day four which is followed by a fringe rise from 5th day to 8th day (Fig. 13.2). Net reduction of DO was considerable in the pre-monsoon season from the initial value (maximum). The DIC and TAlk trends were almost same throughout the all seasons (Fig. 13.2). DIC and Talk gradually declined in monsoon and post-monsoon months. In the pre-monsoon, both

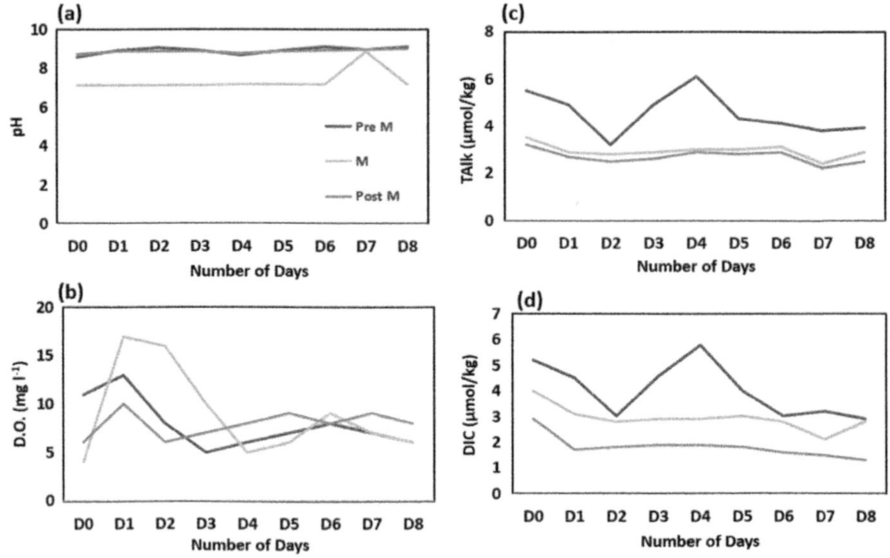

Fig. 13.2 The day-to-day disparity of **a** pH, **b** dissolved oxygen, **c** total alkalinity, and **d** dissolved inorganic carbon during the 08 days microcosm study during monsoon, pre-monsoon, and post-monsoon season

DIC and TAlk exhibited an elevating tendency from 3rd to 5th days followed by a clear decrease during pre-monsoon season (Fig. 13.2).

Positive values indicates reduction and negative values indicates increase.

13.3.3 Temporal Trend of pCO_2(Water) Along with Chlorophyll and Nutrient During the 08 days (a Microcosm Study)

We have tabulated the % reduction of pCO_2(water), Chl-*a*, and nutrients during the eight (08) days microcosm study in Table 13.2. A preliminary reduction in pCO_2(water) till 2nd day was detected in all the three (03) seasons. During monsoon, pCO_2(water) were more or less constant from 2nd to 6th days followed by a reduction in 7th day, whereas in pre-monsoon a sharp increase in pCO_2(water) was detected between 2nd and 4th days followed by a strong reduction. An alike rise in pCO_2(water) during the middle of the research was detected in post-monsoon. But it was low compared to pre-monsoon (Table 13.2). Chlorophyll-*a* in pre-monsoon reduced from 0th day to 4th day. During post-monsoon season, chl-*a* declined from 1st day to 3rd day followed by a fringe rise till 5th day and it again reduced till 8th day. Chl-*a* in monsoon season reduced gradually from the start till the end.

Table 13.2 The percentage reduction of pCO_2(water) and nutrients in the microcosm setups

	% Reduction	0th–2nd Day	2nd–5th Day	5th–8th Day	0th–8th Day
Pre-monsoon	ΔpCO_2	85	−143	83	90
	Δ Fe	77.2	−51	15	70.51
	ΔPO_4^{3-}–P	32	1.21	38	60.44
	ΔNO_3^-–N	41	50	70	90.11
	ΔNH_4^+–N	9	−4.39	21	24
	ΔSiO_3^-	22.1	−40.11	16.1	3.4
Monsoon	ΔpCO_2	62.3	−11.66	30.1	70.3
	Δ Fe	44.2	29.2	40.4	72.7
	ΔPO_4^{3-}–P	−20.8	43	80.1	85.3
	ΔNO_3^-–N	55	−88	45.2	56
	ΔNH_4^+–N	4.2	−30.2	56.2	55
	ΔSiO_3^-	25.2	60	−122	33.65
Post-monsoon	ΔpCO_2	56	35	20	72
	Δ Fe	53	46	8	73
	ΔPO_4^{3-}–P	69	18	8	76
	ΔNO_3^-–N	33	55	65	90
	ΔNH_4^+–N	66	70	−98	80
	ΔSiO_3^-	8	53	3	55

Nitrate (aquaculture pond water) reduced gradually in the pre-monsoon and post-monsoon. During monsoon, an increase was noticed between the 3rd and 4th days. The net decrease in nitrate was ~88% in both post-monsoon and pre-monsoon. In monsoon, the net reduction was only 59% (8th day) (Table 13.2). A net reduction in ammonium was noticed only in the post-monsoon (~81%), whereas in pre-monsoon and monsoon, the net reduction was less (Table 13.2). The maximum net reduction of phosphate was found in the monsoon (~86.5% by the end of the 8th day) though it showed a rise during the initial days. In the pre-monsoon and post-monsoon, phosphate concentration decreased. However, a lower reduction was noticed in the pre-monsoon (Table 13.2). A net decrease in Fe value was almost the same in all three seasons (70–75% reduction by 8th day). In contrast to other nutrients, the use of SiO_3^{2-} was noticed less (pre-monsoon ~4% and post-monsoon ~50%).

13.3.4 Link Between pCO_2(Water), chl-a and Nutrients

All the nutrients of this microcosm experiment revealed a positive relationship with chl-*a* in pre-monsoon, monsoon, and post-monsoon season. Most of all were statistically significant (exceptions are there). Nitrate showed a positive (strong) correlation ($R^2 = 0.96$) during monsoon. There was no significant relationship in the post-monsoon. On the contrary, ammonium displayed a strong correlation in the pre-monsoon. It was not significant during monsoon and post-monsoon. SiO_3^{2-} value also showed *a* similar relationship like ammonium concentration with chl-*a* value. Iron concentration revealed a significant (positive) relationship with chl-*a* in all the three seasons. Alike, ammonium and silicate, phosphate also showed a significant (positive) correlation with chl-*a* only during pre-monsoon. Comparison to chl-*a*, pCO_2(water) also showed a positive correlation with the respective nutrients. But the relationship was not statistically important all the time. The R^2 values were much less than the detected chl-*a* value.

13.4 Discussion

There are several studies on the coastal water or estuarine water of Indian Sundarbans regarding CO_2 emission or uptake. But there is no study on the carbon dynamics of the aquaculture pond. The present study is the first approach to reveal that the brackish pond water acted as a source of CO_2 in the months of pre-monsoon and monsoon in respect to pCO_2(water).

During the post-monsoon, the surface waters showed lower pCO_2(water) values (under-saturated concerning atmospheric CO_2 level). Bhattacharyya et al. (2018) also observed the same trend while working the aquaculture ponds of East Kolkata Wetland. During pre-monsoon and monsoon season, the pCO_2(water) level was reduced (drastically) during the eight-day experiments (even below the atmospheric CO_2 level). The present study demonstrated that the photosynthetic potential of the aquaculture pond of Indian Sundarbans is high. That means the pond system must be in a net autotrophic state. But in reality, the system is net heterotrophic. Moriarty (1997) stated that there is a source of organic matter to make the pond water system net heterotrophic in this type of situation. While pCO_2(water) showed decreasing trend throughout the microcosm, chl-*a* increased in corresponding. But the chl-*a* was decreased steadily in post-monsoon and monsoon season. Lindsey et al. (2010) also observed the same trend and explained the lower life cycles of phytoplankton standing stock in the aquaculture pond system. During pre-monsoon months, the chl-*a* reduced until the fourth day, it rose significantly at the final stage of the microcosm study. Higher surface temperature and lesser light penetration in the water column are the common features during pre-monsoon months, thus lower concentration of chl-*a* present in the surface water. Roy and Pal (2015) also revealed a similar trend during pre-monsoon months. They have found a lesser amount of phytoplankton in the same season.

During monsoon season, surface water of brackish water pond act as a source of CO_2 under in-situ and microcosm experiment till the 7th day. The higher amount of organic load and heterotrophic decomposition are the principal reason behind the above-mentioned observation. Boyd and Tucker (2012) described that a higher amount of DIC produced during monsoon months in the surrounding areas of aquaculture ponds and makes this pond net heterotrophic. They also identified diatoms as a controller of photosynthetic activity during monsoon. But DIC makes the aquaculture pond a source of CO_2 in-situ conditions. According to Giordano et al. (2005), the present research experienced sufficient CO_2 consumption and nutrient removal during the microcosm experiment.

During post-monsoon, because of the higher amount of sunlight and lower turbidity of pond water, the photosynthesis occurred at its higher level and reduced the pCO_2(water). In most of the case, liming activities are performed in the aquaculture ponds of Indian Sundarbans during post-monsoon months. Due to these activities, the pH values of the pond water are high in this season. According to Wurts and Durborow (1992) nutrients are more available at higher pH. All the above-mentioned statements explained the higher photosynthesis rate of the aquaculture pond water during post-monsoon in Indian Sundarbans. Thus, pond water act as a sink of carbon-di-oxide.

The key finding of the present research is the higher pH value of the aquaculture pond water during post-monsoon. This is a traditional practice of local fishermen to perform lime treatment in their pond to lower the turbidity in Indian Sundarbans. This traditional practice is the key reason behind the low pCO_2(water) values in this season. Finally, it is understood that nutrient removal make the aquaculture pond as a sink of carbon di-oxide in all three seasons from the microcosm experiment.

In conclusion, it was found that the pCO_2(water) variability of the aquaculture pond was generally dependent on the photosynthetic potential of pond water and the environments which hamper or enhance the frequency of photosynthesis in Indian Sundarbans. For the management purpose, it can be suggested that the lime treatment in the fixed interval could increase the nutrient deduction as well as minimize the CO_2 as a source from the aquaculture pond of the Indian Sundarbans.

References

Aufdenkampe AK, Mayorga E, Raymond PA, Melack JM, Doney SC, Alin SR (2011) Riverine coupling of biogeochemical cycles between land, oceans, and atmosphere. Front Ecol Environ 9(1):53–60. https://doi.org/10.1890/100014

Bhattacharyya S, Chanda A, Das S, Akhand A, Pattanaik S, Choudhury SB (2018) Nutrient removal Vis-a-Vis change in partial pressure of CO_2 during post-monsoon season in a tropical lentic and lotic aquatic body: a comparative study. Earth Syst Environ 2:233–245

Biswas H, Jie J, Li Y, Zhang G, Zhu ZY, Wu Y (2015) Response of a natural Phytoplankton community from the Qingdao coast (Yellow Sea, China) to variable CO_2 levels over a short-term incubation experiment. Curr Sci 108:1901–1909

Boyd CE, Tucker CS (2012) Pond aquaculture water quality management. Springer, Boston. https://doi.org/10.1007/978-1-4615-5407-3

Dickson AG (1992) The determination of total dissolved inorganic carbon in sea water using extraction/coulometry: the first stage of a collaborative study (No. DOE/RL-01830T-H14). Scripps Institution of Oceanography, Marine Physical Lab, San Diego, U.S

Downing JP, Meybeck M, Orr JC, Twilley RR, Scharpenseel HW (1993) Land and water interface zones. Water Air Soil Pollut 70(1–4):123–137

Giordano M, Beardall J, Raven JA (2005) CO_2 concentrating mechanisms in algae: mechanisms, environmental modulation, and evolution. Ann Rev Plant Bio 56:99–131

Holgerson MA, Raymond PA (2016) Large contribution to inland water CO_2 and CH_4 emissions from very small ponds. Nat Geosci 9(3):222–226

Huszar VL, Silva LHS, Domingos P, Marinho M, Melo S (1998) Phytoplankton species composition is more sensitive than OECD criteria to the trophic status of three Brazilian tropical lakes. In: Alvarez-Cobelas M, Reynolds CS, Sanchez-Castillo P, Kristiansen J (eds) Phytoplankton and trophic gradients. Springer, Dordrecht, pp 59–71. https://doi.org/10.1023/A:1017047221384

Kattner G (1999) Storage of dissolved inorganic nutrients in seawater: poisoning with mercuric chloride. Mar Chem 67(1–2):61–66

Lewis E, Wallace DWR (1998) Program developed for CO_2 system calculations, ORNL/CDIAC-105. Oak Ridge, TN: Carbon Dioxide Information Analysis Center, Oak Ridge National Laboratory, U.S. Department of Energy

Lindsey R, Scott M, Simmon R (2010) What are Phytoplankton. NASA's Earth Observatory (online). http://earthobservatory.nasa.gov/Library/phytoplankton. Accessed 22 October 2020

Mitsch WJ, Bernal B, Nahlik AM, Mander U, Zhang L, Anderson CJ (2013) Wetlands, carbon, and climate change. Landsc Ecol 28(4):583–597

Moriarty DJ (1997) The role of microorganisms in aquaculture ponds. Aquaculture 151(1–4):333–349

Natchimuthu S, Selvam BP, Bastviken D (2014) Influence of weather variables on methane and carbon dioxide flux from a shallow pond. Biogeochemistry 119(1–3):403–413

Panneer Selvam B, Natchimuthu S, Arunachalam L, Bastviken D (2014) Methane and carbon dioxide emissions from inland waters in India–implications for large scale greenhouse gas balances. Glob Change Biol 20(11):3397–3407

Parsons TR, Maita Y, Lalli CM (1992) A manual of chemical and biological methods for sea water analysis. Pergamon, New York

Roy AS, Pal R (2015) Planktonic Cyanoprokaryota and Bacillariophyta of East Kolkata Wetlands Ecosystem, a Ramsar Site of India with reference to diversity and taxonomic study. J Algal Biomass Util 6(3):47–59

Song C (2011) Carbon dynamics in aquatic ecosystems in response to elevated atmospheric CO_2 and altered nutrients availability. Doctoral dissertation, University of Kansas, U.S

St. Louis VL, Kelly CA, Duchemin E, Rudd JW, Rosenberg DM, (2000) Reservoir surfaces as sources of greenhouse gases to the atmosphere: a global estimate: reservoirs are sources of greenhouse gases to the atmosphere, and their surface areas have increased to the point where they should be included in global inventories of anthropogenic emissions of greenhouse gases. AIBS Bull 50(9):766–775

Wurts WA, Durborow RM (1992) Interactions of pH, carbon dioxide, alkalinity and hardness in fishponds. Southern Regional Aquaculture Center publication no. 464. Liming Fishponds 3. Auburn University

Printed by Printforce, the Netherlands